CORRESPONDENCE, INVARIANCE AND HEURISTICS

BOSTON STUDIES IN THE PHILOSOPHY OF SCIENCE



VOLUME 148

Heinz Post, summer 1992. (Photograph by Ginny Post.)

# CORRESPONDENCE, INVARIANCE AND HEURISTICS

## *Essays in Honour of Heinz Post*


*Edited by*

STEVEN FRENCH

*Southeast Missouri State University, Missouri, U.S.A.*

*and*

HARMKE KAMMINGA

*University of Cambridge, Cambridge, United Kingdom*



KLUWER ACADEMIC PUBLISHERS

DORDRECHT / BOSTON / LONDON

**Library of Congress Cataloging-in-Publication Data**

Correspondence, invariance, and heuristics : essays in honour of Heinz Post / edited by Steven French and Harmke Kamminga.
p. cm. -- (Boston studies in the philosophy of science ; v. 148)
ISBN 0-7923-2085-9 (alk. paper)
1. Correspondence principle (Quantum mechanics) 2. Symmetry (Physics) 3. Heuristic. 4. Post, Heinz. I. Post, Heinz. II. French, Steven. III. Kamminga, Harmke. IV. Series.
QC174.17.C65C67 1993
530.1'2--dc20 92-40985

ISBN 0-7923-2085-9

Published by Kluwer Academic Publishers,
P.O. Box 17, 3300 AA Dordrecht, The Netherlands.

Kluwer Academic Publishers incorporates
the publishing programmes of
D. Reidel, Martinus Nijhoff, Dr W. Junk and MTP Press.

Sold and distributed in the U.S.A. and Canada
by Kluwer Academic Publishers,
101 Philip Drive, Norwell, MA 02061, U.S.A.

In all other countries, sold and distributed
by Kluwer Academic Publishers Group,
P.O. Box 322, 3300 AH Dordrecht, The Netherlands.

*printed on acid-free paper*



Printed in the Netherlands

*With all good wishes for Heinz, a defender of reason. I greatly value the years when you came, with your excellent students, to my Seminar at the LSE: we all learned much from one another. As ever,*

Karl Popper

*My years in the Department of History and Philosophy of Science, headed by Heinz Post, were the happiest years of my academic life. Now the department is no more – it was over-taken by a world dominated by market forces, which does not value knowledge for its intrinsic worth, merely as an instrument for making profit. But even in this new Dark Age there are enough of us left who will keep alive the subversive flames of free thought and liberal learning, a tradition which Heinz did much to promote.*

Moshe Machover

*To Heinz Post who, from our early meetings in Alpbach when I was still a student to our late encounters in London when I had become a professor, taught me that while science has many holes and while some scientists work hard to conceal them, the attempt to understand scientific knowledge and scientific change in a non-doctrinaire way is still worth undertaking – best wishes for his 75th birthday.*

Paul Feyerabend

# Table of Contents

Acknowledgements xi

HARMKE KAMMINGA, STEVEN FRENCH AND MELVIN EARLES / Introduction xiii

HEINZ R. POST / Correspondence, Invariance and Heuristics 1

ERIC R. SCERRI / Correspondence and Reduction in Chemistry 45

HARMKE KAMMINGA / Taking Antecedent Conditions Seriously: A Lesson in Heuristics From Biology 65

ALLAN FRANKLIN / The Rise of the 'Fifth Force' 83

ELSPETH CRAWFORD / Michael Faraday's Thought: Discovery or Revelation? 105

NORETTA KOERTGE / Ideology, Heuristics and Rationality in the Context of Discovery 125

NEWTON C.A. DA COSTA AND STEVEN FRENCH / Towards an Acceptable Theory of Acceptance: Partial Structures, Inconsistency and Correspondence 137

AVINASH K. PURI / Tales from the Classroom: The See-Saw 159

GIORA HON / The Unnatural Nature of the Laws of Nature: Symmetry and Asymmetry 171

ALAN CHALMERS / Galilean Relativity and Galileo's Relativity 189

CLIVE KILMISTER AND BARRIE TONKINSON / Pragmatic Circles in Relativistic Time Keeping 207

HARVEY R. BROWN / Correspondence, Invariance and Heuristics in the Emergency of Special Relativity 227

JAMES T. CUSHING / Underdetermination, Conventionalism and Realism: The Copenhagen vs. The Bohm Interpretation of Quantum Mechanics 261

ARTHUR FINE / Measurement and Quantum Silence 279

SIMON SAUNDERS / To What Physics Corresponds 295

MICHAEL REDHEAD / Is the End of Physics in Sight? 327

NOTES ON CONTRIBUTORS 343

INDEX OF NAMES 347

INDEX OF SUBJECTS 353

# Acknowledgements

The idea for this *Festschrift* in honour of Heinz Post is due to Steven French. We began to explore the possibilities for such a venture in 1990 and, encouraged by the enthusiastic response we received from potential contributors, we decided to take on the not inconsiderable task of bringing the project to fruition. The warm support of Robert S. Cohen enabled us to place this volume in the Boston Studies series. It was he who suggested a thematic approach and we thank him for thus stimulating us to aim for a coherent collection of papers.

The theme of 'Correspondence, Invariance and Heuristics' was chosen because we regard Heinz Post's classic paper bearing this title as the prime focus on which research in the Chelsea tradition of philosophy of science was centred. We feel that the contributions presented in this volume bear out this proposition. To Heinz Post, then, our thanks for being an inspiring teacher and for providing the inspiration for this volume.

We thank all contributors to this volume for responding so enthusiastically to our invitations, for dealing so graciously with our requests for revisions and for making such valiant efforts to meet the deadlines we imposed (most of you, that is!). We apologise if our patience seemed to wear a bit thin at times and hope that you all feel that our 'nagging' helped to improve the collection.

We are especially grateful to Dr. Janet ('Ginny') Ramage Post for smuggling biographical details concerning Heinz Post's pre-Chelsea days to us and for keeping the secret of our plans for this *Festschrift* from Heinz until we informed him ourselves. We thank Ginny also for the remarkable feat of having persuaded Heinz to allow her to take the photograph for the frontispiece of this volume.

Steven French would like to thank his colleagues in the Dept. of Philosophy at Southeast Missouri State University and, in particular, Hamner Hill, the chairperson, for their support, both moral and material. He would also like to thank Jennifer Rigdon for her help with the initial preparation of the manuscripts and last, but of course by no means least, Dena Golf for her patience and understanding.

Harmke Kamminga should like to thank Rosemary Ward for her help with the transcription of some of the submitted papers. She also thanks her colleagues in the Cambridge Wellcome Unit for a great deal of practical advice and affectionate support, especially Andrew Cunningham and Perry Williams.

We are grateful to Professor Nicholas Jardine and Dr Andrew Cunningham, editors of *Studies in History and Philosophy of Science*, and to the journal's publishers, Pergamon Press, for permission to reproduce Heinz Post's paper on correspondence, invariance and heuristics in this volume.

Finally, our thanks are due to Annie Kuipers and her staff at Kluwer Academic Publishers for guidance and advice – and for her generosity in allowing us to extend our initially somewhat overoptimistic deadlines.

*Steven French, Cape Girardeau* *Summer 1992*
*Harmke Kamminga, Cambridge*

HARMKE KAMMINGA, STEVEN FRENCH AND MELVIN EARLES

# Introduction

This volume is presented in honour of Heinz Post, who founded a distinctive and distinguished school of philosophy of science at Chelsea College, University of London. The 'Chelsea tradition' in philosophy of science takes the content of science seriously, as exemplified by the papers presented here. The unifying theme of this work is that of 'Correspondence, Invariance and Heuristics', after the title of a classic and seminal paper by Heinz Post, published in 1971, which is reproduced in this volume with the kind permission of the editors and publishers of *Studies in History and Philosophy of Science*. Described by Paul Feyerabend in *Against Method* as "brilliant" and "... a partial antidote against the view which I try to defend" (1975, p. 61, fn. 17), this paper, peppered with illustrative examples from the history of science, brings to the fore some of Heinz Post's central concerns: the heuristic criteria used by scientists in constructing their theories, the intertheoretic relationships which these criteria reflect and, in particular, the nature of the correspondence that holds between a theory and its predecessors (and its successors).

The appearance of this volume more than twenty years later is an indication of the fruitfulness of Post's contribution: philosophers of science continue to explore the issues raised in his 1971 paper. Given the resurgence of interest both in the rationale of discovery and in the actual practice of scientists, we feel that the collection of papers presented here represents an important contribution to philosophy in the context of discovery and thus a fitting tribute to Heinz Post.

The contributors to this work include former pupils and colleagues of Heinz Post, as well as a number of philosophers who used to come to Chelsea as visiting scholars, in some cases year after year. As reflected in this volume, the Chelsea tradition has made its impact all over the world, through the appointment of Chelsea Ph.D. graduates to university posts in many different countries and through the international network of 'Chelsea sympathisers' and friends of Heinz Post.

Heinz Paneth (later Post) was born in Vienna on 26 June 1918, a member of an intellectually distinguished family. His father was Professor A.F. Paneth,



*S. French and H. Kamminga (eds.), Correspondence, Invariance and Heuristics*, xiii–xxiii.

who devised and developed the radioactive tracer technique and who was later Director of the Max-Planck-Institut in Mainz from 1953 until his death in 1958. Heinz's grandfather was the physiologist Joseph Paneth, who first observed the tissue cells that still bear his name, and his maternal grandfather was the eminent historian Ludo Hartmann. Heinz spent his first fifteen years in Austria and Germany. In Königsberg he attended the *Gymnasium* where Immanuel Kant had been a pupil, a fact that Heinz would drop proudly into the conversation from time to time.

In 1933, his parents, recognizing the gravity of the Nazi threat, moved to London where Heinz attended St Paul's School. In 1937 he was awarded a Millard Scholarship in Natural Science to study chemistry at Trinity College, Oxford, where C.N. Hinshelwood was his tutor. Heinz Post's studies at Oxford were interrupted by a period of internment as an 'enemy alien' (Category C) by the British Government during the course of 1940 and 1941. The significance of Category C is that it comprised what were officially designated 'friendly' enemy aliens, who were allowed to live a relatively normal life, in so far as that is possible under conditions of enforced isolation. During his internment, Heinz attended courses on quantum mechanics and on set theory and lectured on chemistry to interned school-certificate students. He graduated from Oxford in 1941 and, after taking a course in radio engineering, was engaged for the rest of the war in scientific activities connected with the war effort, first with the Admiralty Signal Establishment and then with the British Atomic Energy Project in Montreal.

In 1946, Heinz Post moved to the University of Chicago to investigate neutrino recoil at the Institute for Nuclear Studies. He was an Instructor of Physics from 1948 to 1950 and did theoretical work under C. Zener at the Institute of Metals in Chicago, leading to the award of his Ph.D. in 1950. Throughout his time in Chicago, Heinz attended Rudolf Carnap's courses and seminars and developed a profound intellectual passion for philosophy of science.

Returning to Britain in 1950, Heinz Post worked in the Department of Mathematical Physics at the University of Birmingham. In 1952 he was appointed Lecturer in Theoretical Physics at Chelsea Polytechnic which was shortly to be designated a College of Advanced Technology. In 1966 the College was admitted as a School of the University of London to be known as Chelsea College. It was in this developing institution that Heinz Post became instrumental in establishing a department for the teaching of history and philosophy of science.

After organising a successful course of lectures on 'Atomism' during 1963, Heinz Post, by then a Senior Lecturer in Physics at Chelsea, drew up proposals for a full-time postgraduate course in history and philosophy of science, which was submitted to the Academic Board of the College in early 1964. The circumstances were favourable for the project. For a long time, the only department devoted to teaching and research in history and philosophy of science in the United Kingdom was at University College, London, where a Chair in the subject had been established in 1924. In the post-war period, however, teaching posts in history and philosophy of science were created in other colleges and universities and, by 1964, the subject had been recognised for the award of research grants by the Departments of Education and Scientific and Industrial Research of the British Government.

The course proposed by Heinz Post was accepted and the first postgraduate students were admitted in October 1964. The success of the course soon led to the establishment of a separate Department of History and Philosophy of Science at Chelsea, with Heinz Post as its head. He was appointed a University Reader in the subject in 1966 and was awarded a Personal Chair in Philosophy of Science by the University of London in 1972. To complete this brief summary, in 1978, Heinz Post was appointed to the newly created Chair in the Philosophy of Natural Science at Chelsea College. Upon his retirement in 1983, he continued to take an active part in the Department's teaching and research pursuits as one of its Honorary Research Fellows.

In the early 1960s, when Post first proposed his course, several of the staff at Chelsea were inclined to regard the subject of history and philosophy of science in the context of concerns at that time with overspecialisation in science education and with the 'Two Cultures' debate initiated by C.P. Snow. They saw the subject as one that might serve to broaden the education of undergraduates locked into highly specialised science and technology courses. Heinz Post, on the other hand, wished to promote the subject as an intellectually demanding and highly technical pursuit for people well versed in the sciences. He was, therefore, opposed to teaching philosophy of science to undergraduates and the course he argued for successfully was a postgraduate course for people with a good degree in one of the natural sciences or mathematics.

The aim of the course was to work on problems in philosophy of science arising out of a study of the natural sciences, and to maintain the closest links with these sciences. The subject was to be approached through technical problems in science with which only graduates in a natural science or mathematics could be expected to be familiar. In his own teaching, Heinz Post encouraged students to test methodological principles against some area of science that they knew thoroughly, steering them away from the grand metaphysical questions and from purely abstract (or, at least as bad in his view, sociological) approaches. This close contact with the content and practices of science was the distinguishing mark of the course and, later, of the Chelsea Department of History and Philosophy of Science.

The technical rigour of the course was exemplified by the demand that all students attend the series of lectures on mathematical logic that was introduced with the appointment of Moshe Machover in 1968. Mathematics graduates could subsequently take a separate M.Sc. course on the foundations and philosophy of mathematics, the lectures on mathematical logic forming the common link between the two courses. Making the demanding course in mathematical logic compulsory for his students in history and philosophy of science was one means whereby Heinz Post aimed to ward off what he called 'refugees from science'.

The course introduced by Heinz Post was intended to prepare students for research in history and philosophy of science, with an emphasis on methodological and foundational questions, and assist them in addressing the problem (in Post's words) "How is it that science 'works'?" Heinz was later to express disappointment when a research student, privileged to spend some time at the research station of the eminent ethologist Konrad Lorenz, claimed to have spent much of his time arguing philosophy with his host. Post wrote "What he should have done, as a philosopher of science, was, of course, to observe the behaviour of the scientist just as the scientist observed his grey lag geese."

In arguing for the course, Post was at pains to point out that it was intended to avoid misleading teaching on an authoritarian basis, something to which he believed philosophy of science to be particularly vulnerable. In a memorandum he wrote:

> Science graduates may be expected to stand up to some inevitable bias in the presentation of generalizations about science. Moreover, without discouraging bias on the part of lecturers, the sheer multiplicity of views expressed by different lecturers (departmental or invited) should force the students into exercising increasingly mature judgement.

In his Inaugural Lecture as Professor of Philosophy of Science in 1974, he returned to the subject of bias. The lecture, entitled *Against Ideologies* (Post 1974), was characteristically addressed to students with the comment "Professors are past hope". The students were told not to take ideology too seriously: "By all means try any ideology you like for heuristic purposes but *keep your critical powder dry*."

A special feature of the course was the weekly seminar open to members of the public, academics and students from other institutions. This seminar was compulsory for all students on the course, it being one of the means of confronting them with a 'multiplicity of views' on matters relating to the history and philosophy of science. The seminars continued in an unbroken succession from October 1964 and hosted distinguished scientists, mathematicians, philosophers and historians of science from all over the world. The seminars were renowned for their intense question and answer sessions and the vigorous discussions they engendered.

'Heckling' was the standard feature of these seminars and even speakers who had begged for a clear run (an instruction to the audience that Heinz Post would pass on earnestly in introducing the speaker) were lucky if they could talk for ten minutes before being interrupted (usually by Heinz). On one famous occasion, a speaker uttered two sentences before the first interruption came and members of the audience started arguing amongst themselves – the speaker never got another word in, but simply sat at the front of the lecture hall, looking bewildered. These were exciting occasions, especially for the students, who were gradually initiated into debate of the highest intellectual order, even if their first impression of the seminars might be one of mayhem. The suffering speakers also seemed to appreciate the sessions: the vast majority of them eagerly accepted repeat invitations to undergo the same treatment, in several cases again and again.

Heinz Post firmly put his stamp on the style of the Chelsea seminars. At the same time, he was closely, but by no means uncritically, associated with the Popperian school of rationalist philosophy of science in its heyday at the London School of Economics, joining in heated and sparkling debates with his friends Popper, Lakatos and Feyerabend. The passionate intellectual sparring that philosophers of science in London enjoyed in those days was to survive longest at Chelsea under Heinz Post.

In 1969 Professor Sir Karl Popper, in a letter in support of the department, wrote

> In my opinion the best course of training for a philosopher of science is to get first a science degree and then follow it up by a post-graduate course of the type which is offered at Chelsea.

In the decade that followed, the Department began to build a reputation for distinguished research carried out by what Heinz Post described as "a new generation of historians and philosophers who continue to be technically concerned with their science". Examples of the fruits of this style of research are presented in this volume which, incidentally, includes contributions from Heinz Post's first and last research students: Noretta Koertge obtained her Ph.D. under Post's supervision in 1968 and Eric Scerri in 1992.

As Heinz Post used to point out to colleagues in the College, some of his research students were appointed to Professorships before he himself was so honoured. The Chair to which Heinz Post was appointed in 1972 was a personal one and would, therefore, have been strictly coeval with his own tenure. In order to guarantee the continuation of the work of the Department on the principles now firmly consolidated, Post campaigned in the 1970s for the establishment of a University Chair in the Philosophy of Natural Science. It came about in 1978 and, as mentioned earlier, Heinz Post was appointed the first incumbent. This was to be the culmination for the subject at Chelsea College. Within a short time, the College began to suffer from the cuts in university finance imposed by the Thatcher Government. The Department of History and Philosophy of Science survived, but when Heinz Post retired in 1983, the established Chair was 'frozen'. It was not reinstated until 1990, by which time Chelsea College had merged with King's College, London. In 1992 the Department, by then much changed, was absorbed into the Department of Philosophy at King's College.

In an institutional sense, then, the Department created by Heinz Post has ceased to exist, but, intellectually, the 'Chelsea tradition' continues to flourish in many academic centres worldwide. Some of those 'carrying the banner' are represented in this *Festschrift*. The methodological priorities and intellectual impacts of Heinz Post's teaching are visible in every contribution that follows. This is, emphatically, not to say that Post's own views, whether they concern the philosophy of atomism, simplicity, symmetry, experiment or heuristics, are simply being propagated uncritically here. Above all, Heinz Post instilled a critical attitude in his students and he would be the first to agree that what is sauce for the goose is sauce for the gander. At the very least, however, the papers collected here have been inspired by Post's teaching and writings in the philosophy of science, notably his work on 'Correspondence, Invariance and Heuristics'.

In the paper which provides both the starting point and the theme for this *Festschrift*, Heinz Post set out to challenge a consensus that had grown among twentieth-century philosophers of science, according to which the heuristic process is too subjective to be open to philosophical analysis and the procedures of theory construction in science cannot be defined. Post proceeded to argue forcefully that both the motivation for formulating a new theory and the procedures of theory construction are underpinned by a rationale of discovery that it is the business of the philosopher of science to uncover.

What Heinz Post uncovered was a set of restrictions that serve as theoretic guidelines in the processs of theory construction. We shall now outline briefly Post's characterisation of the rationale of discovery, although we in no way pretend to be able to capture the full richness of his paper in a few paragraphs.

Post's first claim is that any new theory grows out of internal problems in the old theory; more precisely, it is an internal analysis of available theory that exposes formal flaws (such as logical inconsistency or incompleteness) in the theory and it is the attempt to overcome these particular formal flaws that both motivates and guides the construction of a new theory. Once identified, such flaws serve as 'neuralgic' or growing points, where theory modification is most likely to bear fruit. From a heuristic perspective, these flaws may be regarded as 'footprints' of the new theory on its predecessor. Among the examples of such footprints, Post mentions the equality of gravitational and inertial mass in the transition from Newtonian mechanics to general relativity.

Post characterises the procedure of theory change as an inductive one, in the sense that it leads from a weaker to a stronger theory: the successful part of the old theory is invariably less precise or less general than the successful part of the new theory. The process is also inductive in the sense that the successful part of the old theory is *conserved* in the new theory. The latter claim about the relation between successive theories is both descriptive and prescriptive. It is descriptive in that Post backs it up with a wealth of illustrative examples, drawn primarily from the history of physics. From this case material, Post draws the powerful conclusion that there have been *no* Kuhn losses in modern science, putting up a vigorous challenge to Feyerabend's radical incommensurability thesis in passing (Post 1971, Appendix II).

The claim becomes prescriptive when Post demands that the relation between any new theory and its predecessor must satisfy what he calls the General Correspondence Principle. This principle requires that a new theory must account for the success of its predecessor by 'degenerating' into the old theory under those conditions under which the latter has been well confirmed by tests. Meeting the requirements of the General Correspondence Principle entails the conservation of the successful empirical consequences of the old theory and the conservation of those features of the old theory that gave it its explanatory power and that are conceptually fruitful. However radical the restructuring at the 'higher' theoretical levels may be, the entire lower-level structure of the old theory must be retained within its confirmed range of validity for the successor theory to be acceptable.

With respect to the relation between successive theories, Post draws particular attention to (well-confirmed) invariance requirements, such as those concerning symmetry and conservation principles. The emergence of such principles as important heuristic criteria represents one of the more striking developments in modern physics – Lorentz and gauge invariance being two notable examples – although they are clearly not infallible.

After a thorough discussion of the evaluative procedures that constrain and guide theory construction, Post summarises his "recipe for constructing new, better theories" as follows:

(i) Look for a flaw or footprint in existing theory. Treat any flaw as a footprint, and build your L-theory up from it.
(ii) Try to retain all present confirmed universal invariance principles.
(iii) Give a tentative interpretation to hitherto uninterpreted parts of the formalism of the theory, or try to take one of the available models of present theory seriously (this, of course, also implies (ii)).
(iv) Of all the candidates satisfying the first two requirements, reject any not satisfying the correspondence principle.
(v) Choose the simplest of the surviving candidates. (Post 1971, p. 249)

The explication, testing and exploration of this "recipe" form the subject matter of this volume.

The fifteen contributions that follow Post's paper have been organised into three groups, corresponding to the themes of 'Heuristics', 'Invariance' and 'Correspondence', in that order. Because of the strong interrelationships between these three categories, this grouping is somewhat rough and ready. Nevertheless, there are differences of emphasis in the individual contributions that have guided this arrangement.

Eric Scerri opens the collection with a discussion of the border area between chemistry and physics, a territory of long-standing fascination to Heinz Post. Scerri's analysis of atomic orbital theory shows that oft-heard claims that atomic chemistry has been reduced to quantum mechanics are unfounded. He shows that the correspondence relations between the two fields are of a much weaker form and that, in the context of discovery, physical chemistry has a great deal more autonomy than is generally appreciated by philosophers of science – and by educators in chemistry, for that matter.

Harmke Kamminga presents a case study from biology which illustrates progressive theory change that satisfies the requirements of Post's correspondence principle, but which is unusual (for philosophers) in that the most consequential theoretical changes were made at the level of non-universal statements. Discussing the implications of this case for heuristics in general, she advocates a re-evaluation of the status of antecedent conditions as parts of the structure of theories. As an antidote to the emphasis on the physical sciences in this volume, she suggests that a concern with fields such as biology can bring into view general practices in science that tend to remain hidden if philosophers only examine the abstract formalisms of theoretical physics.

Discovery and pursuit are typically regarded from the point of view of the theorist. However, experimentalists may also have 'experimental' reasons for investigating a hypothesis. Allan Franklin, who shares a deep interest with Heinz Post in the epistemology of experiment, presents a detailed account of the strategies of pursuit followed by those investigating the possible existence of a 'Fifth Force'. The original suggestion and subsequent investigation of this force, although now discredited, followed a sequence of reasonable and plausible steps according to Franklin, thus adding further support to the claim that both heuristics and pursuit possess a well-defined structure.

Elspeth Crawford tackles the very difficult question of whether philosophers can legitimately say anything about the subjective processes that guide and constrain discovery in science. Taking her cue from psychoanalytic insights due to Wilfred Bion, she argues that Bion's notions of truth and lie serve to illuminate the account of discovery that Michael Faraday left in his Diary. Crawford's account of Bion's thinking may not fully convince the hard-nosed philosopher, but her case study demonstrates that the records left by scientists themselves can give us important insights into the subjective process of creative thinking.

Noretta Koertge examines the much discussed demarcation between the context of discovery and the context of justification. In Reichenbach's version, this demarcation was commonly interpreted as one between the domains of psychology and epistemology, respectively. Heinz Post showed how evaluative criteria and theoretic guidelines can legitimately enter the context of discovery. On the other side of the divide, social and psychological factors have been imported into the context of justification by social constructivists and sociologists of knowledge. Given the apparent 'leakiness' of the demarcation, Koertge asks: if we can use any ideology we like for heuristic purposes, is it possible to keep ideology out of the domain of justification or evaluation, or, in Post's terminology, is it possible to 'keep one's critical powder dry'? She answers this question in the affirmative, by drawing a strict line, not between psychology and epistemology, but between the (ideological) desirability of a theory and its (objective) plausibility.

Newton da Costa and Steven French also consider the discovery/justification distinction, but from a somewhat different perspective. Their focus is on theory acceptance in science and they develop a model of such acceptance, drawing on their previous work on 'partial structures'. Their model introduces a notion of 'partial acceptance', which can accommodate the 'acceptance' of both empirically false and inconsistent, yet heuristically fruitful, theories, such as Newtonian mechanics and Bohr's theory of the atom, respectively. On this account, the discovery/justification distinction is under pressure from two directions: the domain of heuristics has a structure that is open to analysis in accordance with Post's correspondence principle, and the domain of justification must allow pragmatic commitments to theories that are known to be empirically false or inconsistent.

Avinash Puri recounts a tale of discovery from the classroom, an intellectual space that has always intrigued Heinz Post. He reconstructs the heuristic process whereby a group of students with little schooling in science were gently led to 'discover' Archimedes' principle of the lever. Puri's account is a witty illustration of Post's 'recipe' in action, not in the higher spheres of Nobel Prize science, but in the 'nether' regions of an inner-city classroom. But there is a philosophical twist in the tail, in the form of a thought-provoking counter-example to Post's dictum of correspondence that could fruitfully be explored further in other regions.

Giora Hon also draws upon the example of Archimedes' law, but he uses it to illustrate how symmetry considerations can lead the scientist astray. This argument forms part of his general thesis that current concerns with symmetry and invariance principles have effectively divorced science from Nature, which we experience as inherently asymmetric. Hon concludes with a call for a new physics that takes asymmetries in space and time seriously. In accordance with the correspondence principle, this new physics would conform to the asymmetric and irreversible features of actual processes, features that may be regarded as footprints pointing the way to more 'natural' theories.

Two symmetry principles that have played a fundamental role in the development of modern physics are, of course, Galilean invariance and Lorentz invariance. In the first of a trio of papers examining the nature and heuristic power of these principles, Alan Chalmers argues persuasively that Galilean relativity is not, in fact, Galileo's relativity, although the latter can be regarded as a 'footprint' of the former. On the basis of careful textual analysis, Chalmers points out the Aristotelian features of Galileo's physics which prevented him from formulating the Newtonian concept of inertia and, correspondingly, what is now called Galilean relativity. In line with Post's general thesis, Chalmers argues that Galileo's physics grew out of internal problems in Aristotelian theories of motion and that, for this very reason, the 'conservative' physics of Galileo is conceptually much closer to Aristotle – and much further away from Newton – than is generally appreciated.

Clive Kilmister and Barrie Tonkinson then discuss the transition from Newtonian to relativistic physics. In the context of Post's notion of 'inconsistent correspondence' between special relativity and classical mechanics as $v/c$ approaches zero, they compare Newton's and Einstein's concepts of time. In particular, they pose the question of what one means by a 'good' clock.

The obvious answer to this question, namely that good clocks are defined as those which measure proper time when at rest in an inertial frame, immediately gives rise to another question: when is a frame inertial? Kilmister and Tonkinson call this trap 'the pragmatic circle'. They argue that such circles are common in physics and are to be dealt with in a progressive manner by invoking an iterative process in which both theory and experiment are variously definers of and defined by the terms involved.

Harvey Brown continues the theme of special relativity, discussing the emergence of Einstein's theory in the light of Post's scheme of heuristics. He argues that the postulates presented by Einstein in 1905 have larger footprints in classical physics than is commonly appreciated. Thus Brown notes that the basis of Einstein's light postulate is to be found in those aspects of Lorentz's theory of the electron which he thought were robust enough to survive future developments. Such footprints have, however, been obscured by recent interpretations of the 'geometric' approach to special relativity, in response to which Brown presents a penetrating critique of Friedman's derivation of relativistic kinematics. He shows that, by eschewing an appeal to the principle of relativity, Friedman's 'dynamics-free' derivation is rendered physically mysterious. Questioning the status of the assumption that arbitrarily constructed clocks act as hodometers of Minkowski spacetime, Brown argues that it is precisely the principle of relativity, understood to apply to all dynamical laws, that constrains the phenomena of length contraction and time dilation – a point that was well understood by Pauli and by Einstein himself.

Moving on from special relativity, James Cushing opens another trio of papers which examines the general correspondence principle in the context of quantum mechanics. Starting from the proposition that the standard Copenhagen interpretation and Bohm's theory of quantum mechanics are empirically equivalent, he presents the following counter-example to Post's thesis: the Copenhagen version is generally preferred to Bohm's quantum theory; yet the latter has many more formal and conceptual features in common with classical mechanics than does the standard theory and should, therefore, be preferred on the grounds that it better satisfies Post's correspondence principle. The pervasive underdetermination in this case presents obvious difficulties to the scientific realist and Cushing goes on to argue for a crucial role played by historically contingent factors in theory construction and selection.

Arthur Fine notes the importance of correspondence to Einstein, who invoked it to rule out Bohm's theory. Fine argues, however, that Bohm's theory does satisfy a weaker principle of 'observational' correspondence: where classical mechanics is well confirmed, a theory should 'degenerate' into the classical account of what one is expected to observe. Rejecting Bohm's theory on other grounds, he examines the prospects for reconciling the standard version of quantum theory with this principle of observational correspondence. Fine effects this reconciliation by way of his proposal of a solution to the measurement problem: the pure state of an object system interacting with the measurement apparatus should be replaced by the appropriate mixed state, from the very start. The underlying physical rationale is that measurements are 'nonholistic' in the sense that the apparatus only interacts with part of the object system.

Simon Saunders pays special attention to quantum mechanics in his vigorous defence of the cumulative, progressive view of the history of physics that is part and parcel of Post's general correspondence principle. His detailed consideration of the mathematical structures of theories that retain their identity across the revolutionary divides poses a strong challenge to the relativist and incommensurabilist. Saunders' thesis is that the entire development of dynamical physics is characterised by the gradual evolution of a 'plastic' mathematical and conceptual structure that has maintained its overall integrity through time. With respect to quantum mechanics, Saunders concludes that it is the very fact that quantum theory is not autonomous with respect to pre-existing mathematical structures that makes it so difficult to define clearly its relations to classical mechanics.

Michael Redhead concludes this collection, appropriately, with a critical discussion of the claim that the 'end of physics' is in sight, in the light of current attempts to formulate a Theory of Everything. The linearly progressive and unifying view of theory change that is at the root of Post's correspondence principle might reasonably (but not necessarily!) be taken to promise the possibility of some ultimate theory. Surveying a range of obstacles on the way to this goal, Redhead concludes that the end of physics is likely to be a receding horizon. We believe, with Redhead, that Heinz Post will interpret this as an optimistic conclusion and it is, therefore, a fitting note on which to end this *Festschrift* in honour of that critical optimist, Heinz Post.

*Wellcome Unit for the History of Medicine,*
*Dept. of History and Philosophy of Science,*
*University of Cambridge, Cambridge, U.K.*
and
*Dept. of Philosophy,*
*Southeast Missouri State University,*
*Cape Girardeau, MO, U.S.A.*
and
*Dept. of History and Philosophy of Science,*
*King's College, London*

## BIBLIOGRAPHY

Feyerabend, P. (1975), *Against Method*, New Left Books.

Post, H.R. (1971), 'Correspondence, Invariance and Heuristics: In Praise of Conservative Induction', *Studies in History and Philosophy of Science*, 2, pp. 213–255.

Post, H.R. (1974), *Against Ideologies*, Inaugural Lecture, 28 November 1974, Chelsea College, University of London.

*H. R. POST*

# CORRESPONDENCE, INVARIANCE AND HEURISTICS:

## IN PRAISE OF CONSERVATIVE INDUCTION

THE principles which constituted the triumph of the preceding stages of the science, may appear to be subverted and ejected by the later discoveries, but in fact they are (so far as they were true) taken up into the subsequent doctrines and included in them. They continue to be an essential part of the science. The earlier truths are not expelled but absorbed, not contradicted but extended; and the history of each science, which may thus appear like a succession of revolutions, is in reality, a series of developments.

W. Whewell, *History of the Inductive Sciences*, 3rd edition (London, 1857), 8.

. . . each theory passes on to the one that follows it a share of the natural classification it was able to construct.

P. Duhem, *The Aim and Structure of Physical Theory*, 2nd edition (1914, translation, Princeton, 1954), 32.

. . . and the continuity of our science has not been affected by all these turbulent happenings, as the older theories have always been included as limiting cases in the new ones.

M. Born, 'Physics in the Last Fifty Years' (1951), reprinted in *Physics in My Generation* (New York, 1969), 122.

The history of these two principles indicates the importance of the study of the relationship between theories. It shows how one can be guided by such considerations toward a guess of a yet unknown covering theory.

F. Rohrlich, *Classical Charged Particles* (New York, 1965), 5.

When once apprehended, it [the doctrine of the conservation of energy] furnishes to the physical enquirer a principle on which he may hang every known law relating to physical actions, and by which he may be put in the way to discover the relations of such actions in new branches of science.

J. Clerk Maxwell, *Matter and Motion* (London, 1877), 60.

It is now natural for us to try to derive the laws of nature and to test their validity by means of the laws of invariance, rather than to derive the laws of invariance from what we believe to be the laws of nature.

E. P. Wigner, 'Invariance in Physical Theory', *Proc. Amer. Phil. Soc.*, **93** (1949), 521–6.



*S. French and H. Kamminga (eds.), Correspondence, Invariance and Heuristics, 1–43.*

*Kluwer Academic Publishers. Printed in the Netherlands.*

# INDEX


I Introduction 215

II Heuristic Criteria 221

1 The Flaw 221

2 The Footprint 225

3 Conservation 226

4 (*a*) The General Correspondence Principle; exact ('consistent') and approximate ('inconsistent') correspondence 228

(*b*) Criticisms of the General Correspondence Principle 235

5 Simplicity 240

6 Adding to the Interpretation 241

7 Taking Models Seriously 241

8 Enlarging the Domain 242

III Relations between Criteria 243

What happens to symmetry under correspondence? 243

The 'normal' case: *S* degeneracy 243

The 'anomalous' case: *L* degeneracy, found only in cases of 'inconsistent' (approximate) correspondence 243

IV Some Present Problems 244

V Conclusion 248

Appendix I 251

Appendix II 252

# I

## Introduction

Philosophers of science should be concerned with Science; that is, with the activity of scientists, whether the concern be descriptive, prescriptive or both. The business of scientists is to construct science, to build and extend coherent systems, simple and unifying theories with predictive power describing the world of phenomena, increasing our understanding thereof;[1] in short, to find new, better theories.

Philosophers of science of this century (as distinct from philosophically inclined scientists) have neglected analysis of that activity. They have thrown out the baby heuristics with the bathwater of naive inductivism. They have concentrated largely on developing a methodology of *post-hoc* criticism.[2]

This paper is concerned with the better part of philosophy of science, namely heuristics. It is intended to combat two common views:

(1) That it is impossible or improper to define any procedure for the constructing of new theories.[3]

(2) That the most one can say of the heuristic process is that it is a trial-and-error procedure.

A variant of the first view relegates an enquiry into heuristics to psychology, as if that disposed of it.[4]

[1] 'For the whole burden of philosophy seems to consist in this—from the phenomena of motions to investigate the forces of nature' (Newton).

[2] This part of philosophy of science is covered by that stock character, the logician with the dictionary, at best further equipped with a relative-simplicity sieve to compare existing theories.

[3] Just as I do *not* believe that it is healthy to confine philosophy of science to linguistic analysis of statements about science, but *do* believe that philosophy of science should, amongst other things, present a truthful account of what actually goes on in science, so I believe it would be a pity to treat heuristics as a matter outside philosophy whereof we must be silent.

[4] 'There are historical problems, both as to what causes the individual scientist to discover a new idea, and as to what causes the general acceptance of scientific ideas. The solution of these historical problems involves the individual psychology of thinking and the sociology of thought. None of these questions are our business here. What we are concerned with are the straight logical problems of the internal structure of scientific systems and of the roles played in such systems by the formal truths of logic and mathematics, and also the problems of inductive logic or epistemology concerned with the grounds for the reasonableness or otherwise of accepting well-established scientific systems' (R. B. Braithwaite, *Scientific Explanation* (Cambridge, 1953), 20–1).

'Another issue customarily connected with the study of scientific method is the quest for "rules of induction". Generally speaking, such rules would enable us to "infer", from a given set of data, that hypothesis or generalization which accounts best for all the particular data, in the given set. Recent logical analyses have made it increasingly clear that this way of conceiving the problem involves a misconception: While the process of invention by which scientific discoveries

To deal with this last point immediately: I do believe that there is a demarcation of the legitimate field of the philosopher of science, but that this line runs through the field of psychology.[5] This demarcation is provided by a criterion of relevance: the line is to be drawn between what is intersubjectively relevant, and what is at best of private relevance (*e.g.*, Köstler's hypothetical, and probably jocular, psychoanalytic explanation of Pythagoras' preoccupation with triangles; the effect of maternal neglect on Newton; the fact that Poincaré had an inspiration while stepping on a bus). The cut occurs between the private circumstances leading the scientist to acquire a prejudice on the one hand, and the rational connections between the prejudice in question and further developments in science on the other hand. It might be considered the business of the historian of science to discover precisely those deep-seated prejudices of a particular scientist which (in many cases) are found to have guided his work throughout his life (for example, Planck's absolutism).

While I think further philosophically interesting work could and should be done in discussing the psychological factors in heuristics (*e.g.*, analogy between the tactics of searching one's memory and the tactics of problem solving), the field of heuristics I wish to discuss is even firmer than intersubjectively relevant psychology: it is concerned with the structure of the heuristic objects.

To return to the major point: the object of this exercise is to point out some steps towards a rationale, if not a 'logic', of scientific discovery.

are made is as a rule *psychologically guided and stimulated* by antecedent knowledge of specific facts, its results are *not logically determined* by them; the way in which scientific hypotheses or theories are discovered cannot be mirrored in a set of general rules of inductive inference. . . .

'. . . there can be no general rules of induction in the above sense; the demand for them rests on a confusion of logical and psychological issues. What determines the soundness of a hypothesis is not the way it is arrived at (it may even have been suggested by a dream or a hallucination), but the way it stands up when tested' (Carl G. Hempel, *Mind*, **54** (1945), 4 (c)).

'. . . for the act of conceiving or inventing a theory, seems to me neither to call for logical analysis nor to be susceptible of it. The question how it happens that a new idea occurs to a man—whether it is a musical theme, a dramatic conflict, or a scientific theory—may be of great interest to empirical psychology; but it is irrelevant to the logical analysis of scientific knowledge' (Karl R. Popper, *The Logic of Scientific Discovery* (London, 1959), 31).

'We pointed out in the beginning of our enquiry (§1) the distinction between the context of discovery and the context of justification. We emphasized that epistemology cannot be concerned with the first but only with the latter; we showed that the analysis of science is not directed toward actual thinking processes but toward the rational reconstruction of knowledge. It is this determination of the task of epistemology which we must remember if we want to construct a theory of scientific research' (H. Reichenbach, *Experience and Prediction* (Chicago, 1961), 381–2).

[5] Thus I consider Hadamard's *The Psychology of Invention in the Mathematical Field* (New York, 1954) relevant for the philosopher of science. One of the messages of this book is the importance of non-conscious processes in scientific work. I have experienced myself the phenomenon of failure at the conscious level to connect theories that subconsciously have been constructed jointly.

Clearly, criteria for *a posteriori* assessment of theories are included amongst the guiding lines in the further, and more interesting, field of heuristics, of *a priori* recipes for work in progress. This is summarized by the ambiguity in the phrase 'construction of theories': if there are any criteria of merit which we accept in judging the structure of given theories, then these criteria may presumably be carried into the field of building new theories. The study of the structure of existing houses may help us in constructing new houses. This does not imply mere copying of designs,[6] nor does the importation of these posterior criteria into heuristics exhaust the list of heuristic criteria.

First of all, we have to refer to the criterion 'agreement with experience'. This pragmatic criterion does play a role heuristically, though we might here introduce a usually unimportant distinction between 'experiment' and 'observation'. The former plays a role within a given theory in deciding between two or more possibilities within the theory (frequently at a low level of generality: an extreme example is measurement deciding on a numerical value out of a continuous infinity of possibilities) or, rarely, in deciding between rival available theories (some 'crucial' experiments become so in retrospect, when the new theory has become available).[7] Experiment by itself does not lead to a theory so new as to break out of the theoretic framework in which the experiment is conducted, although general observational experience provides some basis which limits the choice of theories.[8] We may reinterpret observations, but only in the light of another theory already available (and arrived at on other grounds, such as the ones outlined in section II below).

I propose to concentrate on 'theoretic' guide lines to new theories, for four reasons:

(1) It does not seem very startling from the philosophic point of view

[6] *Cf.* the role of parody in the development of the novel. On the other hand, different contemporaneous fields of physics developed distinct forms of treatment: in classical physics the picaresque novel of electromagnetism, the neo-classical poetry of thermodynamics and the dead language of mechanics.

[7] Though the lack of cylindrical symmetry in certain experiments with beams of particles might rationally have been detected within the theoretic framework in which these experiments were conducted (and perhaps was: R. T. Cox, C. G. McIlraith and B. Kurrelmeyer, 'Apparent Evidence of Polarization in a Beam of β-rays', *Proc. US Nat. Acad. Sci.*, **14** (1928), 544), no conclusive test of cylindrical symmetry was in fact carried out till Lee and Yang suggested non-conservation of parity in their paper (to account for a flaw in elementary particle theory). The two-component theory of Weyl, proposed in a different context (*Zeitschrift für Physik*, **56** (1929), 330), was probably not influential in this development, though it was then revived.

[8] An example of such a general result of observation is the fact of the occurrence of line spectra, calling for appropriate theories to account for them.

that observation may be relevant to the discovery or invention of new theories.

(2) As a matter of historic fact, observations not already expressed within the body of available explicit theory have contributed only rarely to progress towards new theories. We do consider, as one of the 'theoretic' guide lines, the important guide line provided by a regularity implicit in, but not derived by, available explicit theory.[9]

(3) It is particularly interesting to point to progress in science towards new, better theories, entirely guided by factors internal to the old theory.

(4) We use theories as well as events (observations, experiments) to refute a theory. In the case of experimental 'refutation' we require, in general, an indefinite number of ancillary hypotheses (such as absence of disturbing fields, *etc.*) to establish the *relation* of refutation to the theory. Confronted with clashing theories, on the other hand, the relation of refutation not only is definite, but (as discussed below) also provides hints in which direction one or the other theory is to be amended. We set a theory to amend a theory. We need theories to suggest or derive a theory.

The *General Thesis*, then, is four-fold:

(*a*) There *is* a rationale of scientific discovery, over and above mere trial and error. There is a series of restrictions (some of which are listed below) which render the activity of the scientist constructing new theories essentially different from that of a clueless rat trying one trapdoor after another (a remark probably also applying to any actual rat). 'In der Beschränkung zeigt sich erst der Meister.'

(*b*) These restrictions are 'theoretic', that is, based on internal analysis of available theory.

(*c*) The procedure is inductive, in the sense of leading from a weaker to a stronger theory ('weaker' at least in the sense that the successful part $S^*$ of the old theory is less precise or less general than the successful part of the new theory $L$).[10]

(*d*) It is also essentially inductive in retaining the old theory in a certain sense: it is conservative (as every good scientist is).

Apart from rational arguments, such as the one on importation of criteria of merit applicable to existing theories into the field of heuristics,

[9] *E.g.*, the identity of inertial and gravitational mass in Newton's mechanics (see 'footprint', p. 225).

[10] Again, $L$ may be 'deeper' in the sense that the weaker $S$-theory of first quantization of atoms determines correctly the frequencies of the line spectrum emitted, but, unlike the stronger, deeper $L$-theory of quantum electrodynamic field theory, has no theory of the process of emission and fails to predict intensities.

and the requirement that the new theory should explain the success of the old,[11] the argument in support of the thesis will be meta-induction from empirical historic evidence.

In dealing with the methodology of advancing science I shall have in mind primarily science of the present century, though N. Koertge[12] has investigated the validity of one of the criteria discussed below in case studies including eighteenth-century chemistry. Recent science is an easier field of study for the testing of any theory of theories, because we have here the most ramified and extensive example of theoretic structure, and because it includes the two most radical revolutions since Newton, thus constituting the most severe test of those of our criteria which imply a feature of continuity.[13]

By the same token, I shall deal largely with physics, not because it is particularly 'successful' as a science, as is frequently claimed. It is not; it is remarkably unsuccessful.[14] It is, however, the most formalized (into an abstract, ramified system) and therefore, in a neutral way, most suitable for analysis (whatever theory of inter-theory relations we may entertain).[15]

I wish to press (*d*) as strongly as possible, dealing with the most interesting, most challenging cases, namely the cases of revolution, and point to the continuity provided by the guide lines right across even revolutionary breaks. I shall refer in what follows particularly to the four revolutions starting from classical physics (Newton–Hamilton mechanics plus Maxwell–Hertz electromagnetism):[16] the Special Theory of Relativity, the General Theory of Relativity, Quantum Theory (Planck, de Broglie, Heisenberg),

[11] This is a plausible requirement, which goes beyond the mere requirement that the new theory should agree with the old on factual data that are successfully explained by the old theory. It is the requirement of an explanation in terms of the new theory of the success of the *explanation* by the old theory. This requirement is invariably met by good theories, in which insight into why the old theory worked is never lost by the new. This almost trivial empirical historical point is so neglected in most discussions that I consider it worth emphasizing. It is of philosophic interest, like any non-tautological general truth.

[12] 'A Study of Relations Between Scientific Theories: A Test of the General Correspondence Principle' (Doctoral Thesis, London University, 1968).

[13] Incidentally, I do not subscribe to the 'volume' view of science that sees a monotonically accelerating accumulation of science in recent times. The period 1900–35 (Planck to Yukawa) was particularly rich. I consider the subsequent stagnation (at least in the field of physics) to call for an explanation.

[14] Thus classical electromagnetism, for instance, is unable to answer the simplest empirical question meaningful in its field of application, namely the question as to the motion of an electron in a uniform electrostatic field. On the other hand, botany has fair predictive power in its own field of application.

[15] See also note 72, paragraph 2 following.

[16] As is usually the case, the prevalent theory cannot be associated with any one man. It is the world of textbooks and secondary publications (review articles and reports on applications) that

and Theories of Symmetry (Pauli, Parity, 'Elementary Particle Theory').

With regard to (*b*), I shall not here discuss the problem of the relation of an observed event to a statement referring to or describing that event in some language (including mathematics). Nor shall I discuss the less difficult but equally important question of the role and status of such basic or observation statements in our formal system of theory. I am here concerned entirely with the problem concerning the relationship between successive theories in the history of science.

The reference to successive theories is taken here to imply a roughly linear view of history of science, in the sense that at any one time there is usually only one dominant general theory (and not the multiplicity of rival theories dearly beloved by philosophers of science)[17] covering a particular field, and that such a theory *S* is followed chronologically by a theory *L* that is regarded as the 'successor' of the theory by any neutral observer, that is to say, the historian of science, irrespective of the particular philosophic commitment of that historian.

The very notion of a 'predecessor' theory *S* implies that we have two theories *S* and *L* which refer (in their statements) to at least some events or facts which are identifiably the same. This implication is regarded as absurd in the Feyerabend view of 'incommensurability of theories', which holds that for some competing theories there are no common 'facts' whose description is translatable from one theory to the other. This absurd view, which denies the possibility of any tentatively agreed starting level, makes any discussion or comparisons of theories impossible. In particular, it fails to distinguish between the case of theories which claim to cover the 'same' data (*e.g.*, wave *versus* particle theories of light) on the one hand, and the case of theories that claim to cover disparate data (*e.g.*, hydrodynamics and optics) on the other hand.[18]

However, it is not the purpose of this paper to enter into the discussion of Feyerabend's and related views, but rather to show that the opposite view, that theories can and should be compared, and are in fact related rationally as well as historically, leads to interesting detailed problems, such as the relations between correspondence and invariance criteria (section III below). These criteria are 'internal' in the sense that textual

expresses the consensus. Original fundamental papers are interesting not only by virtue of the way in which they introduce a theory (and frequently discuss problems ignored by followers), but also by virtue of their deviations from the ultimately accepted form of their theory.

[17] Here I agree with Strauss, 'Inter-theory Relations', in *Induction, Physics and Ethics*, ed. Weingartner and Zecha (Dordrecht, 1970).

[18] For a discussion of Feyerabend's view, see Appendix II.

criticism is 'internal' to literature. We are concerned with inter-theory relations,[19] and their application in the heuristic situation. Heuristic criteria and their interrelation will be discussed, and attention drawn to cases that should be studied further.

## II

## Heuristic Criteria

We shall now discuss, with examples, some heuristic guide lines provided by analysis of the internal structure of the given *S*-theory. Counter-examples are also given to elucidate what is meant by the respective criteria. 'Exceptions prove the rule' not only in the usual sense of indicating the limits of validity, but, in this case, are cited as counterexamples to indicate what the (fallible) rule asserts.

(1) A formal '*flaw*' in the old theory, a neuralgic point, is the starting point for the new. We may regard this as an importation of the criteria of consistency and coherence into the field of heuristics. The flaw may range all the way from a downright logical inconsistency, *via* various more or less formal paradoxes, to mere failure to unify existing theory (taken together as *S*) into a single monolithic integrated theory.[20]

[19] I have come across two publications in this field, by Tisza, *Reviews of Modern Physics*, 35 (1963), no. 1, and Strauss (*loc. cit.*), respectively. They are written from points of view somewhat different from that of the present paper, and concentrate on inter-theory relations in present-day physics. While welcoming an implied point of Tisza's paper, *viz.*, the (within certain limits) arbitrary character of the choice of theories integrating the system, I disagree with several points of detailed application, in particular with Tisza's choice of integrating system, which I attribute to his paying insufficient regard to criteria of simplicity. Philosophically, I would disagree with Tisza on many grounds, particularly with his instrumentalist view (theories agreeing within experimental error appear to be treated as identical).

I have learned in conversation with Professor Tisza that one of his grounds of preference for a thermodynamic-type theory as integrating theory is precisely the paucity of the (phenomenological) parameters required for a complete description (*cf.* Einstein's 'Theories of Principle'). For his distinction of 'dynamic' and 'chemical', *cf.* Dürr, 'Approximate Symmetries in Atomic and Elementary Particle Physics', in *Properties of Matter Under Unusual Conditions*, ed. H. Mark and S. Fernbach (New York, 1969).

Strauss's 'partial formal anticipation' is not a 'footprint' in my sense, but rather one of a set of equivalent formulations of *S*, which is only singled out later from the standpoint of *L*. I am grateful to Professor Strauss for sending me an advance copy of his 'Die Entwicklungsdialektik in der Physik und das Dilemma in der büergerlichen Philosophie' (*Deutsche Zeitschrift für Philosophie*, 1971). It appears that extensionally his term 'dialektische Aufhebung' coincides with my term 'inconsistent correspondence'. I flatly disagree with him when he says (*ibid.*, p. 5) 'Es zeigt sich nämlich, dass die dialektische Aufhebung einer Theorie $T_1$ durch eine Theorie $T_2$ in allen untersuchten Faellen mit einer Begriffsaufspaltung verbunden ist. . . .' The most interesting case discussed in the present paper is precisely *L*-degeneracy, which refutes Strauss's statement.

[20] The instruction to attempt to unify might be taken as a normative heuristic rule, which may fail: 'What God has put asunder, let no man join together' (Pauli).

Examples of mathematical inconsistencies in a theory include the infinities of quantum field theory (which lead to 'renormalization' programmes, but suggest more radical new theories, such as the inclusion of gravitational effects).

The ultra-violet catastrophe in the application of classical statistical mechanics to black-body radiation is a strong guide to quantum theory. This unacceptable 'catastrophe' of energy contributions increasing towards infinity with shorter wave length is a strict consequence of classical theory.

Boltzmann claimed that the removal of the reversible mechanics–irreversible thermodynamics paradox constituted the greatest success of his statistical mechanics.

Paradoxes, apparent formal contradictions in our system, are useful hints of a higher and wider system in which these contradictions are eliminated. Progress in physics is frequently analogous to the steps from paradox (the liar) *via* incompleteness (Gödel) to truth (at a meta-level) in logic. 'The knowledge of privation, like all other knowledge, in the end is positive. You cannot speak of the absent and lacking unless you assume some field and some presence elsewhere.'[21]

There is the possibility of detecting the incompleteness of the *S*-theory. This incompleteness[22] of theories, or their openness to growth beyond themselves, is a stronger and more interesting case than just filling the gaps in the data covered by a theory, such as missing elements in the periodic system. Just as the system of real numbers is not closed, in the sense that one can formulate equations in terms of real numbers only whose solutions point *outside* the system, so a physical theory may show its lack of closure. The points at which a theory is open, or anomalous, provide a strong indication of the way in which the future larger theory is to be built[23] (see also 'Enlarging the domain of the theory' below).

An example of the flaw of incompleteness is the failure of thermody-

[21] F. H. Bradley, *Appearance and Reality* (Oxford, 1962), 457.

[22] We are not, of course, referring to the fact that any theory that contains the arithmetic of natural numbers, is incomplete in the sense that there are statements in the language of the theory such that neither they, nor their negation, can be logically derived from the axioms. This Gödelian difficulty can be overcome, for instance, by joining to a physical theory all true statements of the mathematics used, *i.e.*, usually at least those of arithmetic.

[23] I was very pleased by Professor Dingle's example of a school arithmetic problem allowing the solution 'number of workmen = minus 3', and his point that a modern physicist would conclude that there *are* three negative workmen. Whilst Professor Dingle intended this as a warning caricature, I welcome it as a fair example. An actual example is provided by Bonnor, who points out that Maxwell's theory allows charges to move freely with the speed of light, and suggests a physical interpretation (W. B. Bonnor, 'Solutions of Maxwell's Equations for Charge Moving with the Speed of Light', *International Journal of Theoretical Physics*, **2** (1969), 373–9). This amounts

namics (before 1906) to allow calculation of the absolute[24] values (beyond the mere temperature dependence) of chemical equilibrium constants from thermal data. This led Nernst to postulate his third law, making the value of the missing integration constant zero.

The most revolutionary example of a flaw leading to a new fundamental physics is the transition from classical physics to special relativity. Classical physics consisted of two disparate parts: Newtonian (Hamiltonian) mechanics on the one hand, and Maxwell's electromagnetic theory (including optics) on the other. Every physicist hoped, and many worked towards, a unification of these two parts (most attempted a reduction of electromagnetism to mechanics, following Maxwell). Now Newtonian mechanics is, of course, Galileo-invariant. On the other hand, Maxwell's theory is not. It is invariant under a different transformation, the so-called Lorentz transformation (though Lorentz did not fully appreciate this point).

Now it is obviously impossible to unify two theories with different invariance properties, since unification in one coordinate system would be destroyed in another. One or both theories had to be altered. Einstein[25] describes how he was led to his theory of relativity by a thought experiment showing that a Galileo transformation on a light wave would falsify Maxwell's theory. Einstein was so imbued with Maxwell's theory (in particular, with the law that any curvature in the field implied time-dependence) that he was led to reject Galileo-invariance. He then reformulated mechanics in a Lorentz-invariant way (special theory of relativity). We shall show under the heading 'Correspondence' that for a unified physics the choice in favour of Lorentz- as against Galileo-invariance was not arbitrary.

An analogous example is Maxwell's argument leading to the unifying theory that currents are the sole cause of magnetic fields:

> It is impossible, by any arrangement of magnetized matter, to produce a system corresponding in all respects to an electric circuit, for the potential of the magnetic system is single valued at every point of space, whereas that of the electric system is many-valued.

to giving physical significance to hitherto uninterpreted elements of the theory (*cf.* criterion 6).

We should, however, distinguish sharply between the fruitful policy of tentatively extending interpretation to regions where a theory is genuinely open, and the heresy of misplaced concreteness when a special interpretation is given to certain approximations within a closed calculus (such as the 'non-conservation of energy' in second-order perturbation).

[24] But see the amusing polemical remark in Nernst: *Theoretische Chemie* (11th–15th edition, Stuttgart, 1926), 805.

[25] *Albert Einstein: Philosopher–Scientist*, ed. P. A. Schilpp (New York, 1951), 53.

But it is always possible, by a proper arrangement of infinitely small electric circuits, to produce a system corresponding in all respects to any magnetic system, provided the line of integration which we follow in calculating the potential is prevented from passing through any of these small circuits. This will be more fully explained in Art. 833.

The action of magnets at a distance is perfectly identical with that of electric currents. We therefore endeavour to trace both to the same cause, and since we cannot explain electric currents by means of magnets, we must adopt the other alternative, and explain magnets by means of molecular electric currents. . . .

This mode of explaining magnetism requires us also to abandon the method followed in Part III, in which we regarded the magnet as a continuous and homogeneous body, the minutest part of which has magnetic properties of the same kind as the whole.

We must now regard a magnet as containing a finite, though very great, number of electric circuits, so that it has essentially a molecular, as distinguished from a continuous, structure.

If we suppose our mathematical machinery to be so coarse that our line of integration cannot thread a molecular circuit, and that an immense number of magnetic molecules are contained in our element of volume, we shall still arrive at results similar to those in Part III, but if we suppose our machinery of a finer order, and capable of investigating all that goes on in the interior of the molecules, we must give up the old theory of magnetism, and adopt that of Ampère, which admits of no magnets except those which consist of electric currents.[26]

Here again, as in the previous example, we start with the flaw of incoherence of two theories existing side by side. Our motivation is towards unification. We attempt to let one of the already existing theories override the other, but demand a 'correspondence' in the sense that the overriding theory should degenerate (within experimental accuracy) into the other theory within that range of cases in which that other theory has proved successful. It turns out that this is possible, using one theory as the overriding one, but not the other.

A curious example of a flaw is provided by the Pauli principle in quantum theory prior to quantum (or wave) mechanics. The formulation 'no two electrons may be in the same state' renders the Pauli principle vacuous, since no electron, strictly speaking, is in a well-defined state of its own (owing to interaction with other electrons). It is interesting to see how Pauli got over this difficulty in his 1925 paper.[27] He cunningly referred to quantum numbers obtained in the case of a very strong external applied

[26] J. C. Maxwell, *A Treatise on Electricity and Magnetism* (reprinted New York, 1954), vol. 2, 274–6.

[27] *Zeitschrift für Physik*, 31 (1925), 765. This also has a footnote referring to statistics, making the major error of failing to apply his principle. *Cf.* Einstein's failure to use 'Bose–Einstein' statistics in 1905, and Planck's use of Bose–Einstein statistics in 1900.

field (which decouples the electron–electron interaction to a good approximation). He then referred to an extraordinary, almost metaphysical theoretic device (now considered to be false) invented by Bohr and entitled 'Postulate of Invariance of Quantum Numbers'[28] to go on to the pragmatically more important case of absence of strong external fields. The fact remains that a (strictly speaking) vacuous principle played an extremely useful role in the early quantum theory of spectra before being reformulated two years later by Heisenberg in an unexceptionable way in terms of the new wave mechanics.

In all these examples the flaw in the *S*-theory was removed in an *L*-theory that profited from the limited success of the *S*-theory. Construction of the *L*-theory starts from *S*. Alterations or additions are made to remove the flaw. These alterations may be profound, leading to an *L*-theory fundamentally inconsistent with *S*. But the genesis of *L* is such that we can see that the successes of *S* are inherited by *L*.[29]

(2) A related heuristic criterion is '*the footprint*' of the *L*-theory on *S*. The *S*-theory (taken to include operational interpretation) may contain specific well-delineated features or elements that do not fit into the rest of the theory, though not inconsistent with it. Thus, the statement of the existence of point masses with electronic charge may be added to Maxwell's equations of electromagnetism in order to permit the theory to yield operational consequences (electrodynamics) at all. Electrons are clearly aliens in the theory. Historically speaking, they are inconsistent with the spirit of Maxwell's programme. Thus the footprint may shade over into a flaw. The existence of electrons, whether point masses or of extended size, is inconsistent with the present theory (relativity) and clearly points to a new theory of elementary particles.

A more subtle example of a footprint, this time of general relativity within Newtonian mechanics, is the equality of gravitational and inertial mass. It is an interesting historical paradox that the separation (necessary

[28] '... A mechanically inexplicable invariant character ...' (Bohr, *Annals of Physics* (1923), 225–6); a totally *ad hoc* principle designed precisely to justify the theoretic treatment of electrons in a many-electron atom in the only way then available.

[29] There are, of course, flaws which have not been removed (see also section IV). Indeed, some have been inherited by successive theories.

The inability of classical electromagnetic theory to deal exactly with the motion of a charged particle in a field shows its incompleteness. This flaw has never been removed. In my view quantum mechanics is unsatisfactory because many such classical flaws are compounded in it. Incompleteness may be masked by the use of an undue number of independent axioms: the mere use of Lorentz's electrodynamic equation of motion (however unsuccessful) as an axiom independent of the Maxwell field equations points to a non-linear *L*-theory in which the equation of motion is derivable from the field equations, as is the case for the gravitational field.

to give meaning to Newton's definitional laws of motion) between force, as determined by the external situation, and mass as an intrinsic feature of the body acted upon, is destroyed precisely in Newton's paradigm application to gravitation.

The existence of two separate extremal principles, Fermat's and Maupertuis', in classical physics ($S$), epitomizing optics and mechanics respectively, might suggest a stronger $L$-theory providing a unifying relation between mechanical and phase velocity.[30]

The existence of parity as a good quantum number has consequences expressible even in classical physics (zero electric dipole of nuclei in ground state), though parity itself is not expressible in classical physics. The last two examples are footprints of quantum mechanics in classical (non-quantum) physics.

(3) *'Conservation'*. A third criterion of heuristics is that the new theory should conform to those general laws, invariance or conservation principles, that have been confirmed without exception either in the $S$ theory, if any, or in other theories, not necessarily overlapping the new theory.[31] Strictly speaking, if the new theory does not at least interact with the other theories (*e.g.*, interconversion of electrical and mechanical energy, allowing *extension* of the principle of conservation of energy to electromagnetic theory) or share, say, a common physical space (allowing application of a previously confirmed symmetry principle), any importation of laws from non-overlapping theories would be 'analogical' in the sense discussed in the fourth paragraph of section V below.

We certainly wish to conserve[32] charge and energy in any theory of electromagnetism or elementary particle theory. Conservation of energy is the starting point, for example, of Planck's presentation of Maxwell's theory. Again, conservation of momentum applied to electrodynamics led

[30] Though Hamilton applied the same formalism to both optics and mechanics, this did not lead to a revolutionary advance by itself. It was only after the wave theory of light had become accepted, and Fermat's principle of least time rehabilitated, bringing out a sharp distinction between optics and mechanics, that de Broglie pointed out the relation between Fermat's and Maupertuis' principles that led to a revolutionary unification of wave and particle physics. 'And though, among these problems of mathematical optics, it is not here intended to include investigations respecting the *phaenomena of interference*, yet it is easy to perceive, from the nature of the quantity which I have called the characteristic function, and which in the hypothesis of undulations is *the time of propagation of light from one variable point* to another, that the study of this function must be useful in such investigations also' (J. Hamilton, 'On a View of Mathematical Optics', *Brit. Assoc. Report* (1831–2), 545–7).

[31] *Cf.* Newton's third rule of reasoning.

[32] No pun is intended in the use of the word conservation, (i) in a context such as 'conservation of energy', and (ii) at a meta-theoretical level referring to the retaining of certain laws which may themselves be conservation laws.

to the attribution of momentum to the field. This idea proved very fertile indeed. Conservation of momentum together with Lorentz invariance was used by Tolman to derive the special theory of relativity in its present formulation. This is an example of a combination of invariance requirements leading *uniquely* to the new $L$-theory. Dirac was forced into a four-component spin-positron representation by the joint requirements of linearity, Lorentz invariance and conservation of probability.

We would, at present, reject any field theory that is not locally Lorentz-invariant. Gauge invariance (which is closely related to charge conservation) is used as a guide in constructing new field theories.

Again, without quantum theory it can at least be asserted that *if* there is a functional relationship between energy and frequency only, this relationship is necessarily a linear one, $E \propto \nu$. This follows at once from the theory of relativity which states that both $E$ and $\nu$ vary covariantly with time.[33] Chronologically, this fact was not, of course, a footprint, but it at least illustrates the strength of a requirement of conservation (in this case of relativity in quantum mechanics).

The criterion of conservation, like the laws of science themselves, is not infallible. It was long considered an obvious symmetry requirement that physics should be invariant under reflection in space. This invariance requirement, corresponding to conservation of parity, was valid within all theories, but is now considered invalid at least for neutrino fields. We *might* restore its validity by a deeper theory of structure of the neutrino.

Nor are these very general requirements in general sufficient to render the construction of the new theory unique. Following Wigner, we may look forward to the day when a theory is uniquely given by a conjunction of invariance requirements. In most cases at present we are faced with a looser situation. We have tentatively accepted certain requirements, and are faced with the task of making the theory unique. Starting with his law of motion defining force, and the requirements of interaction being central forces and depending on a single power of the distance, Newton showed that adding the fact of planetary orbits being elliptic sufficed to deduce the inverse square law.[34]

[33] This kind of argument is analogous to the dimensional argument, based on scale invariance, which again only allows us to assert that *if* there is a functional relationship between two quantities, it must be of a certain form.

[34] Similarly, laws may be determined uniquely by the empirical determination of a finite number of parameters, provided we assume some general requirements to start with. Such 'deduction from the facts' is the empirical analogue to the mathematical procedure of 'proof by a finite number of instances'; we may prove Pythagoras' theorem by first establishing that the expression for the square of the hypotenuse is necessarily bi-linear in terms of the lengths of the

The policy of retaining the law of conservation of energy (against some evidence and the revolutionary opposition of Bohr) bore fruit, for example, in the simple strong theory of the neutrino, postulating a particle to carry no charge, but energy, momentum and spin, so as to conserve all four simultaneously in $\beta$-decay.[35]

(4) (*a*) The most important heuristic restriction is the *General Correspondence Principle.*

Roughly speaking, this is the requirement that any acceptable new theory $L$ should account for the success of its predecessor $S$ by 'degenerating' into that theory under those conditions under which $S$ has been well confirmed by tests. We shall denote the well-confirmed part of $S$ by $S^*$.

We may divide a theory 'vertically' into well-confirmed working parts and others. Thus the phlogiston theory 'worked' in that it assigned *consistent* levels of phlogistication (explaining many features) to chemical substances related in more than one way by reactions. On the other hand it tried to establish a connection between colour and phlogistication, and this part of the theory was not successful even at the time.[36]

We may distinguish three pragmatic situations *before* the advent of the $L$ theory:

(i) With respect to some of its aspects, the $S$ theory is unrefuted and appears to be highly confirmed throughout the claimed range of validity of $S$.

(ii) With respect to some aspects, the $S$ theory is unrefuted, but we notice that the confirmed range is smaller than the range of validity claimed. A major example occurred recently when Yang and Lee pointed out that conservation of parity was confirmed for strong interactions, but not for weak interactions, suggesting a 'vertical slicing' of the $S$ theory to whittle it down to $S^*$.

(iii) The $S$ theory is already refuted in some respects beyond a certain range.

There are also horizontal divisions in a theory, separating various levels

other two sides, and then 'determining' the coefficients of the bi-linear terms by reference to a small number of particular instances, which may be particularly simple (in this case, the single instance of the isosceles triangle).

[35] We may compare this with the discovery of Neptune on the basis of Newton's law of gravitation. Less rigorous are such cases of 'in-filling' as Mendeleev's postulation of the existence and properties of the 'missing elements' to conserve his periodic table, and the prediction of the $\Omega^-$ particle in recent elementary particle theory.

[36] See Koertge, *op. cit.*, note 12, as against T. S. Kuhn, *The Structure of Scientific Revolutions* (Chicago, 1970), 107 and 148.

of generality or deducibility. The well-confirmed part of a theory extends only up to a certain level. The *S*-theory is stripped of dispensable, not independently confirmed superstructure. (The high-level *interpretation* of the total coherent pattern of data is changed by a revolution.)

The manner in which this well-confirmed part $S^*$ is inherited by the new theory is such as to preserve *coherence* within $S^*$; it is taken over wholesale. It is not the case that we rearrange the data within $S^*$ into a new pattern. The pattern is preserved. No insight into why the old theory worked is lost by the new. We have sliced off some very high levels and substituted new ones, but the low (particularly classificatory) levels remain undisturbed within their confirmed ranges of validity. For example, we do not believe in immutable chemical elements any longer, but we have taken over the whole pattern of the periodic system in the low-energy range.[37]

$S^*$ may consist of several disjoint parts. I claim that it is the duty of any successor theory to explain $S^*$. I shall even claim that, as a matter of empirical historical fact, *L* theories always explained the *whole* of $S^*$; that, contrary to Kuhn,[38] there is never any loss of successful explanatory power.

The General Correspondence Principle not only demands agreement of the *L*-theory with the *S*-theory in singular statements relating to certain events, but claims that any *L*-theory will in fact embody a good deal of the (lower) theoretical structure of the *S*-theory. Only in this way is the success of *S* *explained*.

Actually, this agreement on singular events, and the preservation of

[37] The polemic concerning the use of the term 'element' following Soddy's discovery of isotopes is instructive. It was suggested by leading physical chemists (*e.g.*, K. Fajans, 'Radioaktivität und die neueste Entwicklung der Lehre von den chemischen Elementen', 2nd edition (Braunschweig, 1920), 88–103) that the term element now had to be applied to each isotope separately. These chemists were, in fact, following Mendeleev's (and Dalton's) philosophy attaching primary significance to the combining mass or atomic weight of an element as a source of its individuation. If this policy of retaining Mendeleev's ontology had been pursued radically it would have meant the break-up of the periodic system. It was fortunate for chemistry that the policy advocated by the other side in the polemic was adopted. This other side (*e.g.*, F. Paneth, *Naturwissenschaften*, **43** (1920), 839–42) advocated changing the definition of the relevant term 'element' in an *ad hoc* fashion precisely to preserve the simple pattern of the periodic system. A change of definition was preferred to a change of structure. Ontology followed theory.

[38] Kuhn, *op. cit.*, 169: 'new paradigms seldom or never possess all the capabilities of their predecessors'.

Paradoxically, the lower-level (more specific) theories seem to provide a relatively more promising hunting-ground for Kuhn losses than the most general theories. The chronological sequence: Dulong–Petit theory of atomic heat applied to metals—theory of free electrons in metals—theory of free electrons plus Fermi statistics, appears to be an example of a temporary Kuhn loss at the intermediate stage, when the previous agreement between theory and specific heat data was lost, to be recaptured on the application of Fermi statistics. Here, the temporary failure was in a particular field requiring the application of a conjunction of theories.

part of the theoretical structure, is not necessarily exact. We shall come across cases where the agreement is only approximate ('inconsistent correspondence'). In such cases, too, the $L$-theory explains the success of the $S$-theory, the approximate agreement extends over several deductive levels.

To establish the correspondence between $S$ and $L$ we need, in general, a system of translation $T$, from the language of $L$ into that of $S$, with the necessary restriction on $T$ that it should carry the statement of a given event in $L$ into the statement of the same event (see Introduction) in $S$. Moreover, $T$ is, in general, simple in the technical sense ('linguistic simplicity').[39]

We now formulate in the language of the $L$-theory conditions $Q$ which define the range of validity of $S$.[40] This validity may be quantitatively approximate only ('inconsistent correspondence').[41] At the other extreme we may have a $Q$ condition so generous as to incorporate the whole of the $S$-theory ('reduction' of $S$ to $L$). Now the thesis of the General Correspondence Principle is that applying the $Q$ condition on $L$ and then applying the translation $T$ yields $S^*$. This is the strong thesis of zero 'Kuhn-losses'.[42]

The General Correspondence Principle can be tested. Take any two theories $S$ and $L$, parts $S^*$ and $L^*$ respectively of which give successful explanatory accounts of their respective range of data well-confirmed by experience. Let $S^*$ and $L^*$ intersect with regard to their reference. Let $S^{**}$ be that intersection. ($S^{**}$ may be identical with $S^*$; this is the case of zero Kuhn losses.)

Then we may have the case $S^{**} \subset L^*$, *i.e.*, $L^*$ entails $S^{**}$ without translation; or it may be that we need a translation $T$ between the languages of $S$ and $L$. This translation may be trivially a one-to-one mapping of terms. This case is more trivial even than a translation from French into

[39] See H. R. Post, 'Simplicity in Scientific Theories', *Brit. J. Phil. Sci.*, **11** (1960), 32, and H. R. Post, 'A Criticism of Popper's Theory of Simplicity', *ibid.*, **12** (1962), 328.

[40] We identify any theory with the collection of all theorems provable in it. $(L|Q)$ is a subtheory of $L$ consisting of those theorems of $L$ satisfying condition $Q$.

[41] This case of 'approximate correspondence' may be subsumed under the general treatment of correspondence given here: a theory $S$, one of whose axioms may be 'the mass of the electron is $m$' includes amongst its theorems 'the mass of the electron is $m \pm 1\%$'. It is this latter statement which might be in $S^*$. Of course, it is preferable to confine one's discussion to the axioms rather than to the unmanageable richness of the set of all theorems of a theory. But this move, again, presents no hardship. We may stamp any publication 'axiom system', declaring every sentence (except possibly the expression of gratitude to the Office of Naval Research) an axiom. In the case of approximate correspondence we are well advised to confine ourselves to publications that are not frauds, in the sense that they honestly state conservative estimates of accuracy.

[42] See note 38 above.

English. Or $T$ may be more complicated: for example, 'Affinity' in the $S$ theory may split into 'Free Energy' and 'Valency' in the $L$ theory. $T$ may be more complicated still. However, there are restrictions on $T$:

(i) For any fact which is identifiably the same in $S$ and $L$ (*cf.* p. 220 f.), $T$ must transform a statement of that fact in $L$ into a statement of that fact in $S$.

(ii) $T$ must be reasonably simple in precisely the sense in which scientific theories are required to be reasonably simple in order to allow one to speak of them as theories at all.[43] $T$ is not expected to be less simple than the less simple of the two theories $S$ and $L$.

(iii) There is at least a local homomorphism as $T$ maps at least a part of $L$ into $S^*$. Roughly speaking, we preserve structure under $T$.

The correspondence principle may be stated as asserting that $S^{**} \subset TL$. The successful theorems of $S$ having common reference with $L$ form a subset of the theorems of $L$ as translated. The zero-Kuhn-loss postulate allows us to write $S^* \subset TL^*$ making $S^*$ the 'smaller' of the two theories: chronologically, $S$ is the 'predecessor' theory. This must be the case in any rational development of science.

When $S$ is truncated horizontally to $S^*$, terms in $S^*$ are not fully reduced to the highest, most abstract, explanatory terms of $S$, and $T$ maps into $S^*$ only. The correspondence principle is non-trivial to the extent that $S^*$ extends to explanatory levels beyond the 'lowest' level of facts both $S$ and $L$ are called upon to explain.

The catch in using the correspondence principle heuristically is of course that the true extent of $S^*$ is only conjecture at any one time. We do not know the exact limits of validity of $S$ for certain (as, indeed, we do not know the truth of any statement of universal empirical content in science for certain; this paper does not defend the introduction of an infallible 'proto-science'). What *is* claimed here is that any well-confirmed $S$-theory always has a non-empty part $S^*$, large enough to cover all its successes, which will be wholly in $TL$. Indeed, it is frequently the advent of $L$ that leads to the discovery of further refutations of $S$ shrinking $S^*$ still further.[44] On the other hand, it is precisely $L$ that makes this further

43 See note 39 above.

44 P. K. Feyerabend, 'Problems of Empiricism', in *Beyond the Edge of Certainty*, ed. R. G. Colodny (New Jersey, 1965), 175–6, makes a beautiful point by drawing attention to the case of Brownian motion 'refuting' thermodynamics. It is not just that this refutation is originally suggested by the $L$-theory. The refutation is dependent on $L$. The refutation establishes limits of

limitation on $S^*$ intelligible. We may formulate a condition $Q$ in $L$ that explains the (reduced) success of $S$.

This $Q$ condition may be a straightforward discrete classificatory one (for example, not all particles obey the Pauli principle, but only those of odd spin); or it may limit the value of a parameter (hidden in the $S$ theory but explicit in $L$) ranging over a continuum: for example, the $L$-theory may explain that the rate of acceleration under gravity is dependent on latitude, and only at a particular latitude takes a single value as demanded by Galileo. For a given accuracy of experiment, then, a band of latitude (explained by $L$) is allowed.

In some cases, this $Q$ condition is a condition limiting strict validity of $S$ to an asymptotic value. For example, according to special relativity theory, classical mechanics is correct only in the limiting case $v/c \to 0$, but yields successful predictions at low values of $v/c$ to an extent depending on the accuracy of the experiment. Here we have a case of 'inconsistent correspondence'.[45] But again, the correspondence rescues the (approximate) validity of *several* levels of $S$.

We may, then, formulate the General Correspondence Principle in the form $S^* = T(L \mid Q)$, where $S^*$ may be smaller than the part of $S$ historically (but mistakenly) accepted as successful before the advent of $L$.

$Q$ again is a simple condition in the language of $L$. Certainly the correspondence principle claims informally a simple explanation of $S^*$ by $L$, at least as simple as $S$ itself was.[46]

acceptability on the concepts of the $S$-theory, provided we accept the interpretation in terms of the $L$-theory. Thus, in this case, the 'refutation' cannot be done within the resources of the $S$-theory at all. We need an $L$-theory for this criticism of $S$.

But this $L$-theory has to draw strength from somewhere else in order to be taken sufficiently seriously to refute the $S$-theory. This strength is not parasitic on $S$, but must be drawn from elsewhere, an established mechanics plus atomic theory ($L$). This $L$-theory has to be enriched to $L'$ by a set of factual requirements (see also note 53). It is the bridge theory $L'$–$L$ which is at issue, and this bridge theory must achieve '(inconsistent) correspondence' between $L'$ and $S$.

Historically, the $L$-theory (statistical mechanics) was produced by Boltzmann precisely to reduce thermodynamics to atomism. Boltzmann refrained from going beyond 'relative atomism' to 'numerical atomism' and from pushing the consequential aspect, fluctuations, which may be regarded as instances refuting thermodynamics (H. R. Post, 'Atomism 1900', *Physics Ed.*, 3 (1968), note 44). In some views these instances are not even regarded as refutations of (a suitably formulated) thermodynamics. But the point I wish to make here is that historically this case is an example of the heuristic benefits of a reductionist programme.

[45] An analogous example is Nernst's third law explaining how the second law of thermodynamics degenerates to Thomson's rule at low temperatures.

[46] In certain cases the $Q$-relativized $L$ cannot be 'translated' (in the intensional sense) into $S^*$, not because of the richness of $L$, but on the contrary because $L$ may fail to refer to restrictive conditions peculiar to $S$: $S \vdash$ all animals have trunks; $S^* \vdash$ all animals on this island have trunks; $L \vdash$ not all animals have trunks; $(L \mid Q) \vdash$ all grey-skinned animals have trunks. We may, then, enrich $L$ by a remark concerning extension, of the form 'All animals on that island are grey-skinned', and this enriched $L$-theory is then translatable, after applying $Q$-relativization, into $S^*$.

Even in the case of 'inconsistent correspondence', the (asymptotic) $Q$ condition applied to laws of $L$ renders laws in $S$ according to $T(l|Q) = s$, where $l$ and $s$ are laws in $L$ and $S$ respectively. This is a far from trivial heuristic restriction on $L$.

The new theory must correspond to the old theory, in the sense of coinciding with $S^*$, the working part of the old theory in the range in which $S$ has been found to work. According to the General Correspondence Principle, this is a necessary requirement. It is, of course, sufficient that the $L$-theory correspond to more than $S^*$, such as the $S$ theory with enlarged domain (see section II.8 below), or richer models transcending $S$ (see section II.7 below).

The new theory may be, and frequently is, inconsistent with the old theory. It is frequently the case that the new theory interprets the inherited low-level part, $S^*$, of the old theory by a new high-level superstructure that is formally inconsistent with higher levels (not in $S^*$) of the old theory $S$. Moreover, as discussed above, even the agreement of the two theories in $S^*$ may be only an approximation. We may formally take care of this by specifying a certain accuracy in our formulation of $S$. This accuracy may (uniformly) converge to exactness by a definite limiting process.

The 'conservation' criterion discussed in section II.3 above was found to be fallible. Paradoxically, the only counterexample we have been able to find to the General Correspondence Principle is the paradigm example of the relation of quantum mechanics to classical mechanics. Contrary to the impression that may be given in some textbooks, it is not possible to reduce quantum mechanics to classical mechanics except 'locally', *i.e.*, with respect to certain sub-theories such as some of those involving angular momentum. Ehrenfest's theorem establishes a correspondence between the motion of the centre of a quantum mechanical distribution and the motion of the corresponding classical mass point. No correspondence with respect to higher momenta of the distribution has been found.[47]

This additional 'law', 'animals on that island are grey-skinned' (a fact not necessarily noted by the $S$-islanders), may in turn be explained by a further theory (*e.g.*, climatic conditions). Actual examples of such *ad hoc* bridging theories being required include the modern account of the periodic system in terms of elements consisting of 'contingent' statistical proportions of isotopes, and the present attitude to the fact of prevalence of protons (rather than antiprotons) in our universe. In both cases, some cosmological theory would have to be invoked in turn to explain the low level bridging laws.

[47] See G. Fay, 'On the Correspondence Principle', *Acta Physica Hungaricae*, **29** (1970), 2–3.

One way out of the difficulty might be to postulate *ad hoc* particular initial conditions, somewhat analogously to Boltzmann's way of accounting for the irreversibility of thermodynamics. In this case of quantum mechanics such initial conditions are, in fact, postulated ('collapse of wave function') as the result of measurement–observation. But no quantitative derivation of the

This failure to establish a general correspondence between quantum mechanics and classical physics (in the limit where $\Delta P \Delta Q \gg h$) should in our view be regarded as a shortcoming of quantum mechanics (at least, of the theory of 'first quantization') in its claim to the status of *L*-theory, rather than as a breakdown of the General Correspondence Principle. This failure of quantum mechanics,[48] together with its incompleteness,[49] indicates clearly the need for a new, better *L*-theory. Quantum mechanics has a flaw of incompleteness not found in classical mechanics. It is precisely because quantum mechanics is not 'incommensurable' with classical physics that we are able to point out these flaws in quantum mechanics.

The general idea of correspondence is, of course, older, at least as old as the relativity revolution. Einstein was faced with the following problem: in *some* approximations, both classical electromagnetism and classical mechanics must be valid. He rightly refused to accept as a coincidence, or rather a long series of coincidences, that these theories worked. On the other hand, there was reason to believe that these theories were not both true (see p. 223 above). He was thus faced with the problem of constructing a new, better theory, that, amongst other things, would explain why these older theories worked, and explain it in the framework of a new, larger, better theory which degenerates respectively into the old theories in certain classes of cases.

Einstein's decision to let Lorentz override Galileo was not arbitrary, since the Lorentz transformation degenerates into the Galileo transformation (for $v/c \to 0$), but not *vice versa*. This is an example of 'inconsistent correspondence'. I shall come back to the general criterion of deciding which invariance must apply to the new, better, theory *L* in other contexts, but we already have one instance here in which there is no ambiguity: the *L*-theory must be Lorentz- rather than Galileo-invariant.

The task, then, is to reformulate mechanics, rather than Maxwell theory, and to do so in a Lorentz-invariant way. This task does not lead of itself to a unique answer. We might reformulate mechanics in such a way that mass is conserved. It was Tolman who derived the special

validity of, say, Kepler's laws within observational error has been achieved. It is possible that such quantitative agreement could be achieved by postulating *ad hoc* a suitable probability distribution consistent with (initial) observation which would not show appreciable dispersion within cosmic times. Even if this could be done, it would be an unsatisfactory explanation.

[48] See also E. L. Hill, in *Mind, Matter and Method*, ed. P. K. Feyerabend and G. Maxwell (Minneapolis, 1966).

[49] See H. R. Post, 'Incompleteness of Quantum Mechanics', in *Quantum Theory and Beyond*, ed. E. W. Bastin (Cambridge, 1971).

theory of relativity in its present form from the alternative requirement of conservation of momentum. Certain further invariance requirements thus had to be invoked to make the task unique.

The correspondence relationship between the new larger theory *L* and the old theory *S* can, and must, apply in one direction only: it is a requirement on the theory *L* that it degenerate into *S* in that range of cases in which *S* has been found to work. This relationship is frequently misused. It is, for example, just as ill-conceived to try to explain the lack of individuality of particles, or to construct an atomic theory on a dimensional basis, as it is to explain quantum mechanics on a classical basis, or diffraction on the basis of geometrical optics. Classical mechanics should be explained on a quantum mechanical basis; we explain individuality on the basis of non-individuality.[50] In explanation, we go from *L* to *S*.

In heuristics, going from *S* to *L*, the correspondence principle is at least a guiding constraint. In general, it is to be used *post hoc* (*i.e.*, after formulation of *L*) to eliminate certain candidates for *L*-theory. We shall consign to the wastepaper basket any *L*-theory, the brainchild of a careless night, when we realize in the morning that this candidate does not fulfil the general correspondence principle, does not explain why the previous *S*-theory worked. The principle is not, in general, a sufficient constraint to make the transition $S \rightarrow L$ unique.

(*b*) *Criticisms of the General Correspondence Principle.* We said that we would not go into the difficult question of the relation of empirical data to their formulation in a theory, or into the problem of the relation of such observation statements to the rest of the theory, but it is apparently necessary to defend the general correspondence principle against several charges in that field:

(i) *The charge of triviality.* This is refuted by emphasizing that the principle claims far more than mere 'agreement' on individual instances of factual data found to be successfully covered by the '*S*-theory'. The General Correspondence Principle claims that the *L*-theory inherits a coherent pattern of theorems, including some higher levels, from the *S*-theory.

(ii) *The charge of absurdity* in the spirit of Feyerabend, mentioned in the Introduction, is dealt with in Appendix II.

(iii) *The charge of falsity*, levelled, in effect, by Kuhn and others, who

[50] Thus, the flaw in classical statistical mechanics represented by Gibbs' paradox points to a radical theory of non-individuality such as Bose's (H. R. Post, 'Individuality and Physics', *The Listener* (10 October 1963), 534–7). See also G. Weinreich, *Nature*, **184** (1959), 1825–6.

insist not only on the occurrence of 'losses' in the $S \rightarrow L$ transition, but on the radical change in world picture in the case of revolution. This calls for case studies; but some general remarks may not come amiss.[51]

Any $L$-theory is necessarily stronger in *some* way than the part of the corresponding $S$-theory taken to be successful, otherwise it would not be rational to replace $S$ by $L$. Sometimes the strength of the $L$-theory in some other respects causes us to cast doubt on a part of $S$ hitherto taken to be successful.

$L$ may be weaker than $S$ in some respects: thus a law in $S$ may turn out to be (the consequence of) a boundary condition (relating to $S$) in $L$. We certainly must consider a theory using a boundary condition in the place of a law to be weaker, precisely because our present definition of boundary condition is a condition not laid down by the theory.[52] We do not consider the status as boundary condition unalterable: $L$-theories may replace a boundary condition in $S$ by a law.

We have these possibilities in decreasing order of strength of the relation $L \rightarrow S$:

(i) Reduction: no horizontal, no vertical slicing. $S$ is taken over into $L$ as a whole.[53]

(ii) Consistent correspondence with no explanatory losses: cutting off the top, retaining, however, several levels of $S$; no vertical slicing beyond $S^*$, the well-confirmed core.

(iii) Consistent correspondence with no explanatory losses, but requiring *ad hoc* bridging statements: these bridging statements are matter-of-

[51] These charges against the General Correspondence Principle are dealt with at length by N. Koertge, *loc. cit.*, note 12.

[52] 'The laws of nature are regularities, that is, conditions between events, whereas the initial conditions are the unforseeable events' (Hontappel, Van Dam and Wigner, 'The Conceptual Basis and Use of the Geometric Invariance Principle', *Rev. Mod. Phys.*, **37** (1965), 596).

Our remarks concerning the inferior status of boundary conditions apply also to so-called initial conditions. We are not concerned with the difference between these two classes of conditions which have to be added to the theory to yield definite predictions. Nor are we referring to such conditions as the vanishing of Schrödinger wave functions towards infinity, which are, in fact, laws, and only misnamed 'boundary conditions'.

[53] This may involve the aid of factual linkages (rather than coordinating definitions) which are added to $L$ to form that theory $L'$ which is to supersede $S$. An example is provided by the relationship: mechanics ($L$)—statistical mechanics ($L'$)—thermodynamics ($S$). *Cf.* E. Nagel, *The Structure of Science* (London, 1961), 354–5, on connectability and derivability. Nagel seems to go further and to claim that 'temperature' is not reducible to 'mean kinetic energy', just because the meaning of these two terms is not identical. While admitting the dubious character of the various 'disorder hypotheses' (see also below, p. 246), I claim in this particular case that an operationally defined 'temperature' can be shown to be identical with the mean kinetic energy derived from statistical mechanics with the aid of such disorder hypotheses.

fact statements about empirical 'boundary' conditions peculiar to $S$, whose accuracy can be checked.[54]

(iv) Approximate or inconsistent correspondence with no explanatory losses.

(v) Correspondence with Kuhn losses: vertical slicing beyond $S^*$.

(vi) No correspondence: no part of $S$ (other than the 'lowest' agreed factual level) inherited by $L$.

Possibilities (v) or (vi) would be counterexamples to our thesis.

Quite generally, the thesis may be put this way: no theory that ever 'worked' adequately turned out to be a blind alley.[55] Once a theory has proved itself useful in some respects, has shown its semantic simplicity or explanatory power, it will never be scrapped entirely. Even the phlogiston theory had features that were useful scientifically in its day, and those features translate smoothly into present theory.

From the point of view of present-day philosophy, the fact that there is continuous progress in science is a problem, while the fact that there are occasional revolutions is not. There is no logical reason why a scientific theory should not be replaced by a theory different from the old at every level; but this is not what has happened so far. The most radical revolutions have destroyed the top levels and give a somewhat different interpretation to the lower levels, but they have not destroyed the whole of the lower-level structure. The internal connections, the pattern, are preserved. The periodic system is the basis of inorganic chemistry. This pattern was not changed when the whole of chemistry was reduced to physics, nor do scientists ever expect to see an explanation in the realm of chemistry which destroys that pattern. The chemical atom is no longer strictly an atom, yet whatever revolutions may occur in fundamental physics, the ordering of chemical atoms will remain.

[54] See note 46, and *cf.* note 53.

[55] Interesting case studies are presented by the following apparent counterexamples: (i) Boerhave's theory (which surpassed the present theory in simplicity, but lacked the fertile concept of heat capacity leading to a non-trivial concept of temperature). (ii) Arrhenius' theory of weak electrolytes. It is not at all clear that the qualitatively successful predictions (variation of electrical conductivity with concentration) of this basically false theory (partial disassociation in the case of salts) can be accounted for by present computational methods (Debye–Hückel, *etc.*) available to deal with interaction between ions. (iii) First-order perturbation in the case of divergence in higher orders. In so far as first-order perturbation yields qualitatively correct results, the fact of divergence in higher orders must be regarded as a flaw, to put it mildly. This flaw would have to be removed in an $L$-theory which corresponds (probably inconsistently) to the first-order perturbation theory. (iv) Sommerfeld's theory of the hydrogen atom. In certain respects any one-dimensional ('orbit') model can only be a poor approach to the wave-mechanical treatment. But Sommerfeld's theory was successful in giving the same energy values as Dirac's theory did later, though it took no account of either wave-mechanics or spin, and is therefore now abandoned.

It is true that the interpretation of most of the theoretic terms changes as a consequence of a revolution. The constitutional character of many an office of state changes when a monarchy is overthrown and a republic replaces it. But the function of that office may well remain almost the same as it was under the monarchy. Again, in government the nominally higher functionaries may play a lesser, mainly representative role. There is a certain loss of content as we climb higher and higher in a theory to increasingly abstract, more general laws. The higher levels usually are uninterpreted to an increasing extent. Many of the most general laws at present are of a rather formal character.[56] Just as Newton's laws of mechanics gain content by inserting into them a particular force-law, the law of gravitation, so many present-day super-laws (to borrow Wigner's phrase) such as the uncertainty principle of quantum mechanics, really only acquire empirical content after particular laws of interaction, or of fields, have been injected. The emperor, the super-law, may well be overthrown, but then the emperor not only lacks clothes (that is, the statements of initial conditions without which present-day laws cannot be applied to actual situations), but is only a skeleton anyway (largely of uninterpreted terms). In countries with unstable governments there may yet be a stable Civil Service carrying on the machinery of government at lower levels. Similarly, I claim that even the most radical revolutions in science have saved not only the phenomena, the population of facts, but a good deal of the lower parts of the old theory as well.

My claim is not merely that there is an element of continuity in change, which is necessarily the case in all human activity, but that in science part of the content of the old theory is preserved, as far as one can tell, for ever; not just in the next theory, but throughout all future theories. Astronomy, for example has retained a core of kinematics observed in the heavens while attributing this to gods, holes in crystal spheres, or gravitating masses. Again, chemistry has retained the central fact (and problem) that materials acquire entirely new properties when mixed together in certain circumstances, and that the original properties can be recovered, whether this be explained in terms of principles, elements, 'atoms', or 'elementary particles'. Successive good theories do not merely exhibit family resemblances: $S^*$, though never defined exactly, does not shift. It is

[56] *Cf.* Weyl, *Raum, Zeit, Materie* (Berlin, 1919), 263. 'Maxwell's theory and analytic geometry are practically interchangeable in their mathematical constitution.... Physics has no more significance for reality than formal logic for the realm of truth.' Maxwell's equations can be derived from the assumptions (i) that the electrostatic field is a vortex-free vector field; (ii) that its sources are conserved; (iii) Lorentz invariance.

pinned down by facts. The true core may be found to be smaller than $S$ at any one time. On the other hand, the advent of $L$, while possibly shrinking $S$, will, in general, add to the total core of truth. As Nernst says:

This conception is not without importance for the question of the certainty of scientific progress altogether. It is common to speak of the inadequacy of human inquiry, simply because many a long-recognized law has had to undergo revision to meet the requirements of the progress of knowledge. If we consider the matter more closely, it is obvious that the law in question has retained its validity over a wide range, but that the limits of its applicability have been more sharply defined. It can even be said that since the development of the exact natural sciences, there is scarcely one law established by an investigator of the highest rank which has not preserved for all time a wide range of applicability, *i.e.*, which has not remained a serviceable law of nature within certain limits. We cannot say, for example, that the electromagnetic theory of light has completely overthrown the older optical theory put forward by Fresnel and others. On the contrary, now as formerly, an enormous range of phenomena can be adequately dealt with by the older theory. It is only in special cases that the latter fails; and further, there are many relations between optical and electrical phenomena which certainly exist, but of which the older theory takes no account. Hence the electromagnetic theory implies a great advance, but by no means nullifies the successes of the older theory.

So scientific theories, far from dropping off like withered leaves in the course of time, appear to be endowed under certain restrictions with eternal life; every famous theoretical discovery of the day will doubtless undergo certain restrictions on future development, and yet remain for all time the essence of a certain sum of truths.[57]

Science may well be converging ultimately to a unique truth.[58] In any case, the progress of science appears to be linear. There is at least one profound difference between biological evolution and the evolution of

[57] W. Nernst, *Theoretical Chemistry* (London, 1911), 4–5.

[58] As Einstein says: 'It is similar to a man engaged in solving a well-designed word puzzle. He may, it's true, propose any word as the solution; but there is only *one* word which really solves the puzzle in all its forms. It is an outcome of faith that nature—as she is perceptible to our five senses—takes the character of such a well formulated puzzle. The successes reaped up to now by science do, it is true, give a certain encouragement for this faith' (Einstein, *Out of My Later Years* (Westport, 1950), 64). Or again: 'The historical development has shown that among the imaginable theoretical constructions there is invariably one that proves to be unquestionably superior to all others. Nobody who really goes into the matter will deny that the world of perceptions determines the theoretical system in a virtually unambiguous manner, although no logical way leads to the principles of the theory.' (Address for Planck's sixtieth birthday, quoted by Weyl, *Philosophy of Mathematics and Natural Science* (New York, 1963), 153.) Or Jean Rostand on the particular example of Mendel: 'Mendelism has been established once and for all in biological thinking and nothing will oust it, for it expresses a considerable part of the real basis of life

scientific theory: a biological line may develop along any of a vast number of *divergent* paths. The path taken depends in part on external influences, and development may, indeed, be reversed under certain changes of external conditions. Contrary to Kuhn, I believe that scientific theory *converges* towards a unique truth; this process is only helped by external influences, however one-sided, such as scientific fashions, and prejudices of individual scientists.

(5) *Simplicity*. Existing theories must meet three criteria of acceptability: (i) internal logical consistency; (ii) agreement with experience; (iii) simplicity.[59] In some sense these criteria are met in the heuristic situation respectively by: (i) the removal of flaws; (ii) correspondence; (iii) the all-pervasive requirement of simplicity that gives bite to the other criteria: it is only in conjunction with a requirement of simplicity that Einstein's requirement of covariance amounts to a restriction on possible theories.[60]

The criterion of simplicity is of more use as a *post hoc* eliminator of potential *L*-theories than for *a priori* construction. This is due to the fact that we would need a basic language[61] in order to arrive at an absolute, rather than merely a relative, measure of simplicity, and that such a language is rarely given *a priori*, except possibly in the natural numbers.[62] Such rules as 'simple questions have simple answers' are, like the method of dimensions, only applicable within a given theoretic framework. However, even a relative measure of simplicity is sufficient for use within a given language, *e.g.*, to choose a 'dominant theory' in the sense of Tisza,[63] or to prefer a single straight-line interpolation to a complicated curve. The criterion of simplicity is probably the most powerful means of eliminating crank-theories, for instance totally *ad hoc* numerological 'theories' of elementary particles.[64]

itself.' While I think that this statement is too strong in claiming *finality* achieved in a finite number of steps, I do agree that a well-confirmed theory will never be scrapped entirely.

59 See Popper, *Logic of Scientific Discovery* (London, 1965), chapter 7.

60 *Op. cit.*, note 25, 69.

61 See note 39 above.

62 See G. N. Lewis, 'Ultimate Rational Units', *Philosophical Magazine*, **49** (1925), 750.

63 See L. Tisza, 'The Conceptual Structure of Physics', *Reviews of Modern Physics*, **35** (1963).

64 In the absence of better rival theories, theories claiming absolute numerological simplicity cannot be rejected on grounds of relative simplicity. They can be rejected by showing that there is a high probability that randomly distributed data would fit formulae of the same *form* and simplicity (within the limits of experimental error, where applicable), in other words, by showing that agreement of the data with the theory does not pass the significance test as against a null hypothesis of randomness. Such tests might be used to reject such theories as Stoney's or E. E.

The criterion is certainly fallible (though circles are simpler than ellipses, planets do not move in circles).

(6) *Adding to the Interpretation.* We take a hitherto incompletely interpreted part of the abstract formalism of the theory, and give it a tentative interpretation of our own, at some level. This further interpretation may serve predictively or as an explanation of already observed effects. Examples include (i) Born's interpretation of the Schrödinger $\psi$-function, allowing an account of the phenomenon of scattering; (ii) the interpretation of Dirac's four-component equation in terms of spin and the positron; and (iii) interpreting singularities in the gravitational field to be massive particles, an interpretation confirmed by the fact that these singularities follow the appropriate equations of motion. Another example is given in note 23.

(7) A closely related guide line is the procedure of *taking models 'seriously'*. In this case we adopt a model already available which may initially have been offered as an arbitrary articulation of the formalism only. This criterion is highly fallible. It is a case of borrowing a model of the *S*-theory which contained features not *essential* for the modelling of the *S*-theory ('neutral analogy'[65]), and assigning physical significance to such extra features (which may be regarded as features imported as the result of tentative union of *S* with another theory). This is the opposite of 'stripping': we should only strip after persistent failure to confirm. It is frequently the case that the old theory contains the seed of the new theory, not as a flaw, but by way of a suggested model. We should distinguish between the strict logical consequences of a theory, its 'dividends', and such further suggestions, which might be called a 'bonus'.[66] To give an example:

Witmer's (*Physical Review*, **71** (1947), 125). They have been used in support of Prout's (C. Marignac, 1860, reprinted in *Alembic Club Reprints*, **20** (Edinburgh, 1932), 56 f.) and Hudson's hypotheses. 'It has been seen that if the ring forms on the $\gamma$-carbon there are twenty-four agreements, no disagreements and none in doubt. As the chances that an event which can happen in two equally probable ways will happen in exactly the same way twenty-four times out of twenty-four trials is only one in seventeen million it seems certain that the lactonic ring in these monobasic sugar lactones forms on the $\gamma$-carbon atom' (C. S. Hudson, 'A Relation between the Chemical Constitution and the Optical Rotatory Power of the Sugar Lactones', *Journal of the American Chemical Society*, **32** (1910), 346).

We see here that we can apply such significance tests even to theories covering an essentially finite set of data only, and without reference to experimental error.

65 M. Hesse, *Models and Analogies in Science* (Indiana, 1966), 10. The policy of taking models seriously should also be implemented in replacing vague metaphors by definite models. What is the point of introducing terms such as 'pressure' into psychology? We should make clear just what we wish to commit ourselves to when using this term,

66 See, for instance, H. R. Post on 'Pragmatic Simplicity' in 'Simplicity in Scientific Theories', *loc. cit.*, note 39, 32.

chemical formulae were only gradually accepted as actually denoting the configuration of atoms in space (as in the work of van 't Hoff). Similarly, crystallographers had been using regular arrays of units as models of crystals for a century before physicists took this model sufficiently seriously to identify the units with physical atoms (Laue). Taking seriously a model suggested by a theory is often pragmatically successful. This is only one form of the general phenomenon that scientists 'builded better than they knew'. Planck introduced light quanta *ad hoc* to account for the spectral distribution of radiation; later Einstein suggested that these photons also showed up in the process of emission and absorption, predicting the photochemical and photoelectric effects. Emerson's poem applies to scientific theories as well as to abbeys. The criterion applies to the invention of what Einstein called 'constructive theories'.[67] The programme of atomism is an apotheosis of a constructive theory.

We do not necessarily have models with additional features transcending the *S*-theory available. Nor are such models, if available, by any means infallible (mechanical models of the ether, for example, were not).

(8) *Enlarging the domain of the theory*. This bears the same relation to the procedure of building the closure of a theory, referred to under criterion 1, as criterion 7 to criterion 6. It may take many different forms.

Examples include the method of analytic continuation, extrapolation, and such examples as Rydberg's generalization of Balmer's formula. The first step in astronomy was to account for the appearance and disappearance of heavenly bodies in terms of continuous motion in space and time. Embedding discontinuous manifestations in an otherwise continuous *S*-theory into a wholly continuous *L*-theory, a procedure far transcending straightforward interpolation or filling in, may, in many cases, be considered an advance.[68] Certainly we find a theory attributing a discontinuous change in one parameter to a continuous change in another parameter unsatisfactory. *Natura non facit saltus*. Parity is an invariant associated with a discontinuous transformation in a continuous space. Such discontinuous

[67] A. Einstein, *Out of My Later Years*, *op. cit.*, note 58, 54.

[68] 'Certainly, in our own intellectual behaviour we rarely solve a tricky problem by a steady climb toward success. I doubt that in any one simple mechanism, *e.g.*, hill-climbing, will we find the means to build an efficient and general problem-solving machine. Probably, an intelligent machine will require a variety of different mechanisms. These will be arranged in hierarchies, and in even more complex, perhaps recursive, structures. And perhaps what amounts to straightforward hill-climbing on one level may sometimes appear (on a lower level) as the sudden jumps of "insight".' M. Minsky, 'Steps Toward Artificial Intelligence', *Proc. I.R.E.*, **49** (1961), 10.

transformations may correspond to continuous transformations in a still richer *L*-theory.

It might be satisfactory to explain away quantum jumps in a continuous theory. Even a strictly discrete *S*-theory may be transformed fruitfully into an *L*-theory embedded in a continuous space: a radical *explanation* of the existence of elementary particles would presumably be based on a (non-linear) continuous theory.

## III

## Relations between Criteria

We have indicated some overlap of heuristic principles above; some flaws are clear footprints of the *L*-theory, some are not (*e.g.*, mere failure to unify physics). Some heuristic criteria may conflict: we may be faced with the choice of either maximizing simplicity amongst available theories (resulting, for example, in an ultimate discrete elementary particle theory) or enlarging the domain of the theory (by explaining the discrete values in an embedding continuum theory).

There is one particularly interesting relationship: that between correspondence and invariance.

In all cases of 'consistent correspondence', the *S*-theory must be at least as symmetric as the *L*-theory.[69] Even if we have a *Q*-condition limiting the range over which correspondence occurs, that is (according to the General Correspondence Principle), shrinking *S* to the core $S^*$, $S^*$ must be invariant under at least all the transformations under which *L* is invariant. Thus, if *S* is classical mechanics, which is symmetric under time reflection $t \rightarrow -t$, $S^*$ may be a degenerate form of a strictly irreversible *L*-theory, whose footprints show up (as flaws in *S*) in the many irreversible processes in nature. Very commonly, certain degeneracies in the *S*-theory are removed by the more precise *L*-theory.

There is, however, the interesting case of 'inconsistent correspondence' ('controlled inconsistency' in Tisza's terminology) where $S^*$ agrees only approximately with *L* (for example, *L* asymptotically converges with $S^*$ for some parameter $x/m \rightarrow 0$), and *S* asserts (falsely) a proposition $f(x)$ which only agrees with *L* for $f(0)$. There is then an $S^*$, defined by the *Q*-condition $x/m < \alpha$, in which $f(x)$ agrees with the prediction of *L* in that range, within

[69] *Cf.* Curie, 'Symmétrie dans les phénomènes physique', *Journal de Physique*, 3 (1894), 393–415.

experimental error. In such cases of inconsistent correspondence, $S^*$ may be less symmetric than $L$, may have fewer invariance restrictions. However, $S^*$ may contain elements of $L$ that have the full-blown symmetry of $L$ ('footprints' of $L$). Indeed, the invariance requirement explicitly characterizing $L$, but not $S$, may appear implicitly in $S$, or at least in $S^*$ (examples of this are the equality of the effect of moving a magnet towards a conductor, and the effect of moving a conductor towards a magnet even according to Maxwell's theory; these effects are treated asymmetrically in Maxwell's theory, and only receive a symmetric treatment in the later Einstein theory. A further example is the relation $E = h\nu$ which is natural in four-dimensional special relativistic representation, but not in classical mechanics). Again, Fermi statistics corresponds (in the 'non-degenerate' case of dilute filling of levels, or in the case of very large numbers of possible values of distinguishing parameters[70]) to the non-symmetric Boltzmann $S$-theory. However, in a metal (a 'low-temperature' degenerate case) or in an atom (in the ground state) the Pauli exclusion principle projects into the $S$-theory. The Pauli symmetry restriction appears, as we have seen, as a flaw in pre-wave-mechanical quantum theory.

Thus it is a fallacy to assume, as Hamilton does implicitly (Appendix I), that in a theory $S^*$ limited by the $Q$-condition 'gravitational effects quantitatively negligible' we may ignore symmetry restrictions which are peculiar to the $L$-theory (in this case general relativity). As discussed in Appendix I, there are many examples in physics of effects that are qualitatively essential in a theory, and yet may be neglected quantitatively in certain calculations. In cases where this approximation brings about inconsistent correspondence, the effect may appear as a footprint.

Thus we may find 'symmetry' footprints of the $L$-theory in the $S$-theory, either in the form of an inexplicable element of asymmetry in an otherwise symmetric $S$-theory, when $L$ is less symmetric than $S$, or in the form of an inexplicable symmetry restriction, when $L$ is more symmetric than $S$. The latter can only occur in the case of inconsistent correspondence.

## IV

### Some Present Problems

Doubt is sometimes cast on criteria which have been claimed here to be

[70] See second reference in note 50.

guidelines in the construction of theories, on the ground that most examples of application are taken from the past. This might suggest that these criteria can only be pointed out with the wisdom of hindsight. We can point to historical evidence that such criteria as 'flaws' (p. 223) and 'correspondence' (note 78 below) were used consciously. Inherited invariance criteria are being used every day in the construction of further elementary particle theories.[71] However, the general thesis might be considered more persuasive if its applicability to present *S*-theories were indicated. Some indications of this have already been given (note 29, and the beginning of the section on 'footprints').

Looking at over-all developments in science, we find that the concept of evolution plays a curious role in converting biology into a historical science. Any reductionist programme for biology (*S*) into present physics (*L*) will require *ad hoc* boundary conditions defining *S* in *L*, and probably factual linkages as well, enriching *L* to *L'*. The reduction of genetics to chemistry (and hence to physics) in the field of molecular biology seems to be proceeding smoothly (except, perhaps, for quantitative short-fall in error in genetic transmission[72]).

[71] The use of mathematical desiderata by themselves in the construction of new theories is illustrated by such publications as: D. Atkinson, L. A. P. Balazs, F. Calogero, P. di Vecchia, A. Grillo and M. Lusignoli, 'Toward A Unitarization of the Veneziano Model', *Physics Letters*, **29B** (1969), 423; A. Salam, R. Delbourgo and J. Strathdee, 'The Covariant Theory of Strong Interaction Symmetries', *Proc. Roy. Soc. London*, Series A, **284** (1965), 146; M. L. Thiebaux, 'Crossing-Symmetric Amplitudes with Arbitrary Regge Behaviour Based on Elementary Symmetric Functions', *Physical Review*, **188** (1969), 2283. The implication seems to be that symmetry, unitarity and analyticity between them completely define the physical situation. For a still more optimistic programme anticipating uniqueness on the basis of analyticity and unitarity, see particularly G. F. Chew, *The Analytic S Matrix* (New York, 1966): 'Historically, the recommendation of analyticity in the theory of sub-atomic particles developed as much from studies of field theory as from experimental observations.' See also W. Heisenberg, *Introduction to the Unified Field Theory* (London, 1966), 20–1.

I am not referring here to extreme a-prioristic theories, such as Eddington's or Atkin's (*Quantum Theory and Beyond, op. cit.* note 49, 191), which start from scratch with epistemological principles and attempt to construct a physical theory guided solely by mathematical considerations, though I see no binding philosophical reason why such an attempt could not one day prove successful.

Physicists live from hand to mouth as far as mathematics is concerned. The timely discovery of Hilbert space, before the advent of quantum mechanics, is sometimes cited as a minor miracle performed by the Zeitgeist, or whatever. The problem may be turned round: perhaps Hilbert space was assigned its role in quantum mechanics partly because it was there, conveniently constructed in a university with exceptionally good liaison between mathematicians and physicists. (See also C. Reid, *Hilbert* (London, 1966), 183–280.)

'One of the most remarkable discoveries in elementary particle physics has been that of the existence of the complex plane' (Eden *et al.*, *The Analytic S-Matrix* (Cambridge, 1966), vii).

[72] H. H. Pattee, 'The Recognition of Hereditary Order in Primitive Chemical Systems' in *The Origins of Prebiological Systems and of their Molecular Matrices*, ed. S. W. Fox (New York, 1965), 385–405. There does not seem to be sufficient evidence for the absence of 'error' at the detailed biochemical level: error at the biochemical level need not lead to 'error' at the macroscopic level,

Within physics itself, we again have a profound flaw splitting physics as a whole into general relativity theory and quantum mechanics.[73] It seems clear that no unification can be achieved without abandoning some of the highest-level concepts of one theory or the other. As has been pointed out, quantum mechanics itself is seriously flawed by incompleteness resulting from its probabilistic basis. Classical physics has not been reduced to quantum mechanics. Presumably the first step is to mend that flaw so as to allow the precise prediction of single events.

Nor is the join of thermodynamics to mechanics (in statistical mechanics) satisfactory even yet. Irreversible thermodynamics has been developed little beyond a perturbation extension of thermostatics (of the four laws), but the irreversibility paradox in statistical mechanics has been with us ever since the days of Boltzmann and his H-theorem. We may take a weak or a strong line. The weak line is the present orthodox one of regarding thermodynamics as being in inconsistent correspondence with mechanics. Thermostatics is considered to be refuted by the phenomenon of fluctuations. To obtain either thermostatics or irreversible thermodynamics, *Q* (boundary or initial) conditions[74] are required in mechanics, and even within the domains satisfying the *Q*-conditions the agreement of mechanics with thermodynamics (using suitable translations and factual linkages) is only approximate (statistical), there being exceptions of measure zero. I call this line weak because it establishes a relatively weak (inconsistent) correspondence between mechanics and thermodynamics, and because it replaces a law in thermodynamics (*S*) by initial conditions

there is latitude in coding; moreover, there may be interesting biochemical repairing mechanisms at work.

Historically, there are not many theories in pure biology that are sufficiently structured to make structural analysis, let alone a comparative study of structures of pairs of theories, interesting. Possible candidates might be found in embryology, genetics and neurophysiology. The absence of highly structured theories in biology is itself worthy of philosophic consideration.

'Diese Tatsache, dass die Interessen der Biologie sich nicht immer mit denjenigen der mathematisch-physikalischen Wissenschaften decken, hat man bisher unbeachtet gelassen; auch von den Biologen wurde ein GALILEI, ein DESCARTES als Begründer der neuen Auffassung des Lebens gepriesen, obwohl an dieselben keine beachtenswertere biologische Idee anzuknüpfen ist' (E. Rádl, *Geschichte der biologischen Theorien in der Neuzeit* (Leipzig, 1913; reprinted Hildesheim, 1970), vol. 1, viii).

'Und die Biologen, die ja in der Regel erst dann etwas akzeptieren, wenn die Physiker es ihnen zurechtgedacht haben, werden diese neue Philosophie erst wieder aus der Physik zu sich herüberholen' (A. Meyer-Abich, in his Foreword to *Geschichte der biologischen Theorien in der Neuzeit*, 1970).

[73] See, for example, M. Sachs, 'Positivism, Realism, and Existentialism in Mach's Influence on Contemporary Physics' in *Philosophy and Phenomenological Research*, **30** (1970). A further complication is that 'general relativity' by itself as developed so far can hardly be considered to be more than a frame.

[74] See, *e.g.*, A. Grünbaum, *Philosophical Problems of Space and Time* (New York, 1964).

in mechanics ($L$), in other words, brushes the difficulty under the carpet of 'initial conditions'.[75] However, the fact is that Newton's mechanics requires more initial conditions than thermodynamics, and *this* fact is not introduced *ad hoc* to resolve the paradox. The strong line is to look for a consistent correspondence between thermodynamics ($S$) and a new super-theory ($L$). In that case $L$ has to be irreversible in time, just as $S$ is. $L$ should, however, degenerate into reversible mechanics under suitable $Q$-conditions.

The greatest problem or flaw in present physics is the absence of a theory of elementary particles, as distinct from the mere group-theoretical listing of their degeneracies.

That parity is conserved in some parts of physics and not in others points to overall non-conservation, degenerating to conservation in some limiting cases. This seems to have been confirmed recently when slight deviations from symmetry were found even in heavy-particle interaction. Perhaps parity may be conserved in the limit $v/c \to 0$ or mass$\to\infty$. In general, the new two-component theory with 100 per cent correlation between direction of spin-axis and velocity can only apply for $v/c = 1$, and hence presumably only for zero rest mass.

The explanation of the connection between spin and statistics at the moment is cumbersome. A richer model in a continuous space might explain the connection in the same way in which Schrödinger's angular momentum representation in real three-dimensional space explains the connection between orbital angular momentum and parity of molecules. Such elucidation of the correlation between spin and statistics might be compared with the elucidation of the identity of gravitational and inertial mass upon adoption of the larger continuous $L$-theory of general relativity.

An elementary particle theory, to be integrated into physics, must be more than a fine-structure classification matching some pages of a book on group theory. Broadly speaking, we have the choice between cosmological theories (embodying existential statements relating to elementary particles) and non-linear continuum theories.[76]

An $S$-theory employing a continuous space may in turn be reduced to an 'atomic' $L$-theory employing a discrete space. This would be indicated by

[75] See section II. 4, p. 236.

[76] This may be based on an interaction whose energy graph shows a minimum when plotted against the product of charge and particle number.

Broadly speaking, we have a perennial conflict between trying to preserve continuity on the one hand and the atomistic programme (preserving linearity) on the other hand.

the phenomenon of dispersion in the *S*-theory. The dispersion of de Broglie waves for free particles may be a footprint of this kind.

## V

### Conclusion

'All origins are obscure' (H. Weyl). There is necessarily a remaining area of obscurity in any account of intellectual advance, however far back we push our account. Nevertheless, there are heuristic criteria to be applied prior and posterior to the act of invention.

We start within some context of theories. These theories may be construed to contain an ontology, or they may not.[77] Theories suggest a new theory, perhaps by the removal of flaws, or by extending interpretation. This new theory may in turn be refuted, not necessarily by experiment, but quite commonly by other theories, either already existing, such as invariance principles, or by new, simpler ones.

Any 'flaw' (*e.g.*, of incompleteness) in the previous theory, and the 'conservation' criterion, are prior, while the *Q*-condition in correspondence may be posterior to the creation of the new theory. Application of the criterion of simplicity may be either constructive (prior to the new theory) or selectively destructive (choosing *a posteriori* between possible *L* theories). The correspondence principle may be constructive.[78]

The above list (section III) of heuristic criteria 1–8 is not exhaustive. In any case, we have confined ourselves to 'inductive' devices starting from the *S*-theory. We do not take account of 'analogical' devices (*e.g.*,

[77] Concerning atomism, this is the paradigm case for philosophers of science going off the deep end about ontology: modern physics has, in my opinion, abandoned any naive notion of existence of elementary particles. We are dealing not with things, but with events.

[78] Examples include the case of Bohr's determination of the allowed transitions by the prior criterion of correspondence to classical physics for large quantum numbers; and Planck's use of correspondence with Wien's law for short wave length. While Planck *might* have used correspondence with Rayleigh's formula for long wave length, he in fact seems to have been guided in *that* direction by little more than the empirical results of Pringsheim and of Rubens indicating the breakdown of Wien's law.

It might be fairer to credit Rayleigh (*Philosophical Magazine*, **49** (1900), 539) with appreciation of correspondence in that part of the spectrum. Against the available experimental evidence he derived his expression on the basis of Boltzmann's statistical mechanics, and suggested that, though it failed in general, equipartition might 'apply to the graver modes'. He was sufficiently convinced of the correctness of some aspects of orthodox thermodynamics that he considered it an argument against Wien's law that, according to it, the intensity of radiation of given wave length would reach a limit as the temperature is raised.

comparing light with water waves) starting from a theory covering a set of data not overlapping with the data covered by the theory under construction,[79] except in so far as such analogical devices extend beyond the mathematical formalism and are used to unify the two theories physically (as, for example, conservation of energy in a new field, or identification of electromagnetic waves with light).

Laws taken over from neighbouring theories, and general invariance requirements, provide heuristic constraints on the construction of any new *L*-theory. The Correspondence Principle constitutes a particularly strong constraint. More detailed hints as to the direction of construction of the new *L*-theory are provided by points within the old theory. These neuralgic or growing points are the traces of the *L*-theory projected on to the *S*-theory. They may be consequences of wider invariance requirements only expressible in the *L*-theory. It is precisely the interaction between correspondence and invariance requirements that provides the interesting problems and restrictions concerning the future *L*-theory.

Thus we arrive at the following recipe for constructing new, better theories:

(i) Look for a flaw or footprint in the existing theory. Treat any flaw as a footprint, and build your *L*-theory up from it.

(ii) Try to retain all present confirmed universal invariance principles.

(iii) Give a tentative interpretation to hitherto uninterpreted parts of the formalism of the theory, or try to take one of the available models of present theory seriously (this, of course, also implies (ii)).

(iv) Of all the candidates satisfying the first two requirements, reject any not satisfying the correspondence principle.

(v) Choose the simplest of the surviving candidates.

These 'rules of procedure' are fallible, but almost universally successful. The policy of 'taking models seriously' is an exception. It is worth trying

[79] 'In many cases, the relations of physical phenomena in two different physical questions have a certain similarity, which enables us, when we have solved one of these questions, to make use of our solution in answering the other. The similarity which constitutes the analogy is not between the phenomena themselves, but between the relations of these phenomena' (J. C. Maxwell, *Elementary Treatises on Electricity* (1881)). See also J. C. Maxwell, 'On the Mathematical Classification of Physical Quantities', *The Scientific Papers of James Clerk Maxwell*, ed. W. D. Niven (Dover, 1965), vol. 2, 257.

to apply it to the neutral analogies[80] of any model available. There may not be any model already postulated, in which case we may create one, by joining another theory to the *S*-theory in question (such as mechanics to electricity). But we are embarrassed by an infinity of candidates for the role of 'serious' model, most of them presumably misleading. The situation is quite different in the case of extending interpretation to a hitherto uninterpreted part of the theory. Here there may be only one possible interpretation within the existing framework.[81] This illustrates the strength of heuristic guidelines provided by mathematics (whether by way of a flaw, by the importation of overall invariance requirements, or by way of a so far uninterpreted part of the theory), as distinct from the weaker guideline provided by explanatory models decried by Duhem. Theoreticians are carried along by their momentum within mathematics in a definite direction. In the heuristic situation, scientists do not just try to save the phenomena—they also try to save the laws, they accept refutation by the calculus adopted. It is this balance of theory and experiment that results in definite guidelines.

The significance of these rules appears trivial to scientists, but they do not fit trivially into many present philosophies of science. It is hoped that this paper will stimulate philosophers of science to study the heuristics of theory construction in particular cases.

Accidental difficulties (totalitarian governments, stupidity of individuals, economic pressure of other duties) apart, there is nothing in the world to prevent people from putting forward, and trying out, new theories, however wild. This paper is not intended to discourage this; its purpose is to show (i) that there are possible procedures for going from a weaker to a stronger theory—these procedures are not mandatory; (ii) that these procedures are rational; (iii) that they describe much actual scientific work, including the most revolutionary achievements; (iv) that they go some way to account for progress in science in the last three hundred years.

Conservative induction is one way of attempting to meet the elusive requirement of agreement with facts; at least as far as the old theory successfully covered facts, and generally further (pragmatic simplicity). It is the most economical way to progress on stepping stones of the dead theories. Psychologically, in science, as in art, work within imposed restric-

[80] See note 65.

[81] Thus, the frequency of occurrence of dislocations (whose existence had not previously been suspected) could be computed on thermodynamic grounds for any particular crystal structure.

tions is more fertile than the mere assertion of the admitted right to be different.

The overall moral is that scientific progress is linear and that heuristics is possible.[82]

*Chelsea College, London*

## APPENDIX I

This investigation was triggered off by a remark (see * below) in a textbook on Elementary Particle Theory[83] which carries, I believe, a mistaken implication and raises the interesting problem of the relationship between one formal heuristic principle, the correspondence principle, and a class of others, the invariance principles.

*Quotation from* The Theory of Elementary Particles *by J. Hamilton*

'In discussing the equations of the fields associated with elementary particles we require consistency with special relativity* because the particles may move with any velocity up to the velocity of light. This requirement turns out to be useful in classifying the types of free field which are possible, and it severely restricts the possible equations of motion.** The relativity requirement is also of value in limiting the types of interactions which are possible between elementary particles. In practice the requirement is satisfied if the field equations are derived from a Lagrangian function which is a relativistic invariant.

'Another important invariance property—rotation invariance—comes from the requirement that the field equations should be independent of the orientation of the (spatial) coordinate system.*** This leads to the angular momentum analysis of the fields which is an important tool for studying the actual interactions between particles.

'* General relativity effects are ignored throughout. At the shortest distances investigated at present ($\sim 10^{-14}$ cm) the gravitational potential of a particle is very weak compared with other interactions.

'** Field equations containing partial derivatives of higher order than the second, or non-linear free-field equation, have not been intensively studied.

'*** The equations of motion of a closed interacting system of particles should be independent of the position of the origin of coordinates. As is well known, this implies that the total momentum of the system is constant.'

Hamilton may or may not be right in ignoring general relativity restrictions for a domain of physics where gravitational effects are indeed quantitatively negligible. However, it is a fallacy to *assume* that because quantitative effects

[82] This paper is based on a talk with the same title delivered at a meeting of the British Society for the Philosophy of Science in April 1965. I wish to thank Dr A. F. Chalmers, Dr Mary Hesse, Professor N. Koertge, Dr M. Machover and Mr P. M. Williams for helpful discussions on various points, though they are not to be blamed for any statements made in this paper.

[83] J. Hamilton, *The Theory of Elementary Particles* (Oxford, 1959).

are negligible, no qualitative footprint of a restriction may be projected from $L$ (in this case, general relativity).

Examples of important qualitative effects accompanied by negligible quantitative effects include the classic example of 'ideal' gas molecules whose interactions are quantitatively negligible, but qualitatively essential in order to establish equilibrium at all. There is also Bondi's splendid example of calculating the differential ageing of the 'twins' by means of special relativity.[84] Here general relativity projects into the degenerate theory of special relativity to account qualitatively for the non-equivalence of the twins' orbits, though it may be ignored in the quantitative calculation of the difference.

Again, Dirac's $L$-theory differs appreciably from the Schrödinger $S$-theory in its quantitative predictions only in the region $v/c \ll 1$, but it has profound consequences, such as the existence of the positron, even in the non-relativistic region. The existence of spin has an important 'pilot' effect in atomic physics even in the absence of spin forces. Identity of charge of electron and $\beta$-particle was established by Goldhaber to far greater than quantitative experimental accuracy by demonstrating qualitatively that an invariance principle (Pauli's exclusion principle) applied between them.

It is for this reason that I am dissatisfied with Hamilton's argument quoted above: the gravitational effect may well be quantitatively negligible, and yet correspond to definite manifestations in the $S$-theory.

## APPENDIX II

Feyerabend's argument[85] is based on no more than the correct realization that observation statements, as indeed all statements, involve *some* theories. It does not follow that two theories differing profoundly are 'incommensurable', that they cannot have observation statements straightforwardly translatable into each other or even in common.

The example given by Feyerabend[86] goes no further than making the point that the 'interpretation' of a measurement is indeed theory-dependent (incidentally, phlogiston theories were not much concerned with weight; reinterpretation, however, is straightforward: loss of phlogiston is gain in oxygen). This does not provide an example of incommensurable descriptions of a measurement itself.

Of course, if we *define* the 'meaning' of a term by the set of all statements true within a theory involving that term, the meaning of that term will necessarily differ in any two theories differing at all in statements involving that term (*e.g.*, in their predictions).

The observation statement 'this pointer coincides with this mark' *is* intended to be invariant under theory-change. Of course, the statement may be in error (the pointer only seems to coincide with the mark while there is in fact con-

[84] For a discussion of this clock paradox and its geometrical analogue, see H. Bondi, *Assumption and Myth in Physical Theory* (Cambridge, 1967), 51 f.

[85] See 'Explanation, Reduction and Empiricism', in *Minnesota Studies in the Philosophy of Science*. ed. Feigl and Maxwell (Minnesota, 1962), vol. 3.

[86] *Ibid.*, 37, lines 6–10.

siderable parallax between them). But again, the meaning of the statement is such that it should then be false in any general theory of physics.

Feyerabend, being a realist, believes in truth, but he does not believe in science (unlike Kuhn, who believes in science, but not in truth). Unfortunately, he is looking for truth at high ontological levels. The present author advocates starting at more modest levels. Ontology may be an aspect of theory, but need not be prior to it.

It is Feyerabend's curious insistence on the primacy of ontology apart from theory (which may be compared to a linguistic approach confining attention to nouns) that leads to the problem of the emperor's clothes: '. . . most of the correspondence rules will be either false or nonsensical. They will be false if they assert the existence of entities denied by the new theory; and they will be nonsensical if they *presuppose* their existence.'[87]

But the statement 'the pointer coincides with the mark' neither asserts nor presupposes the existence of atoms or of a continuous type of matter; such a neutral observation statement may thus well be shared by or critically decide between two theories differing in these fundamental ontological presuppositions. ('Disagreement in ontology involves basic disagreement in conceptual schemes; yet McX and I, despite these basic disagreements, find that our conceptual schemes converge sufficiently in their intermediate and upper ramifications to enable us to communicate successfully on such topics as politics, weather, and, in particular, language.'[88]) To Feyerabend the meaning of 'pointer' is changed in view of these theoretic interpretations of matter. Elsewhere, Feyerabend appears to agree with my view: 'not only is the description of every fact dependent on *some* theory (*which may, of course, be very different from the theory to be tested*) . . .'.[89]

A good part of Feyerabend's message relates to the need for alternative high level theories in certain cases of refutation, and I accept the moral of pluralism in that sense.

It is remarkable that merely because scientists were linked, however indirectly, by communication, the spatio-temporal complex of European science was a single progressive development. There are no barriers of communication between successive theories at least since the sixteenth century.

Acceptance, rather than rejection, is the more remarkable exercise of scientific judgment. The remarkable fact is the transmission of sciences across space and time. The naive model of followers of a scientific theory pictured as islanders, speculative geographers, extending their horizons by exploration, is, perhaps surprisingly, adequate to describe the progress of science. These islanders did not suffer mutations; they preserved the traditions of the past, and they proved able to communicate with the inhabitants of other islands.

The problem of relations between theories, in view of the role of facts, is not covered by the analogy of the problem of the relation between languages, the problem of reference. What is at issue here is not the mystery of language, nor the mystery of learning a language.

[87] *Op. cit.*, note 44, 170.
[88] W. v. O. Quine, *From a Logical Point of View* (Cambridge, Mass., 1964), 16.
[89] *Op. cit.*, note 44, 175; my italics.

Incidentally, the tower of science is the one monolithic structure built in spite of the babel of languages.

Kuhn's position appears to be less extreme than Feyerabend's and historically based. Indeed, Kuhn recognizes the problem faced in the present paper: 'What is surprising, and perhaps also unique in its degree to the fields we call science, is that such initial divergences should ever largely disappear. . . . For they do disappear to a very considerable extent, and then apparently once and for all.'[90]

If I understand Kuhn correctly, this 'bandwagon' phenomenon is not due to anything deeper than the lack of imagination, daring, energy and time of the individual scientist. Once a paradigm has been established, it saves the scientist further discussion at a fundamental level. Nothing succeeds like success. But what constitutes success? It is not, I think, even in Kuhn's account a wholly sociological phenomenon: some theories do give a better account of the facts than others. 'To be accepted as a paradigm, a theory must seem better than its competitors.'[91] Some theories should be rejected. It is difficult to give a rational account of this without appeal to some criterion of truth, though Kuhn wishes to do so: 'We may, to be more precise, have to relinquish the notion, explicit or implicit, that changes of paradigm carry scientists, and those who learn from them, closer and closer to the truth.'[92]

The notion of truth need not be tied to an explicit ontology of entities 'really there', but statements of fact *are* tied to a notion of truth. We need not go into the dubious quarrel between 'instrumentalists' and 'realists' to convict Kuhn of adherence to the notion of truth: '. . . it should be easy to design a list of criteria that would enable an uncommitted observer to distinguish the earlier from the more recent theory time after time. Among the most useful would be: accuracy of prediction, particularly of quantitative prediction. . . .'[93] Even if Einstein's general theory of relativity were closer to Aristotle's than either of them is to Newton's,[94] this might discredit the view of a 'coherent direction of ontological development'—in other words, it would be yet another example of radical revolutions at the highest level of theory, but would not discredit the view of a continuity of structure at the more modest lower levels. Newton's particles of light suffering fits were moribund at the time and are not rehabilitated by the modern theory of photons.

In the heuristic situation of empirical science, truth does not descend deductively from the (highly fallible) most general laws, but ascends inductively from the more modest lower reaches of the deductive structure of the theory held at any one time. Even that infra-structure has to be pruned, and the remaining core is never true for certain, but the modifications forced on us still allow us to refer to that part of the theory as having a likeness to truth.

It may well be that I am so far blind to the potentially healthy branches of the tree of science being ruthlessly pruned away by the scientific community. (I do

[90] Kuhn, *op. cit.*, note 36, 17.
[91] *Ibid.*
[92] *Ibid.*, 169.
[93] *Ibid.*, 2nd edition (1970), postscript, 205–6.
[94] *Ibid.*, 207.

not, of course, deny that the scientific establishment, being human, obstructs extreme originality.) I do, however, see links and commensurability in all theories so far investigated that ever had any success at all. By success we mean here, of course, not just popular acclaim, but measurable pragmatic success (*e.g.*, correct predictive power), though these usually go hand in hand.

Of course, Feyerabend may well deplore prescriptively the stupid, blind, follow-the-leader attitude of scientists shown up by this description. I would, of course, agree with him on the desirability of playing several rival theories, but only as long as these theories are not too different in merit according to such criteria as agreement with experience and simplicity.[95]

[95] To take an often-cited earlier revolution, Copernicus *vs.* Ptolemy: 'Therefore in this ordering we find that the world has a wonderful symmetry and that there is a sure bond of harmony for the movement and magnitude of the orbital circles such as cannot be found in any other way.' (N. Copernicus, *De Revolutionibus*..., Book I, 10.) Here there is agreement on observations, but an appeal to simplicity.

ERIC R. SCERRI

# Correspondence and Reduction in Chemistry

## I. INTRODUCTION

In his classic paper Heinz Post makes several references to the field of chemistry[1]. For example, he writes:

> The chemical atom is no longer strictly an atom, yet whatever revolutions may occur in fundamental physics, the ordering of chemical atoms will remain. (Post 1971, p. 237)

The paper begins with the claim that the general correspondence principle consists in the requirement that any acceptable new theory *L* should account for its predecessor *S* by 'degenerating' into that theory under conditions in which *S* has been well confirmed by tests. In a later section Post states:

> The well-confirmed part of a theory extends only up to a certain level. The *S*-theory is stripped of its dispensable, not independently confirmed superstructure. (The high level *interpretation* of the total coherent pattern of data is changed by a revolution.) (*ibid.*, p. 229)

What I believe is intended by the last sentence is that *only* the high level or most fundamental interpretation is changed by a revolution. Similarly, he writes:

> There is no logical reason why a scientific theory should not be replaced by a theory different from the old at every level; but this is not what has happened so far. The most radical revolutions have destroyed the top levels and give a somewhat different interpretation to the lower levels, but they have not destroyed the whole of the lower level structure. (*ibid.*, p. 237)

Post claims that there is a rationale of scientific discovery and that the procedure is essentially inductive in retaining the old theory in a certain sense. This leads him to discuss the notion of correspondence as a means of providing continuity across the revolutionary breaks. At another point he draws an analogy with political revolutions and says that while a revolution may occur in a nation, the civil service remains largely unchanged. If Kuhn losses



*S. French and H. Kamminga (eds.), Correspondence, Invariance and Heuristics*, 45–64.

occur anywhere, they occur at the most fundamental theoretical levels and not at the lower levels.

In the present article I would like to examine this view that revolutions in science occur at the most fundamental level by considering, in some detail, a chemical example which Post returns to at several points in his paper. I also examine the influence of revolutions from the point of view of some more general chemical examples. In these discussions the question of reduction of chemistry to physics, on which Post also comments in his paper, is a recurrent theme.

## II. THE FAJANS-PANETH CONTROVERSY

One of the examples Post chooses to illustrate the influence of revolutions in science on lower levels, is the way in which the periodic table of elements has survived the discoveries in fundamental physics. He discusses the Fajans-Paneth controversy which took place in the early part of the present century.

This represents an interesting example in the philosophical study of theory change. The discovery of isotopes provides a strict refutation of the periodic system in the sense that the system does not classify these entities which are the most elementary 'atoms' of the elements. However, the periodic table survives to this day and indeed is of central importance in chemistry. The case also illustrates some aspects of the essential difference between allied fields of science.

The relationship between chemistry and physics, or more precisely the demarcation between the two fields, was addressed explicitly by two radiochemists, Fajans and Paneth, in the course of a heated controversy between the years 1913 and 1922 (van der Vet 1987). The issue began as a debate concerning the nature of isotopes in radiochemistry and rapidly took on a wider significance. In the early part of the 20th century isotopes had been identified (Soddy 1910) but had not been separated by chemical means. One conclusion to emerge from this state of affairs was that isotopes were of no relevance for most chemical effects. Paneth and his collaborator von Hevesey took this view and in 1914 they published a paper claiming that isotopes were completely equivalent chemically (Paneth and von Hevesey 1914). They proposed the notion of replaceability of isotopes, that is, they claimed that the replacement of any isotope with another one of the same species would not produce any noticeable chemical effect.

They set out to verify the correctness of this view through the law of mass action. For any reaction,

$$aA + bB \leftrightharpoons cC + dD$$

it can be shown that an equilibrium constant, $K_c$, at any given temperature, for the reaction is given by

$$K_c = [C]^c\ [D]^d\ /\ [A]^a\ [B]^b$$

where the square brackets denote the concentrations[2] of the species *A-D*.

If the replaceability view is correct, it should be possible to take any of these concentrations as the sum of the concentrations of all the isotopes of any particular species. In order to estimate these concentrations Paneth and von Hevesey used electrochemical experiments and the Nernst equation

$$E = E^{o} + RT/nF \ln c$$

where $E$ is the potential of a half-cell, $E^{o}$ is the standard electrode potential, $R$ the ideal gas constant, $T$ the absolute temperature, $n$ the number of electrons transferred in the course of the half reaction, $F$ the Faraday constant, and $c$ the concentration of the dissolved species.

One of the experiments utilised two isotopes of bismuth, $^{209}Bi^{3+}$ and $^{210}Bi^{3+}$. The observed voltage was found to be constant regardless of the proportion of the two isotopes present in the sample, within the accuracy of the experiment. Fajans objected to the interpretation of these results on the basis of a thermodynamic analysis[3] and put forward the view that isotopes are different, not just physically but also chemically (Fajans 1914). Various other experiments were then devised and carried out in an attempt to settle the issue but the debate continued and indeed was widened to the question of what constitutes a chemical element.

In 1918 Aston discovered isotopes in a number of non-radioactive elements, which led to a strengthening of Paneth's case. He pointed out that elements in Fajans' sense of the word seemed to be multiplying each day and were thus endangering the chemists' picture of the world. It would have been far-fetched for chemists to restructure the foundations of their science on the grounds that 'elements' had turned out to be composite structures. At about this time, Paneth stated clearly what had previously remained an implicit definition of a chemical element:

> An element is a substance of which all atoms have the same nuclear charge. (Paneth 1920, p. 842).

Even at this time, however, isotopes were suspected of consisting of smaller particles. Fajans' suggestion, if taken seriously, would have committed chemistry to research into elementary particles. Chemists were neither prepared for this change, nor were they motivated to abandon their existing successful research programmes.

In 1919 a decisive experiment was conducted by Rutherford who succeeded in artificially decomposing the nucleus by using alpha particles. Fajans was forced to demarcate between decomposition carried out by physical means (Rutherford's experiment) as opposed to chemical means. Paneth, who had advocated such a demarcation from the start, was thus vindicated. Finally, in 1923 the newly formed International Union for Pure and Applied Chemistry (IUPAC) laid down the official definition of a chemical element precisely, according to the criteria suggested earlier by Paneth (Aston *et al.* 1923).

One further interesting development took place in 1932 when Urey discovered isotopes of hydrogen. In these cases the masses of the isotopes differ from each other by as much as 100%[4]. The chemical properties of these isotopes are definitely not the same. For example, water (mostly protium oxide) has a melting point of 0°C whereas the value for deuterium oxide is 3.8°C. Strictly speaking, Fajans had been correct to doubt replaceability, but the problem is only significant in isotopes with very low masses. Urey's discovery did not change the chemical definition of an element in any way.

The conclusion to be drawn from this episode is that chemistry is not obliged to accept every single step towards reduction which may be suggested by research in physics. Re-structuring the subject on the basis of isotopy would have led to the demise of the periodic system.

In an article to commemorate the one hundredth anniversary of Mendeleef's periodic table Greenaway wrote:

> The chief influence of the periodic table on chemistry was to turn it from endless diversification in a search for unknown compounds and elements to a concentration on order and its underlying cause. Chemistry was able to assimilate the new physics – the electron, radioactive transformation etc., without fundamental change. (Greenaway 1969, p. 99)

As Post remarks in his footnote, Fajans was in fact following the Dalton-Mendeleef philosophy of attaching primary significance to the atomic mass or combining weight of an element as the source of its individuation. If Mendeleef's ontology had been adhered to at all costs following the discovery of isotopes, it would have destroyed the periodic table.

## III. THE INFLUENCE OF QUANTUM THEORY AND QUANTUM MECHANICS ON CHEMISTRY

However, one might well argue that the discovery of isotopes does not represent a major revolution in physics and consequently that the lack of any influence of this discovery upon the field of chemistry is not such a good argument for Post's thesis. A more severe test would surely be provided by examining the influence of the major revolution in modern physics, namely quantum mechanics[5].

The question then arises as to where to locate the quantum revolution. Should one consider Bohr's original contribution, sometimes called the old quantum theory, or should one focus on the more mature quantum mechanics of Heisenberg, Schrödinger and others? In order to avoid this question I propose to consider briefly the influence of both theories on the field of chemistry.

According to Bohr's account, the old quantum theory can be used to explain the periodic table in terms of the electronic structure of atoms. This view has been generally adopted, particularly in chemical education, and is seldom questioned. However, even just a casual examination of Bohr's account of chemical periodicity shows that the periodic table cannot be deduced from theory. The alleged explanation is merely an exercise in accounting for the known chemical facts in terms of an alternative representation based on the approximate model of electronic configurations.

As is well known, Bohr entered the atomic physics arena at a time when the Rutherford model of the atom was experiencing a major difficulty. Rutherford's nuclear atom consisted of a central positive charge surrounded by orbiting negatively charged electrons. According to electromagnetic theory, any circulating charged body should lose energy through radiation, and would therefore be expected to spiral into the nucleus. Rutherford's model implied that any atom and consequently all matter should be unstable, contrary to the facts of experience. Furthermore, the model could not explain the discrete nature of the optical spectra of atoms.

Bohr resolved these problems by means of an atomic model into which he introduced Planck's quantum theory (Bohr 1913). He proposed an *ad hoc* stability condition of quantised angular momentum for the nuclear atom by analogy with Planck's quantum theory of energy quantisation, $E = h\nu$. Bohr's quantum condition is

$$L = nh / 2\pi$$

where $L$ is orbital angular momentum, $n$ is a constant and $h$ is Planck's constant. According to Bohr, an atomic system can only exist permanently in certain series of states corresponding to a discontinuous series of values for its energy, and consequently any change of the energy of the system, including emission and absorption of electromagnetic radiation, must take place by a complete transition between two such states. These states are denoted as 'stationary states' of the system.

Bohr also stated that the radiation absorbed or emitted during a transition between two stationary states is 'unifrequentic' and possesses a frequency $\nu$ given by the relation

$$E' - E = h\nu$$

where $h$ is Planck's constant, and $E'$ and $E$ are the values of the energy in the two states under consideration (*ibid.*).

It emerged that Bohr's atomic theory only gave an exact account of the spectrum of atomic hydrogen and an approximate explanation for the spectra of alkali metals. Nevertheless, Bohr was sufficiently confident of his theoretical approach that he set about trying to explain the periodic table of elements on the basis of his quantum theory of atoms. In the same year Bohr produced his first version of an electronic periodic table (Bohr 1913). (See Table I.) Contrary to Bohr's own claims, the configurations were arrived at by intuition and reference to chemical and spectroscopic behaviour rather than quantum mechanical principles. This can be seen from the irregular manner in which Bohr populated the electron shells while trying to maintain agreement with the known experimental facts. For example, the sudden rearrangement in the configuration of nitrogen appears to be an attempt to

*Table I.* Bohr's original scheme for electronic configurations of atoms (Bohr 1913).

| | | | | |
|---|---|---|---|---|
| 1 | H | 1 | | |
| 2 | He | 2 | | |
| 3 | Li | 2 | 1 | |
| 4 | Be | 2 | 2 | |
| 5 | B | 2 | 3 | |
| 6 | C | 2 | 4 | |
| 7 | N | 4 | 3 | |
| 8 | O | 4 | 2 | 2 |
| 9 | F | 4 | 4 | 1 |
| 10 | Ne | 8 | 2 | |
| 11 | Na | 8 | 2 | 1 |
| 12 | Mg | 8 | 2 | 2 |
| .. | .. | . | . | . |
| 18 | Ar | 8 | 8 | 2 |

account for the trivalency of the element. Similar adjustments were made in oxygen and fluorine to ensure the predominant divalency and univalency of these respectively.

Following this work Bohr abandoned the question of periodicity for about a decade, and it was left to various chemists to try to improve upon the electronic version of the periodic table (Langmuir 1919; Bury 1921). Some authors stress the fact that these workers produced improved versions of the electronic periodic table without any firm basis in quantum theory (Hund 1974). This is to say, the work done by chemists seems to be downgraded, although as was argued above, and contrary to his own stated intentions, Bohr's original version was also arbitrary in many respects. MacKinnon has aptly described Bohr's style:

> Bohr's explanation of the periodic table is the crowning achievement of the old atomic theory. It is also one of the clearest manifestations of the type of multifaceted nonlinear reasoning in which Bohr excelled. He was not attempting to deduce the periodic table from his atomic theory. He was, rather, trying to develop a consistent atomic account that fitted the data coming from radically diverse sources; atomic theory, spectral lines, ionisation energies, X-ray spectra, the correspondence principle, the periodic properties of the elements, chemical valencies and information on valent and covalent chemical binding. In blending these together Bohr functioned more like a symphony director than a deductive logician. (MacKinnon 1982, p. 178)

Between the years 1921 and 1923 Bohr returned to the problem of atomic structure and the periodic table. He announced a new improved version of the electronic periodic table, claiming it to be based on an important principle of the old quantum theory, namely the adiabatic principle (Bohr 1923). His method, called the *Aufbauprinzip*, consisted in building up successive atoms by the addition of an extra electron to the previous atom. Bohr developed his new account of the periodic table according to the two quantum numbers *n*, the main quantum number, and *k*, the azimuthal quantum number (see Table II). These numbers emerged from the quantum conditions and served to identify the stationary states of the system. According to this scheme an atom of sulphur, for example, with sixteen electrons, would have an electronic configuration of 2, 4, 4, 6.

*Table II.* Assignment of electrons to shells according to Bohr's 1923 scheme (Bohr 1923).

| *n* | *k* | Total number of electrons |
|---|---|---|
| 1 | 1 | 2 |
| 2 | 1 | 4 |
| 2 | 2 | 4 |
| 3 | 1 | 6 |
| 3 | 2 | 6 |
| 3 | 3 | 6 |
| . | . | . |

As several of the Göttingen physicists who were exposed to these ideas by Bohr's own lectures later commented, the work rested on a mixture of *ad hoc* arguments and chemical facts without any derivations from the principles of quantum theory which Bohr frequently alluded to. According to Heisenberg:

> It could very distinctly be felt that Bohr had not reached his results through calculations and proofs but through empathy and inspiration and it was now difficult for him to defend them in front of the advanced school of mathematics in Göttingen. (quoted in Kragh 1979, p. 61)

Hund later wrote:

> After he had explained a simple spectrum he came to his crucial review of the structure of atoms with regard to their positions in the periodic system. In some respects this turned out to be obscure and not always easy to understand. (Hund 1974, p. 107)

Weisskopf quotes a marginal remark made by Pauli in a book containing Bohr's famous paper on the *Aufbauprinzip* (Weisskopf 1989). In discussing the adding of the eleventh electron to the closed shell of ten electrons, Bohr says: "We must expect the eleventh electron to go into the third orbit." (quoted in Weisskopf 1989, p. 164). Pauli, obviously annoyed by this statement, writes hastily in the margin with three exclamation marks. "No reason to *expect*; you concluded it from the spectra!!!" (quoted in Weisskopf *ibid.*, p. 164).

The notion that the periodic table can be deduced from quantum theory appears to be completely untenable. This position is virtually conceded by Bohr when he states:

> On the general view of the formation of atoms however, and by making use of the knowledge of the properties of the corresponding elements, it will be attempted to . . . obtain indications of what configurations of the electrons may be expected to occur in atoms. (Bohr 1913, p. 478)

As Kragh observes,

> In Bohr's work, where the mechanical and quantum theoretical principles could only fix the configurations very grossly, the chemical-inductive arguments played the major role. (Kragh 1977, p. 208)

The main feature of the building-up procedure was Bohr's assumption that the stationary states would also exist in the next atom, obtained by the addition of a further electron. He assumed that the number of stationary states would remain unchanged apart from any additional states of the newly introduced electron. Bohr thus assumed the existence of sharp stationary states, and their retention on adding both an electron and a proton to an atom.

The basis of this building-up procedure used by Bohr consists of the adiabatic principle introduced by Ehrenfest in 1917:

> Suppose that for some class of motions we for the first time, introduce the quanta. In some cases the hypothesis fixes completely which special motions are to be considered as allowed. This occurs if the new class of motions is derived by means of an adiabatic transformation from some class of motions already known. (Ehrenfest 1917, p. 501)

This adiabatic principle is one of the corner-stones of the old quantum theory. It allows one to find the quantum conditions when an adiabatic change is imposed on a system. It depends on the requirement that it should be possible to derive the new motion from the known one by means of an adiabatic or gradual transformation. The quantities which preserve their values after such a transformation are known as adiabatic invariants. The applicability of the adiabatic principle depends on the condition that the introduction of external forces should not alter the degree of periodicity.

There are further stringent restrictions on the applicability of the adiabatic principle. Ehrenfest himself showed that it was applicable to simply periodic systems (*ibid.*). These are systems having two or more frequencies which are rational fractions of each other. In such systems the motion will necessarily repeat itself after a fixed interval of time. Later Burgers, a student of Ehrenfest, showed that it was also applicable to multiply periodic systems (Burgers 1917). In these more general systems of motion the various frequencies are not rational fractions of each other, and so the motion does not necessarily repeat itself (Goldstein 1980). The hydrogen atom provides an example of a multiply periodic system, with two degrees of freedom being the orbital frequency of the electron and the precession of the orbital frequency.

An even more general class of systems is termed aperiodic, and as far as is known, the adiabatic principle does not apply in those cases. Unfortunately for the field of atomic physics, all atoms larger than that of hydrogen constitute aperiodic systems for which the adiabatic principle does not apply. In the helium atom, for example, the motion of each of the two orbiting electrons varies due to the varying interaction with the other electron as their distance apart changes (in terms of the early Bohr theory). We may no longer speak of a constant period for either of the electrons.

Now, whereas Bohr was claiming in general terms that his *Aufbauprinzip* was founded on the adiabatic principle, he was also aware of the fact that the latter principle did not apply to many-electron atoms. He readily acknowledged this fact in his writings:

> For the purposes of fixing the stationary states we have up to this point only considered simply or multiply periodic systems. However the general solution of the equations frequently yields motions of a more complicated character. In such a case the considerations previously discussed are not consistent with the existence and stability of stationary states whose energy is fixed with the same exactness as in multiply periodic systems. But now in order to give an account of the properties of the elements, we are forced to assume that the atoms, in the absence of external forces at any rate, always possess sharp stationary states, although the general solution of the equations of motion for the atoms with several electrons exhibits no simple periodic properties of the type mentioned. (Bohr 1923, p. 17; English translation in Rud Nielsen 1977, p. 472)

Adopting his characteristic brand of 'mysticism'[6], Bohr simply ignored this problem in the hope that the adiabatic principle would eventually turn out to be valid in aperiodic systems. This was not to be the case. The old quantum theory cannot, therefore, be said to have explained the periodic table. Bohr continued to rely on inductive chemical arguments while at the same time claiming support for his 'electronic periodic table' from the principles of quantum mechanics.

Bohr's hypothesis of the permanence of quantum numbers came under attack from the spectroscopic analysis[7] of the splitting of the lines under the influence of a magnetic field (Landé 1923). Bohr's response was to maintain adherence to the permanence of quantum numbers even in the face of this evidence. He merely alluded to the mystical device which he called a non-mechanical constraint or *Zwang* to save the quantum numbers (Hendry 1984). Bohr's account of the periodic table also came under attack from chemical evidence[8].

On the basis of his theory, Bohr claimed to predict that the element number 72, subsequently named hafnium, was not a rare earth, but a transition element. This claim does not, however, stand up to close scrutiny since, as I have argued elsewhere, many chemists had arrived at the same conclusion from purely chemical arguments (Scerri 1992).

Soon thereafter some of the minor problems in Bohr's scheme were resolved by Stoner (1924) who used not merely two quantum numbers, but also Sommerfeld's recently introduced third quantum number, *j*. Whereas the introduction of a third quantum number suggested additional stationary states in the atom, Bohr did not extend his electronic configuration scheme accordingly, perhaps because he was becoming increasingly concerned with the deeper question of the existence of stationary states for individual electrons in many-electron atoms. Stoner, undaunted by these theoretical problems, suggested that the number of electrons in each completed level should equal twice the inner quantum number of that particular shell.

Pauli then introduced a fourth quantum number for each electron and this was followed by his exclusion principle. In arriving at the exclusion principle, Pauli used an *ad hoc* device whereby a strong magnetic field was imposed on the electrons in order to decouple them. The field was subsequently imagined to be reduced gradually in such a way as to maintain the decoupling even in the absence of the field (Scerri 1991a). These steps seemed to restore the permanence of the quantum numbers which Bohr had sought and for a short period it was again possible to consider the electrons in a many-electron atom as being in individual stationary states. Pauli's *ad hoc* manoeuvre was soon shown to be logically inconsistent with quantum mechanics. The old quantum theory as modified by Pauli's exclusion principle provided a semblance of an explanation for the periodic table, but did not yield any new predictions. Chemistry was provided with an alternative representation in terms of numbers of outer-shell electrons in any particular atom, but was by no means revolutionised by the theoretical discoveries in quantum physics.

According to quantum mechanics, the notion of electrons in individual stationary states or shells, which Bohr had hoped would be retained, was ruled out conclusively. Chemical education seems to ignore this crucial point and continues to operate within the orbital 'lie'.

## IV. *AB INITIO* CALCULATIONS

If developments in physics really had a revolutionary influence in chemistry, the orbital approximation should have been abandoned at this point. In effect it was embraced even more whole-heartedly since quantum mechanics provides an approximate method for calculating the properties of atoms and molecules, whereas the quantum theory could only be applied successfully to the case of the hydrogen atom.

Unfortunately, the use of approximation has led to unfounded claims in theoretical chemistry. The Schrödinger equation can only be solved exactly for the case of the two-particle system, the hydrogen atom. As soon as we turn to the three-body problem, approximations must be introduced in order to proceed. In the crudest of all approximations, interactions between electrons are completely neglected. The wavefunction of the atom is expressed as a single product of one-electron functions. This approximation yields hydrogenic orbitals whose energies may be obtained exactly. Clearly the orbital energies obtained in this way will fail to correspond to observed data on ionisation energies in which inter-electron repulsions are a contributing factor.

The Hartree-Fock method provides a better orbital approximation since it includes the effect of inter-electronic repulsions. The wavefunction now consists of a sum of products to take account of permutation of electrons as dictated by Pauli's principle. The method fails to account for correlation energy, however. The frequently claimed accuracy of the Hartree-Fock method to within 1% of the total atomic energy is not so impressive when it is realised that these errors are of the same order of magnitude as typical chemical bonding energies. In order to correct for these deficiencies the simple orbital concept must be abandoned. The wavefunction takes the form of more products of one-electron functions involving the mixing of many possible configurations. Such more elaborate approximation methods provide an increased level of accuracy. However, an insistence on the part of the *ab initio* community to concentrate on the total atomic energy renders these claims to reducing chemistry less valid. Chemists are more interested in individual energy differences, such as heats of reaction associated with chemical reactions, than in absolute values. Further, a calculation of the total energy does not provide such a severe test for the theory, since the ionisation of the most tightly bound electron (the largest single contribution to the total energy) can be calculated exactly.

An examination of *ab initio* calculations carried out on small atoms shows that, provided that increasingly larger linear combinations of atomic orbitals are taken, the experimental energy of the atom may be approached ever more closely[9]. A true *ab initio* calculation should ideally give some indication of how far the process of expansion must be carried out and should give some indication of the expected errors. In a considerable amount of *ab initio* work the point at which to terminate the expansion is decided upon by reference to the experimental energy which is being calculated. Such mere calculation of already known experimental data does not represent a strong argument for the predictive power of quantum mechanics in chemistry, although it does argue for a certain explanatory power.

The common approach used in both the perturbation and variation methods is one of expansion of the wavefunction for the system as an infinite series of one-electron functions. In using such a procedure it is essential that the series used in the expansion should converge to the function which it is meant to represent.

A number of theoretical chemists appear to be relatively unconcerned about such issues. A survey of the literature shows that most authors adopt the approach of justifying their calculations only by comparison with the known experimental results on the systems in question. To quote a recent textbook on *ab initio* calculations:

The vast majority of contemporary *ab initio* calculations adopt what may

> be termed a pragmatic approach; that is, no error bounds are determined and the accuracy of the calculation is assessed by comparison with quantities derived by experiment. By means of this procedure, the quality of a particular basis set is established and the empirical range of validity for the basis set thus obtained which can be used to make educated guesses at the accuracy of a computed molecular property. The art of selecting a basis set is based on previous experience in treating similar systems using basis sets of comparable quality. (Wilson 1984, p. 169)

There is no guarantee that, although the experimental value is being approached, the next level of accuracy might not show a sudden divergence. Such occurrences are not unknown in mathematical physics[10].

In theoretical chemistry there is another problem which should be borne in mind. The reliance on experimental data to check the validity of calculations presents a certain dilemma, since it is precisely in cases where little or no experimental data are available that calculations would be of the greatest benefit. What is needed in theoretical chemistry is an independent non-empirical method of assessing the accuracy of the calculations.

Although a variational calculation gives an upper bound to the exact solution it does not tell us how close we are to the true value. For many years *ab initio* calculations have yielded a series of results each of which is perhaps better than the previous one, due to the addition of a few parameters. Alternatively, a calculation may be more efficient because a wiser choice of functions gives an equally accurate result with less computational effort. It is usually assumed that with a sufficiently flexible trial function the results will eventually converge to the exact solution. To quote Schwartz,

> The old habit of picking the 'best' (chosen by art) choice of a fixed number of terms is to be discarded if one wants to see how the problem converges. (Schwartz 1963, p. 243)

However, one regularly still finds statements in modern works on theoretical chemistry which make a virtue of being able to guess the correct procedure by a mixture of intuition and past experience. The usual means of proceeding with a calculation appear to be haphazard. As Weinhold (1972) writes, error limits are routinely demanded of experimental results as a matter of course and consequently one should also hope for a corresponding standard of reliability from the theoretical side.

In spite of concentrated work on the question of error bounds in *ab initio* calculations, the situation does not appear to have changed except in some very restricted systems such as those containing two electrons. In 1980, Bunge wrote:

> Rigorous *ab initio* calculations where error bounds are also *ab initio* belong to the future. A few significant calculations have apparently exhausted the present possibilities of current theories of upper and lower bounds to

expectation values. In general, today's rigorous error bounds are too large to be of much use in atomic calculations. (Bunge 1980, p. 328)

A general method of remedying the situation in *ab initio* work would consist in finding a way to calculate both upper and lower bounds to the energy of any particular system under consideration. If this form of 'bracketing' were possible, it would endow quantum mechanics with a genuine power of prediction (Wilson 1965). The problem has been that, whereas the well-known variation method provides an upper bound to the energy (as can readily be proved; see Pauling and Wilson 1935),we do not as yet possess a sufficiently general and tractable method for determining lower bounds in cases applicable to atomic energies. The search for such methods has exercised the best minds in quantum mechanics for over 60 years, but with no sign of a clear solution or even the best way to proceed.

Nevertheless, although until recently predictions made in quantum chemistry were generally unreliable, it has been forcefully argued that since 1970 quantum chemistry has "come of age" (Goddard 1985). This claim seems to be due in part to improvements in technology which have furnished more accurate calculations by virtue of brute computational force. The year 1970 has been designated as the beginning of the new era because of the alleged success obtained in theoretical predictions of the geometry of the methylene molecule, $CH_2$ (Bender and Schaefer 1970).

These authors used theoretical methods which previously had only been applied to atoms and diatomic molecules. The starting point consisted of a set of Hartree-Fock self-consistent orbitals which were used as the basis for an elaborate configuration interaction treatment. The authors report that the calculations were carried out in order to "obtain the most rapid rate of convergence". As a result of this work, a bond angle of 135.1° was predicted in methylene, contrary to previous spectral studies which had concluded that methylene is linear.

At about the same time, a number of electron spin resonance experiments were reported, all of which supported the bent geometry of the molecule (Bernheim 1970, 1971; Wasserman *et al.* 1970). According to these studies the bond angle in methylene lies between 136° and 137.7°.

When the theoretical papers of Bender and Schaefer are consulted it emerges that, whereas the authors claim to have examined their methods for the most rapid rate of convergence, no evidence is given of this aspect in any of the relevant publications. Moreover, a mere examination of the rate of convergence does not imply proof of actual convergence. The authors do not claim to have evaluated both lower and upper bounds, which might provide a reliable *ab initio* estimate of the molecular geometry.

The major reference given by Schaefer with regards to computational methods is to the work of Bender and Davidson (1966). An examination of this source shows several passages which indicate the somewhat *ad hoc* nature of the approach used. For example, the authors state:

> The main difficulty in the selection of configurations . . . was of course the enormous number of possible configurations. In a typical calculation there are billions of configurations which can be formed with the correct symmetry. For an unfortunate choice . . . all of these might be equally important, but for a good initial guess only a few will really contribute to the wavefunction. (Bender and Davidson 1966, p. 2676)

Clearly the procedures seem to be applied in an *ad hoc* manner. Bender and Schaefer also claim that despite having established "beyond doubt" the non-linearity of methylene, experiments were unable to agree on a precise value for the bond angle. The authors therefore " . . . set out to attempt to place error bars on the theoretical prediction of the bond angle" (Schaefer 1986, p. 1102).

As a result of this work employing an even larger basis set than the calculation of 1970, the authors estimated a bond angle of 134 ± 2°. Once again, a detailed examination of the original source reveals a somewhat different picture for this claimed determination of error bars. The article states that

> . . . the agreement with experiment obtained for $H_2O$ will give us an estimate of the reliability for our prediction concerning $CH_2$. (McLaughlin *et al.* 1972, p. 353)

Elsewhere in the same article they say:

> To aid our evaluation of the expected reliability of this 134° $CH_2$ angle, we point to the comparable first order calculations on the ground and excited states of $NH_2$ (Bender and Schaefer 1971), which yielded bond angles differing by 0.6 and 0.7 degrees from experiment. In the light of these results and the $H_2O$ results discussed above we estimate our theoretical bond angle of 134° is accurate to within 2°. (McLaughlin *et al.*, 1972, pp. 356–7)

Table III, adapted from this study, gives the theoretical and experimental values for $H_2O$ and $CH_2$, the theoretical results having been obtained by the same method of approximation.

*Table III.* Bond angles in $H_2O$ and $CH_2$ molecules.

| | $H_2O$ | $CH_2$ |
|---|---|---|
| Bond angle (°) theoretical | 103.4 | 134.0 |
| experimental | 104.5[a] | 136 ± 8[b] |

[a]Benedict *et al.* (1956); [b]Wasserman *et al.* (1970).

It seems odd that only one particular value for the experimental value of the bond angle has been selected for each molecule, in view of the great diversity of available experimental results which these authors themselves discuss elsewhere in their paper.

The approach adopted by the authors represents an extrapolation from the application of the method from one molecule to that of another, and not a rigorous determination of error bars for the $CH_2$ molecule itself, which is implied in the claims made in a review paper by Schaefer (1986).

All in all, the calculations on $CH_2$ represent a perfect example of what Davidson has described as 'calibrated *ab initio*', as opposed to 'true *ab initio*' work:

> For the purpose of this talk, *ab initio* will mean 'from the beginning', i.e. without any input from past experience or any parameters chosen because they worked for other atoms or molecules. With this definition true *ab initio* calculations are always exploratory and of interest mainly to other quantum chemists. If successful they may lead to improvements which can be incorporated into widely used 'calibrated *ab initio*' calculations carried out by applied quantum chemists. Like semi-empirical methods these calculations assume transferability of parameters and methods between molecules and do not check the convergence of every result with respect to quality of wavefunction. (Davidson 1984, pp. 8–9)

This seems to be the approach adopted by Schaefer and his school, since the strict internal criteria demanded by Schwartz (1963), Weinhold (1972) and Bunge (1980) have not been addressed and there is no attempt to calculate both upper and lower bounds.

I conclude that the strong claims made for the coming of age of quantum chemistry are unfounded. Chemistry has not been reduced to quantum mechanics in practice[11].

I would suggest that only such mathematically accessible properties as the total atomic energy and the ionisation energy are approximately reduced. This does not amount to claiming an approximate reduction for the overall behaviour and reactivity of the chemical atoms. Reactivity is not only connected with energetic considerations, but also governed by kinetic factors such as rates of reaction, steric factors and activation energies. There is no available theory which treats both energetic and kinetic factors simultaneously, although some progress has been achieved towards a quantum theory of rates of reactions.

## V. KOERTGE ON CORRESPONDENCE AND THE PERIODIC TABLE

One of the most vocal supporters of Post's views on correspondence has been Noretta Koertge, who also draws on chemical examples. In one of her publications (Koertge 1973) she discusses instances of what she calls "falsification by a striking counter-example" and claims that the discovery of the rare-earth elements was a "massive refutation of the Periodic Law". She goes on to say:

> According to their atomic weights the rare earths should lie in a horizontal

row across the periodic table, but because of their very similar chemical properties they should all lie in one family group. (Koertge 1973, p. 188)

Koertge also mentions the case of reversed pairs in which some elements such as tellurium and iodine were slightly out of order according to the atomic weight measurements. She contends that while Mendeleef acknowledged the difficulties with regard to these pairs, he was not too worried by them.

I believe that it is incorrect to regard the discovery of the rare earth elements as a refutation of the Periodic Law, whilst on the other hand I would argue that the case of reversed pairs does represent a genuine refutation. The main criterion used by Mendeleef for ordering the elements was atomic weight and not the requirement that the periodic table should take any particular form regarding groups of elements. In this respect the classification of the rare earths within the periodic system is completely successful. I suggest that a system of *representation* cannot be refuted in any sense. In the case of the rare earths all that we have is an exception to the notion that a group of elements with similar properties should occupy a column of the periodic table[12]. This is not at all serious in view of the arbitrary choice in which the periodic table may be represented. Literally hundreds of representations have been proposed (van Spronsen 1969).

Koertge also writes that in the introduction of his textbook of 1905 " . . . Mendeleef does not even have the heart to discuss the rare-earths himself" (Koertge 1973, p. 188). In fact, far from regarding the rare earths as a disastrous blow to his periodic system, Mendeleef, as early as 1871, corrected the atomic weight of the rare earth cerium from a value of 92 to 138, which is virtually the currently accepted value. He was also the first to give the rare earth yttrium its correct atomic weight of 88, as compared with the previously held value of 60. As van Spronsen remarks with regard to the atomic weights of these two elements, "Mendeleef's faith in the periodic system prevailed once again" (van Spronsen 1969, p. 264).

Mendeleef was not unduly concerned about the pair reversals because he reasoned that the atomic weights were incorrect. He put more faith in chemical properties than in the values of atomic weights which were available at the time. However, although Mendeleef had successfully predicted and corrected many atomic weights, he was wrong in the case of the pair reversals. In my view the pair reversals do represent genuine refutations of the ordering of elements according to their atomic weights. The question was only resolved following the work of Moseley who showed that the elements should be ordered according to atomic number rather than atomic weight. I agree with Post and many others that, in trying to assess whether certain findings represent refutations or not, we should consider internal factors rather than the perceptions of the protagonists.

## VI. CONCLUSION

To return to Post's paper, he makes the following specific reference to the reduction of chemistry:

> The periodic system is the basis of inorganic chemistry. This pattern was not changed when the whole of chemistry was reduced to physics, nor do scientists ever expect to see an explanation in the realm of chemistry which destroys this pattern. (Post *op. cit.*, p. 237)

Post's apparent acceptance in this passage that chemistry has been reduced seems surprising. I can only suggest that he intends the term reduction in a very weak sense of approximate reduction. In any case, I know that he no longer believes that even the restricted field of atomic chemistry has been reduced to physics, let alone the whole of chemistry[13].

The invasion of chemistry by the orbital approximation may give the superficial appearance of having succeeded in revolutionising chemistry, contrary to Post's thesis regarding the influence of revolutions. However, on closer examination it turns out that although the language of atomic orbitals has taken root in modern chemistry, particularly with regard to the teaching of the subject, quantum mechanics or quantum mechanical approximations do not play a truly explanatory role, let alone a predictive role. This situation has consequences in the field of education, because the current presentation of the subject from a quantum mechanical standpoint gives students the false impression that chemistry is a deductive science[14] (Scerri 1989a, 1989b, 1991b).

After examining the various types of correspondence found in science, Post proposes a classification of the order of strength of the relation of theory *L* to theory *S* (Post 1971, p. 236). He calls the strongest form of correspondence a reduction. In such a case the old theory is completely taken over into the new theory. The weaker forms of correspondence are (i) consistent correspondence with no explanatory losses and (ii) consistent correspondence with no explanatory losses but requiring bridging statements such as matter-of-fact statements about empirical boundary conditions.

I suggest that this form of classification, or a suitably modified analogy, could be applied to the question of the correspondence relationship between different fields of science. As I have argued in this article, the correspondence relationship between chemistry and physics is not one of reduction. This view has now become fairly commonplace and most philosophers would deny that strict reduction of any special science can be said to occur. Reduction has been replaced by approximate reduction. Others seek a solution in the concept of supervenience, that is, the view that higher level phenomena have physical explanations while denying reduction in the usual formal sense (Kim 1982). However, it is not clear that such manoeuvres get us any closer to characterising the relationship between, say, chemistry and physics. I think it might be fruitful to consider which of Post's weaker forms of correspondence might describe the relationship between these two fields. I suggest that the third category mentioned above, involving bridging statements, comes closest to fulfilling this task.

Interestingly, one of the examples Post uses to illustrate his third category is again the question of the periodic table and isotopes referred to earlier:

> Actual examples of such *ad hoc* bridging theories being required include the modern account of the periodic system in terms of 'contingent' statistical proportions of isotopes . . . (Post *op. cit.*, p. 233, fn. 46)

Finally, when the philosopher-chemist Jørgensen was interviewed recently, he was asked which field he would have chosen if he was undertaking graduate work all over again[15]. His reply was that he would wish to work in the border area between chemistry and physics. It has been my privilege to write a thesis on this relatively uncharted territory with the guidance of an outstanding polymath, Heinz Post.

*Dept. of History and Philosophy of Science,*
*King's College, University of London*

## NOTES

1. For his first degree, Post read chemistry at Oxford with C. Hinshelwood as his tutor.
2. Strictly speaking, the terms in square brackets should be activities. These quantities may be taken as approximately equal to concentrations for dilute solutions.
3. Fajans claimed that the experiment represented a non-equilibrium situation for which the Nernst equation did not apply.
4. In the case of $^{1}$H, protium, and $^{2}$H, deuterium.
5. Post himself would not regard this as an omission. He is well known for his view that quantum mechanics is an 'incomplete' theory. He has pointed out that whereas the correspondence principle has become most commonly associated with quantum mechanics, this is one case in which true correspondence with the classical laws does not hold. I leave the discussion of such issues to physicists and have merely accepted the view that quantum mechanics represents the major revolution in modern physics and might therefore be expected to revolutionise chemistry.

   Relativity theory, the other major revolution has not revolutionised chemistry, although it would be incorrect to claim that it has had no influence. For example, in Dirac's relativistic treatment of the hydrogen atom the property of electron spin arises in a natural manner, unlike in non-relativistic quantum mechanics. More specific applications include an explanation of the inert pair effect occurring in elements such as mercury, thallium, lead, bismuth and polonium (Pyykkö 1988). Also, accurate *ab initio* calculations are increasingly turning to relativistic effects (Pyykkö 1978).
6. Many physicists, among them Oseen, Born and Einstein, accused Bohr of mysticism in the sense of holding metaphysical views contrary to the facts of experience (Honner 1982). Post holds a similar view of Bohr and often refers to him in conversation as "that Scandinavian mystic".
7. An atomic core consisting of the nucleus and inner-shell electrons showed a total of $N$ spectroscopic terms in a magnetic field. If an additional electron having an azimuthal quantum number $k$ were to be added, the new composite system would be expected to show $N(2k - 1)$ states since the additional electron was associated with $2k - 1$ states.

However, experiments revealed more terms. In general, the terms split into one type consisting of $(N + 1)(2k - 1)$ components and a second type of $(N - 1)(2k - 1)$ components, giving a total number of $2N(2k - 1)$ components. This represents a violation of the number of quantum states, since a two-fold increase seems to occur in the number of atomic core states on the introduction of an additional electron.

8. The element sulphur, for example, was attributed an electronic configuration of 2, 4, 4, 6, as noted previously. This grouping of electrons suggested that six electrons were more loosely bound than the others and successfully accounted for the formation of compounds such as sulphur hexafluoride, $SF_6$. However, sulphur can also form a tetrachloride, $SCl_4$, and a dichloride, $SCl_2$ (Kotz and Purcell 1977). The formation of these compounds suggests that two or four electrons may be more loosely bound than the remaining electrons, a notion which cannot be accommodated by the configuration given by Bohr.
9. This feature is a result of mathematical completeness. In principle any function can be modelled by expanding it in terms of a complete series of functions.
10. A recent dramatic example occurred in the field of medical tomography scanning, based on the phenomenon of nuclear magnetic resonance. For a number of years, spurious peaks were found in routine scans which were at first taken to indicate tumours. It has since been discovered that these peaks are the result of the failure of convergence in the mathematical methods on which the technique is based. I am grateful to Dr. George Ross of the Department of Physics, King's College, London, for this information.
11. The separate question as to whether chemistry can be reduced even in principle is of some philosophical interest and has been discussed elsewhere (Scerri 1992).
12. Mendeleef produced many different versions of the periodic table, including the so-called short and medium-long forms. A number of these versions successfully incorporated the rare earth elements. The first person to place the rare earths separately was Bassett, in 1892 (van Spronsen 1969).
13. I know this to be the case from frequent discussions on the subject over the past six years, during which Heinz Post supervised my Ph.D. thesis on the question of the reduction of chemistry (Scerri 1992).
14. In recent years chemical educators have become increasingly aware of this problem. For example, the standard text on inorganic chemistry by Cotton and Wilkinson has undergone a major revision with regard to the role of quantum mechanics in the presentation of the subject. Whereas the 945-page second edition (1962) has 105 pages dealing with theoretical principles and 31 pages on quantum mechanics, the fifth edition (1988) has a total of 1411 pages with only a brief section on concepts in stereochemistry and bonding. The most noticeable feature in the latter edition is the complete absence of a review of quantum mechanics.
15. The interview with Jørgensen appeared in the *Journal of Chemical Education* (Kauffman 1985).

## BIBLIOGRAPHY

Aston, F.W., Baxter, G.P., Brauner, B., Debierne, A., Leduc, A., Richards, T.W., Soddy F. and Urbain, G. (1923), 'Report of the International Committee on Chemical Elements, 1923', *Journal of the American Chemical Society*, 45, pp. 867–874.

Bender, C.F. and Davidson, E.R. (1966), *Journal of Physical Chemistry*, 70, p. 2675.

Bender, C.F. and Schaefer, H.F. (1970), *Journal of the American Chemical Society*, 92, p. 4984.

Bender, C.F. and Schaefer, H.F. (1971), *Journal of Chemical Physics*, 55, p. 4798.

Benedict, W.S., Gailar, N. and Plyer, E.K. (1956), *Journal of Chemical Physics*, 24, p. 1139.

Bernheim, R.A. (1971), *Journal of Chemical Physics*, 54, p. 3223.

Bohr, N. (1913), *Philosophical Magazine*, 26, p. 1; 476; 857.

Bohr, N. (1923), *Zeitschrift für Physik*, 13, p. 17.

Bunge, C.F. (1980), *Physics Scripta*, 21, p. 334.
Burgers, J.M. (1917), *Philosophical Magazine*, 514, p. 33.
Bury, C.R. (1921), *Journal of the American Chemical Society*, 43, p. 1602.
Cotton, F.A. and Wilkinson, G. (1980), *Advanced Inorganic Chemistry*, 4th ed., Wiley, New York.
Davidson, E.R. (1984), *Faraday Symposia*, 19, p. 7.
Ehrenfest, P. (1917), *Philosophical Magazine*, 500, p. 33.
Fajans, K. (1914), *Physikalische Zeitschrift*, 15, p. 935.
Goddard, W.A. (1985), *Science*, 227, p. 917.
Goldstein, H. (1980), *Classical Mechanics*, 2nd ed., Addison-Wesley, Reading, MA.
Greenaway, F. (1969), *Chemistry in Britain*, 5, p. 97.
Honner, J. (1982), *Zygon*, 17, p. 243.
Hund, F. (1974), *The History of Quantum Theory*, Harrap, London.
Kauffman, G.B. (1985), *Journal of Chemical Education*, 62, pp. 1002–1005.
Kim, J. (1982), *Philosophical Studies*, 41, p. 51.
Koertge, N. (1973), 'Theory Change in Science', in G. Pearce and P. Maynard (eds.), *Conceptual Change*, Reidel, Dordrecht.
Kotz, J.C. and Purcell, K.F. (1977), *Inorganic Chemistry*, Saunders College, Philadelphia, PA.
Kragh, H. (1977), *Journal of Chemical Education*, 54, p. 208.
Kragh, H. (1985), in A.P. French and P.J. Kennedy (eds.), *Niels Bohr: A Centenary Volume*, Harvard University Press, Cambridge, MA.
Landé, A. (1929), *Zeitschrift für Physik*, 15, p. 189.
Langmuir, I. (1919), *Journal of the American Chemical Society*, 41, p. 868.
MacKinnon, E.M. (1982), *Scientific Explanation and Atomic Physics*, University of Chicago Press, Chicago.
McLaughlin, D.R., Bender, C.F. and Schaefer, H.F. (1972), *Theoretica Chimica Acta*, 25, p. 352.
Paneth, F.A. and von Hevesey, G. (1914), *Physikalische Zeitschrift* 15, p. 797.
Paneth, F.A. (1920), *Naturwissenschaften*, 8, p. 839.
Pauling, L. and Wilson, E.B. (1935), *Introduction to Quantum Mechanics with Applications to Chemistry*, McGraw-Hill, New York.
Post, H.R. (1971), *Studies in History and Philosophy of Science*, 2, p. 213.
Pyykkö, P. (1978), *Advances in Quantum Chemistry*, 11, p. 353.
Pyykkö, P. (1988), *Chemical Reviews*, 88, p. 563.
Rud Nielsen, J. (ed.) (1977), *Niels Bohr, Collected Works*, Vol. III, North-Holland, Amsterdam.
Scerri, E.R. (1989a), *New Scientist*, 1651, p. 76.
Scerri, E.R. (1989b), *Journal of Chemical Education*, 66, p. 481.
Scerri, E.R. (1991a), *British Journal for the Philosophy of Science*, 42, p. 309.
Scerri, E.R. (1991b), *Journal of Chemical Education*, 68, p. 122.
Scerri, E.R. (1992), *The Relationship Between Chemical Periodicity, Quantum Mechanics and the Orbital Model*, Ph.D. thesis, University of London, London.
Schaefer, H.F. (1986), *Science*, 231, p. 1100.
Schwartz, C. (1963), *Methods in Computational Physics*, 2, p. 241.
Soddy, F. (1975), *Annual Reports to the London Chemical Society for 1910*, reprinted in T. J. Trent, *Radioactivity and Atomic Theory*, Taylor and Francis, London.
Stoner, E. (1924), *Philosophical Magazine*, 48, p. 719.
van der Vet, P.E. (1987), *The Aborted Takeover of Chemistry by Physics*, Ph.D. thesis, University of Amsterdam, Amsterdam.
van Spronsen, J.W. (1969), *The Periodic System of Elements*, Elsevier, Amsterdam.
Wasserman, E., Yager, W.A. and Kuck, V.J. (1970), *Chemical Physics Letters*, 7, p. 409.
Weinhold, F. (1972), *Advances in Quantum Chemistry*, 6, p. 299.
Weisskopf, V.F. (1989), *The Privilege of Being a Physicist*, Freeman, New York.
Wilson, E.B. (1965), *Journal of Chemical Physics*, 43, S172.
Wilson, S. (1984), *Electron Correlation in Molecules*, Pergamon, Oxford.

HARMKE KAMMINGA

# Taking Antecedent Conditions Seriously: A Lesson in Heuristics from Biology

## I. INTRODUCTION

Contemporary philosophers of science have, on the whole, treated the process of theory construction with uncharacteristic caution, leaving scientific genius mostly to its own devices. Even the strongly prescriptive methodology of Popper draws back from issuing rules in 'the context of discovery'. While Popper's conjectures and refutations model provides a dynamic for theory change in science and identifies the conditions under which a new theory needs to be formulated (with the focus on experimental anomalies), it is silent on questions of heuristics: the means whereby new theories might be constructed remain a matter of 'scientific creativity', which supposedly belongs in the domain of subjective psychology. It has been Heinz Post's great merit to bring at least part of the heuristic process back into the intersubjective realm where philosophical analysis has its proper place.

In his paper on correspondence, invariance and heuristics, Heinz Post (1971) advocates the policy of treating formal flaws in scientific theories as 'neuralgic or growing points', as the loci where theory modification is most likely to yield benefits. The construction of a replacement theory, far from being a mysterious procedure of conjecture, starts from the existing theory, building on its recognised successes while aiming to overcome its formal flaws. By a formal flaw, Post means analytic features of a theory such as logical inconsistency, generation of paradoxes, failure to unify existing theories into a coherent whole, incompleteness, and suchlike. Post's emphasis on formal flaws, as distinct from experimental anomalies, sets him apart from the traditional falsificationist: he explicitly denies a crucial role for experiment *by itself*, both in the refutation of theories and in the generation of new theories (Post 1971, p. 217).

The advantage of Post's approach is that it not only provides a criterion for the assessment of what, in any particular case, constitutes a progressive move in theory change, but that it at the same time yields an important heuristic principle. Informally, Post's 'general correspondence principle' states that if we are faced with a formal flaw in some theory $T$, we should direct our efforts at formulating a successor theory $T'$ which overcomes this flaw and



*S. French and H. Kamminga (eds.), Correspondence, Invariance and Heuristics*, 65–82.

yet conserves the (empirical, explanatory and predictive) successes of $T$. (As a rule, this would mean that the successor theory $T'$ yields the original theory $T$ under appropriate boundary conditions.) The progressive move here consists in the fact that $T'$ can be said to account both for what are, in a concrete way, understood to be the failures and the successes of $T$. Post's general correspondence principle does not of course determine the transition from $T$ to $T'$ uniquely, but at the very least it places strong constraints on theory modification: if $T'$ does not explain why $T$ worked to the extent that it did, then Post demands that $T'$ should be consigned to the wastepaper basket.

In his paper, Post explicitly chose to deal largely with physics, because it is the most formalised of the sciences (*ibid.*, p. 219) and the networks of abstract theoretical statements that are characteristic of physics are most easily subjected to the kind of formal analysis that Post advocates. Post himself added the rider that physics is, however, "remarkably unsuccessful" at answering empirical questions, comparing it unfavourably with botany in this respect and gaining Feyerabend's approval for doing so (Feyerabend 1975, pp. 60, 61).

My aim here is to discuss theory change in areas of science that are much less well structured at the theoretical level (and much less abstract) and yet have shown marked increases in explanatory force over time – notably the biological sciences. The neuralgic points are indeed less easy to pinpoint here and yet, looking back, it is clear that strategies for modifying theories were developed such that the limited successes of preceding theories were conserved and built upon. What is less obvious is what precisely it was in the existing theories that was modified[1].

In this paper I shall present a case study from biology which illustrates how the detection of formal flaws in the theories in question motivated theory change in a manner that is progressive by the criterion of Post's correspondence principle. The point of interest of this case study is that the primary theoretic changes were at the level of antecedent conditions, not at the level of laws. I shall argue below that this finding has implications for heuristics in general, well beyond the confines of the biological sciences. Indeed, I shall argue that greater weight should be placed on the role of antecedent conditions in all of science if the aim of increasing explanatory success is to be pursued.

## II. SETTING THE SCENE

First it is necessary to introduce a number of themes that will play a role in the discussion of the case material. In particular, some remarks are in order about the neglect of antecedent conditions in philosophical discussions about scientific progress, a theme to which I shall return in the concluding section. I shall then discuss the nature and role of historical reasoning in the biological sciences, in order to introduce some methodological features that are central to my case study.

## II.1. *The Philosophical Neglect of Antecedent Conditions*

Given the fact that most theories in science are complex structures of axioms and laws and boundary and initial conditions, there are several ways in which a flawed theory *T* can, in principle, be modified with a view to improving it:

1. One or more of the axioms of *T* (i.e. core assumptions that are given no justification within the theory itself) can be changed, a move leading in turn to a profound reinterpretation of much of the consequence class of *T*. The move from Newtonian to relativistic mechanics would be an example dear to philosophers of science.
2. One or more of the derived laws of *T* can be changed (or a new law can be added to the structure of *T*), for example the change, in chemical theory, from a law that relates oxidation to the addition of oxygen to a law that relates oxidation to the removal of electrons; or the addition of the law that all genes are located on chromosomes to classical genetics.
3. Some of the antecedent conditions incorporated in the structure of *T* can be changed. Unlike laws, cast in the form of universal statements, antecedent conditions are typically expressed in the form of existential, particular or singular statements. As will be illustrated below, antecedent conditions that are only loosely associated with one theory (and as such often called 'auxiliary statements') can be part of the structure of some other theory.

The three categories listed above are not intended to be mutually exclusive, nor are they exhaustive of possibilities for progressive theory change in science. There is, for example, the further possibility that the known consequence class of *T*, in terms of our understanding of what *T* entails, may change without any change in the axioms, laws or antecedent conditions built into *T*. Hertz's demonstration that radio waves are among the consequences of Maxwell's electromagnetic theory is a well-known example. An exploration of the consequences of a theory as it stands is surely not necessarily the uncreative business suggested by the phrases 'normal science' and 'puzzle solving'. The discussion here, however, will be confined to changes in the structure of a theory as opposed to changes in its perceived consequence class.

In the philosophical literature, discussions of theory change in the context of heuristics have focussed almost exclusively on changes in the laws or axioms of theories. Indeed, scientific theories are often tacitly equated with the set of laws and axioms built into them and antecedent conditions are at best granted an inferior status (see also Post *op. cit.*, pp. 236, 237). Of course, philosophers of science have written at length about modification of auxiliary statements (antecedent conditions) with respect to the *testing* of theories, especially in response to the Duhem-Quine thesis. While Kuhn (1962), Lakatos (1970) and Putnam (1974), among others, acknowledge that auxiliary statement modification is a common and legitimate practice in science, they still tend to associate significant theory change in science with changes at the level of the 'hard core' of laws and axioms, that is, at the level of universal statements. Indeed, the very term 'auxiliary' suggests the ascription of an inferior status.

While Post did not challenge this preoccupation with theory change at the level of universal statements in his paper on heuristics, his statement that the "study of the structure of existing houses may help us in constructing new houses" (*op. cit.*, p. 217) suggests the following analogy. There is a distinction between 'laws' in the form of structural rules of house building (concerning foundations, floors, walls, roofs, rooms, staircases, doors, windows, etc., as well as properties of building materials) on the one hand, and the conditions under which the structures are built (availability of building materials, choice of materials given the type of soil and local climate, choice of building techniques given the specific skills of the construction workers, etc.) on the other hand. When it comes to attempts to build better houses, I suggest that it could prove disastrous to ignore the conditions at the expense of the rules. The analogical situation with respect to theory construction in science has at the very least had the consequence that philosophers have neglected large areas of science where laws do not play as prominent a role as in theoretical physics and where, whether by default or for good reasons, laws do not necessarily have the dominant explanatory import.

The notion of explanatory import is admittedly difficult to capture formally. It hinges on what elements of an explanation are regarded as *informative* and is strongly context dependent. It is unlikely to be informative, for example, to explain Danton's death by guillotine in terms of the law that people die when their heads are cut off. What we are after is an account of the conditions in revolutionary France that led to the condemnation to death of Danton. In other cases the law(s) cited in an explanation may be the most informative element(s), for example the laws of Mendelian inheritance have the dominant explanatory weight in accounting for the distribution of haemophilia in certain families. On the other hand, the explanation of why a specific individual suffers from haemophilia hinges crucially on the particular condition that the X-chromosome inherited from that individual's mother carries the gene responsible for haemophilia. In short, the availability of laws does not guarantee their primacy in explanation: it depends on what is being explained to whom, as well as on the state of knowledge at the time when the explanation is offered.

Although it would be rash to decide *a priori* the precise scientific contexts in which laws and antecedent conditions, respectively, play the dominant explanatory role in science, a number of previous studies have indicated that antecedent conditions do play a prominent role in historical reasoning in the biological sciences[2]. This feature is borne out by my case study, which concerns a set of biological theories of a historical nature. I shall now discuss briefly the nature and role of historical reasoning in biology and introduce some terminology that will enter into the discussion of the case material.

### II.2. *Historical Reasoning in Biology*

The nature of historical reasoning in science, including the biological sciences, has been analysed carefully by Stephen Meyer (1990). He identified three distinctive characteristics of historical theories in natural science. Firstly, the underlying motivation in the construction of such theories is the identification of past causal events to explain present conditions, as opposed to the search for time-invariant laws. Secondly, the mode of reasoning used in identifying past causal events is one of retrospective causal analysis, which typically goes abductively from present conditions to past conditions; these past conditions, plus connecting laws, are then used to explain present (or at least more recent) conditions. In the explanations yielded by this process the primary explanatory work is done by antecedent conditions, the laws in the explanans playing only a supportive role, often tacitly. The theory as a whole will incorporate a set of such explanations, together constituting a causal narrative history. Such causal narrative histories are distinct both from mere chronology and from deductive-nomological explanatory models.

While chronologies are descriptive, causal narrative histories are explanatory in intent: they specify necessary conditions that are jointly sufficient to account for historical change. The difference with deductive-nomological patterns of argument hinges on the dominant explanatory import of particular conditions, as compared with universal laws: the particular conditions specified are crucial elements of the explanatory histories themselves. It is not the case that we start with a set of universal laws, then feed in some particular conditions and work out what the consequences would be; on the contrary, we start with a set conditions and work out how these conditions might have been produced from some earlier set of conditions and it is only when the latter conditions have been specified that the connecting laws begin to play an explanatory role.

Thirdly, Meyer identified a number of conditions of adequacy that such historical theories should meet. To begin with, the *etiological plausibility* of past conditions referred to in the causal narrative history must be established, by providing independent evidence that these conditions, if they were present, could have the required causal efficacy. (This will typically involve appeals to well-understood mechanisms.) Further, it must be established that the postulated causal conditions were in fact present at the relevant time; in other words, the *historical plausibility* of the past conditions must be established, preferably by means of several independent lines of argument leading to the same conclusion. (This will often involve appeals to historical 'records' in nature and also to the consequences of other theories.) Finally, alternative hypotheses about past conditions should be eliminated by establishing their etiological and/or historical implausibility.

In the subsequent discussion, I shall follow Meyer's characterisation of historical theories in science. First I wish to emphasise the important role played by historical reasoning in biology, to ward off any suspicion that this constitutes such a marginal practice in science as to render it unworthy of a philosopher's attention. A brief detour through the overall structure of biological explanation should show that such suspicions are ill-founded.

I have argued elsewhere (Kamminga 1992) that, in general, questions in biology can be interpreted in four different ways, as being concerned with (1) structures, (2) mechanisms, (3) function in the sense of the contribution that an item makes to maintaining the integrity of the system of which it is a part (systemic function) and (4) function in the sense of the selective advantage that possession of this item conferred on the species in question in its past evolutionary history (evolutionary or historic function). I suggested that, at least collectively, biologists aim to answer all four types of question, in a way that maintains overall explanatory coherence. I further argued that, in practice, coherence is maintained by means of a strong interconnectedness of explanations in terms of biological structures, mechanisms and functions: explanations of mechanisms typically refer to structures, explanations of systemic functions typically refer to mechanisms, explanations of historic functions typically refer to systemic functions, and explanations of structure typically refer to historic functions (see Kamminga *ibid.* for further details).

The focus in modern biology on structure-function relationships, mediated by mechanisms, is widely recognised by practising biologists and philosophers alike. What has received less emphasis is that historical reasoning enters crucially into explanations in terms of biological functions. As indicated above, function ascriptions have two interlinked components: a claim about the role that a functioning item has in maintaining the integrity of the system of which it is a part, and a claim about the selective advantage that was conferred by this item in the evolutionary history of the species where it is encountered (see also Mitchell 1989). Without the second component, function ascriptions become teleological in the sense that the overall organisation of constituents of the system is taken as an unexplained given and, in effect, treated as a formal cause. An escape route is provided by supplying a causal narrative history explaining how any particular functioning item was fixed by natural selection. In other words, teleology is avoided by historicising biology[3].

The upshot of this argument is that historical reasoning enters into all areas of biology if the explanatory coherence of biology is to be maintained. Of course, individual research programmes may focus on mechanisms, say, and not enquire into the historical issues, but that does not mean that these issues disappear: the explanatory interconnections outlined above imply that, somewhere along the line, the historical dimension must be taken into account in a concrete way.

The case study that follows is concerned with a set of biological theories that deal directly with historical questions. These theories are about as far away as one can get from the tightly structured formalisms of theoretical physics. What I subsequently hope to show is that the more striking features revealed by this case are by no means unique to scientific theories of an explicitly historical kind, but have implications for heuristics in science in general.

## III. THEORY CHANGE IN BIOLOGY: THE CASE OF THE ORIGIN OF LIFE

The case I shall consider is that of a number of successive theories of the origin of life on Earth that were formulated against the general background of Darwinian evolutionary theory. Soon after the publication of Darwin's *On the Origin of Species* in 1859, a number of evolutionists began to apply the historical thinking that had been imported into biology by Darwin to the question of the origin of life on Earth. In particular, they aimed to bridge theories about early geological history informed by the Kant-Laplace theory of the origin of the solar system on the one hand, and Darwin's theory of biological evolution on the other. The central question they addressed concerned the processes whereby Darwin's hypothetical common ancestor of all extant life on Earth could have come into being on a lifeless Earth. The explicit motivation for addressing this question was that it was seen as a flaw in Darwin's book that he hedged the question of the origin of the common ancestor, writing only in the final paragraph that life had been "originally breathed into a few forms or into one" (Darwin 1859, p. 490). Several contemporary authors commented that this omission left Darwin's theory incomplete[4].

The first point to note is that empirical anomalies played no role whatsoever in motivating theory construction in this case, but the identification of a formal flaw in Darwin's theory, that of incompleteness, did. The notion of incompleteness, one of the formal flaws listed by Post, is not, however, a straightforward one. All extant scientific theories are incomplete, but not all strategies for making them less incomplete are on a par. For a start, theories can be internally incomplete or externally incomplete. For example, Darwin's book, despite its title, had little to say on the subject of *mechanisms* of speciation – a case of internal incompleteness of the theory.

Secondly, Darwin's theory was externally incomplete in that it failed to link up explicitly with a whole range of other theories, but this problem was felt to be acute with respect to only a small subset of these other theories, a subset which included theories of early geological history. In principle, it could simply have been concluded that Darwin's theory was strictly about biology and as such legitimately silent about prebiological events on Earth. However, in the context of 19th-century thought, with its emphasis on the historical dimensions of nature and on the notion of progress, the evolution of life was seen as an important stage in the (supposedly progressive) historical development of matter in the Universe. In this context, it was meaningful to demand explanatory continuity between theories of cosmological, geological and biological evolution. It was seen as one of the *explanatory obligations* of evolutionary theory to provide an answer to the question about life's origins on Earth.

The notion of explanatory obligation[5] expresses the fact that we have certain expectations about what any particular theory ought to explain. Such expectations are bound to depend on the broader theoretical context and need not be stable over time. For example, it would be bizarre in present-day science to demand explanatory integration between astronomy and embryology, but conscientious medieval (and presumably New Age) astrologers might well feel obliged to establish causal links between astronomic events and individual development before birth. Incompleteness in the sense of failure to meet explanatory obligations is always perceived in relation to the problem situation generated within some particular theoretical framework. For this reason, incompleteness is less of a formal flaw than is internal logical inconsistency, for example. It is, nevertheless, a structural flaw in the theory, rather than a flaw exposed through empirical anomalies. And that is how the incompleteness of Darwin's theory was perceived at the time.

### III.1. *Theory Type 1: Evolutionary Abiogenesis*

The first set of theories of the origin of life on Earth motivated by Darwin's *On the Origin of Species* constituted an attempt to overcome the perceived flaw of incompleteness of Darwin's theory while leaving the original theory intact. The successes of the theory were built on by emulating Darwin's explanatory strategies of using gradualistic historical reasoning and excluding teleology in the form of divine design or natural purpose, but extending these strategies to the prebiological domain.

The theories constructed in this way by Ernst Haeckel, Thomas Huxley, Carl Nägeli, John Tyndall, August Weismann and others[6] rested on the following assumptions: (1) the environmental conditions on Earth have been subject to change through history; (2) life on Earth has changed historically in accordance with the principles of Darwinian evolution, having started from very simple beginnings (some sort of primordial cell); (3) vital functions are the outcome of the properties of a physicochemical substratum of complex carbon compounds, notably protein matter. These assumptions were themselves consequences of other theories: the first was derived from the Kant-Laplace theory and geological theory; the second from Darwinian theory; and the third from cell theory and physiological chemistry. In the construction of the theories of the origin of life under discussion, however, these theoretical statements were taken as given and operated as axioms.

The common feature of the theories built on these assumptions was that they presented life as a natural, historical product of inorganic nature: the transition from non-life to life was seen in terms of a series of natural processes, determined by the inherent properties of carbon compounds, acting over prolonged periods. Life emerged as the result of a series of chemical reactions, leading gradually from the simplest carbon compounds present in inorganic nature to increasingly complex organic substances, including eventually protein and protoplasm. Thus, historical continuity between chemical and biological processes under geological conditions in the long distant past was postulated and evolutionary gradualism was extended to prebiological processes.

It is worth emphasising what this set of theories did not provide: they did not incorporate proposals concerning specific mechanisms for the postulated chemical transitions, nor did they specify the environmental conditions under which the chemical processes leading gradually to the formation of protoplasmic bodies would have taken place. In other words, their concrete explanatory content was very limited: they were internally incomplete. What was provided was little more than a rough sketch of a theory, some framework for thinking about the origin of life on Earth, with the explicit aim of integrating the hitherto disjoint theories of early geological and biological history.

### III.2. *Theory Type 2: Pflügers Cyanogen Theory*

The second kind of theory remained within this framework, but moved significantly beyond the previous theories by incorporating accounts of mechanisms whereby supposedly fundamental constituents of protoplasm might have been formed, under specific conditions consistent with theories about the early history of the Earth. This approach was initiated by the physiologist Eduard Pflüger in a long paper on cellular respiration (Pflüger 1875). His theory hinges on a distinction at the chemical level between what he called 'live protein' and 'dead protein', marked, according to Pflüger, by the presence or absence of cyanogen[7]. On the basis of this distinction, he argued that any theory of the origin of life ought to explain the initial formation of cyanogen compounds on the Earth. Pflüger then pointed out that the various ways in which cyanogen compounds can be made in the laboratory have one factor in common, that is, the requirement of large amounts of heat. Hence, cyanogen compounds could have formed readily when the Earth was still glowing hot, as it had been according to the Kant-Laplace theory. He then discussed the formation of hydrocarbons and aromatic compounds under similar conditions and suggested that condensation and polymerisation reactions between these (small) molecules would have given rise to more complex organic matter as the Earth cooled, with the eventual production of 'live protein' and, hence, life.

Pflüger's prime innovation consisted in his proposals of specific chemical mechanisms, consistent with reaction schemes studied in the laboratory, and with specific environmental conditions that were historically plausible in the light of background theory, notably the Kant-Laplace theory. Implicitly, this move suggests an increased awareness of the principles of historical reasoning, in particular of the etiological and plausibility conditions that causal narrative histories should meet. Overall, there was an increase in explanatory content in comparison with the first set of theories.

With respect to the theoretical structure, there are three main differences between the first set of theories and that of Pflüger: (1) he incorporated the condition of a glowing hot, and subsequently cooling, Earth into the theory itself. Haeckel, etc., would not have denied the validity of this condition, but it played no particular role in their theories. (2) He incorporated specific chemical reaction schemes into the theory. Again, there would have been nothing to stop Haeckel, etc., from doing the same[8], but, without the rest of the 'package' presented by Pflüger, there was no criterion for choosing between the myriad organic reaction schemes studied in the laboratory. (3) Pflüger added a new law about the relation between cyanogen and life, making the distinction between life and non-life at the chemical level much more specific than it had been in the first type of theory. (The fact that this law was not widely accepted and was, moreover, soon rejected altogether does not concern me at this point; it is the structure of Pflüger's theory to which I wish to draw attention.)

Pflüger's theory suffered from flaws that are independent of the cyanogen distinction per se. The development of biochemistry in the first few decades of this century led to a rethinking about what a primitive living organism might have been like. The view that all vital behaviour is the outcome of the properties of some single, protein-like substance, was replaced by a perspective where different vital functions were associated with different metabolic reaction sequences, catalysed by sets of specific enzymes. Even a minimal living cell came to be seen as a system of complex networks of coordinated metabolic reactions. This meant that the Pflüger-style theory was at best incomplete: it might explain (in a sketchy way) the origin of protein, but that was not the same as explaining the origin of life.

### III.3. *Theory Type 3: Oparin's Metabolic Theory*

The third type of theory I shall discuss constitutes an attempt to overcome this flaw (and a lot more besides). It preserved the evolutionary thinking and the assumption of chemical continuity between prebiological and biological history that were built into the first two types of theory; and it preserved the concern of Pflüger with plausible mechanisms under historically plausible conditions – in fact, it put this concern at centre stage, with wide-ranging consequences. What was added is an emphasis on colloidal phenomena and metabolic patterns. The prime mover in this respect was the Soviet biochemist Aleksandr Oparin. I shall here focus on the influential theory he presented in his book *The Origin of Life*, first published in Russian in 1936 and subsequently widely translated[9].

The most striking feature of Oparin's theory is the wide range of scientific disciplines that he drew on to present a coherent explanatory scheme, where the proposed chemical and biochemical transformations were made plausible by establishing that they were fully consistent with theories about environmental conditions at each stage of development. In order to get some feel for the ways in which Oparin moved beyond his predecessors, it is necessary to provide an outline of the main features of his theory and of typical arguments that he used to support it.

A consideration of geochemical studies, chemical and mineralogical analysis of meteorites and spectroscopic studies of stars, planets and comets led Oparin to the conclusion that the Earth, like other planets in the solar system, originally had a reducing atmosphere, composed of hydrogen, methane, ammonia and water vapour. Under such reducing conditions, organic compounds would have formed readily as the Earth cooled and would have undergone progressive transformations in accordance with reaction mechanisms well known in organic chemistry, which are presented in detail by Oparin. It is important that under reducing conditions the organic molecules would not have been subject to oxidative breakdown (a problem that had been ignored by Pflüger, among others) and thus the condition of a reducing atmosphere becomes one of the crucial antecedent conditions built into the fabric of Oparin's theory.

As the organic compounds gained in complexity in an aqueous environment, colloidal reactions came into play and organic matter became concentrated in the form of individual colloidal bodies. Laboratory investigations of the phenomenon of coacervation had shown that colloidal bodies of the coacervate type have the property of absorbing and assimilating organic matter from the aqueous environment and of utilising this material in reactions leading to their growth – a feature reminiscent of metabolism in living organisms.

The property of metabolism thus provided Oparin with a possible link between prevital coacervates and primitive living organisms. Drawing on the biochemical traditions of the time, Oparin had a unified view of metabolism, seeing diverse metabolic reactions essentially as a series of coupled oxidations and reductions. At the same time, the diversity of metabolic patterns among extant organisms provided him with a comparative perspective on which he could base a possible sequence for the evolution of the earliest organisms on Earth.

According to Oparin, the first organisms were anaerobic, because there was no free oxygen in the atmosphere, and they were heterotrophic, that is, dependent on organic nutrients for growth. In other words, they were fermenters. Through fermentation processes, however, the environment would gradually have become depleted of organic matter, so that selection pressures in favour of the evolution of autotrophic (including photosynthetic) patterns of metabolism came into operation. The further evolution of photosynthetic organisms resulted in the direct ancestors of present-day blue-green algae, which release oxygen into the atmosphere. Eventually, the reducing atmosphere would have been transformed into an oxygenic one as a result of the photosynthetic activity of living organisms, at which stage conditions were ripe for the evolution of aerobic heterotrophic organisms. Oparin's choice of metabolism as the fundamental characteristic of living organisms and his comparative approach to the early evolution of life on Earth constituted important innovations.

Even after this brief outline, it should be clear that Oparin's theory has a much richer content than the earlier theories discussed above. The theoretical structure is also much more complex – a kind of patchwork of axioms, laws and initial and boundary conditions imported from a range of different, even disjoint, areas of science, yet woven into a coherent causal narrative history. Clearly, Oparin made use of a whole range of new insights, for example in biochemistry, that were not available to, say, Haeckel or Pflüger. But, apart from the detailed content, what were the major changes in the fabric of the theory?

There was, first, a change in the axioms: a move away from the assumption that the origin of life was effectively equivalent to the origin of protein matter, to one that used metabolism as the major operational criterion for any living system, which, as a consequence, had to be multimolecular. New laws about coacervation phenomena and metabolism were incorporated and the scope of the theory was enlarged to take in the early evolution of life, in order to establish the origin of all extant basic patterns of metabolism. Last but not least, new antecedent conditions were built into the theory, notably those concerning the composition of the atmosphere and the aqueous environment in which life was thought to have arisen. The condition of a reducing atmosphere had particularly wide-ranging consequences in terms of the selection of historically plausible chemical reaction mechanisms (quite different from those proposed by Pflüger). Here, a change in antecedent conditions made an enormous difference.

It might still be thought that Oparin's *theory* proper is structured around the assumptions and laws that it rests on and that all the initial and boundary conditions that enter into the causal narrative history as a whole play a peripheral theoretical role. This would be a highly problematic conclusion, because the laws were *selected* as being relevant to the theory, under the constraints of the antecedent conditions thought to be operative. In other words, the antecedent conditions played a central role in the construction of the theory. Each time Oparin presented a mechanism by which one set of conditions could have resulted in another set of conditions, he was of course appealing to laws, be they physical, chemical or biological. But it is the choice of conditions that determined the choice of the mechanisms and, by implication, the laws that were incorporated in the theory.

Oparin did not formulate any new laws – his creativity consisted precisely in formulating new antecedent conditions and only then invoking *appropriate* laws, selected on grounds of etiological and historical plausibility. In short, the antecedent conditions are a crucial part of the theory, not independent elements that are brought in after the fact to be subsumed under a set of pre-selected covering laws.

### III.4. *Summing Up*

The theoretic changes discussed here conform to Post's general correspondence principle. In each case, the motivation for constructing the theory was an attempt, not to overcome some experimental anomaly, but to correct a formal flaw in existing theory (primarily internal and/or external incompleteness) and in each case the successes of the previous theory were conserved. What is striking about the case study is the consequential role played by changes in antecedent conditions that were incorporated in the respective theories. What happens is the following: in the attempt to integrate the original theory *T* with another theory outside its domain, some consequence of the latter is incorporated into *T* as an antecedent condition, which then places strong constraints on the selection of laws that have explanatory relevance in the modified theory *T′*.

For example, in the transition from the Haeckel-type theory to that of Pflüger, the consequence of the Kant-Lapace theory of a glowing-hot earth in the early stages of geological history became an integral part of Pflüger's theory. With Pflüger's additional assumption of a chemical distinction between 'dead' and 'live' protein, appropriate chemical mechanisms could be selected in order to (1) integrate the Kant-Laplace theory with Darwin's theory of evolution and (2) fulfil the explanatory obligations of the bridging theory, by providing what was intended to be an etiologically and historically plausible account of the origin of life.

In the transition from the Pflüger-style theory to that of Oparin, the incorporation of one of the perceived consequences of then current astrophysical theory, that of a reducing atmosphere for the prebiotic earth, came to play a vital role. It enabled Oparin to select a whole range of chemical reaction mechanisms to account for the origin of multimolecular organic systems and their function of metabolism. In doing so, he not only integrated his theory of the origin of life with theories of astrophysics, geology, organic chemistry and evolution, but also with biochemical theory. His additional assumption that the origin of life could not be equated with the origin of protein, but required the formation of metabolising systems, led to an extension of his theory beyond previously established boundaries, thus enlarging its domain and enriching its content.

In summary, the case study as a whole illustrates theory change resulting in progressive increases in explanatory content and in intertheoretic coherence. The changes were motivated by perceived formal flaws and they were driven above all by changes in antecedent conditions.

## IV. IMPLICATIONS FOR HEURISTICS

The heuristic prescription embodied in Post's general correspondence principle does not specifically exclude theoretic change at the level of antecedent conditions, but change at that 'low' level has not exactly been encouraged. In cases where historical plausibility is an important criterion for theory acceptance, the antecedent conditions built into the theory carry considerable weight. It should cause no great surprise, then, that a change in antecedent conditions (together with the consequences of such a change for the selection of appropriate connecting laws) can significantly change the explanatory success of such theories, whether they be theories of the origin of life, of evolutionary change or of geological history. From the point of view of heuristics, all elements of a flawed theory of a historical nature should be considered as possible candidates for change, and the historical plausibility condition may well steer us in the direction of changing particular antecedent conditions before considering any changes in other elements of the theory.

The general point, that a change in the antecedent conditions built into a theory can be a progressive move by the criterion of Post's general correspondence principle, has much broader relevance, however. Consider the much discussed example of the orbit of Uranus and the anticipation of the existence of the planet named Neptune. This case has generally been discussed as a test of Newton's theory of universal gravitation, where a change in auxiliary assumptions not only saved the theory from refutation, but increased its explanatory and predictive success. Putnam (1974) especially used this case as an exemplar in his argument that theories such as Newton's are not highly falsifiable and that much of scientific practice is directed towards finding and changing auxiliary assumptions.

I would go one step further than Putnam and look at the Uranus-Neptune case from another perspective: not as a test of Newtonian gravitational theory, but as a modification of the theory of motion of the solar system (a subtheory embedded within Newtonian theory). It may be more profitable to view the observed orbit of Uranus, not as an observational anomaly for Newton's theory of gravitation, but as an indicator of a formal flaw in the theory of motion of the solar system: the theory is flawed because it is inconsistent with Newton's theory. The addition of a hypothetical planet then constituted, not a change in 'auxiliary' assumptions associated with Newtonian theory, but a change in the antecedent conditions built into the *core* of the subtheory, while all its laws remained unchanged. This change, moreover, did all the explanatory work in accounting for the orbit of Uranus; and by the general correspondence principle it constituted a progressive move in theory change.

Many other examples can be looked at from a similar perspective: Pasteur postulated the operation of an asymmetric physical force to maintain consistency between theories of chemical racemisation and of selective molecular dissymmetry in living things; the existence of messenger RNA was predicted (successfully) by Francis Crick to unify theories of DNA coding for proteins and theories of protein synthesis; and so on[10]. The lesson for heuristics is that, in general, and not just in the historical sciences, Post's correspondence principle can be satisfied by changes in any part of some theoretical structure, whether it be axioms, laws or antecedent conditions. Which parts are to be changed in order to ensure progress cannot be prescribed in blanket fashion *a priori*, but change at the level of antecedent conditions should at the very least be considered seriously as one of the options. In fact, it is plausible that changes at this level generally satisfy the requirements of Post's general correspondence principle in the most straightforward manner.

My emphasis has been on the explanatory success of theories and every philosopher acknowledges that antecedent conditions play an indispensable role in explanation. It still seems to be necessary, however, to stress that there is a trade-off between generalisation and explanatory power that hinges on the role of antecedent conditions. Of course, universal statements have much more *logical* power than do particular (or existential or singular) statements. As Popper has emphasised all along, universal statements (such as laws) are in principle falsifiable, while existential statements (such as antecedent conditions) are not. The greater logical power of universal statements may account for the overwhelming emphasis they have received in the literature on theory change. It is important, however, that logical power is not confused with explanatory import. When it comes to explanation, many universal statements are trivial. 'All men are mortal' is a useless generalisation when it comes to explaining why Einstein died at the age of 76, or why Descartes died of pneumonia rather than a heart attack. (Or, for that matter, why Marie Curie ever died at all.)

Of course, even non-trivial universal statements have no explanatory power on their own. Newton's laws by themselves cannot begin to explain why a particular apple fell on his head when it did, nor can they alone explain the difference between the moon orbiting the Earth and the swing of a pendulum. The construction of explanations is a highly developed skill. It involves the judicious selection of relevant laws and of relevant antecedent conditions, in conjunction. If we did have that great hope of *fin-de-siècle* physicists, just one equation on our T-shirts to explain everything, we would be able to explain precisely nothing. In order to differentiate between diverse things, events, processes, etc., we would still need our countless subtheories and models with distinct boundary and initial conditions built into them. Hence, to preserve any concrete explanatory power at all, we would have to pay more and more attention to these 'auxiliary statements'. Whether antecedent conditions are part of a theory or only brought in when explanations are constructed, they are ignored at our peril if philosophy of science is at all concerned with the explanatory efficacy of scientific knowledge.

My general plea, then, is for philosophers of science to take antecedent conditions more seriously, whether it be in the context of discovery or the context of justification. There has already been some move away from the traditional positivist and logical empiricist emphasis on axiomatisation, formalisation, theory reduction and formal deduction of explananda and predictions. Instead, philosophers have begun to show a greater concern with looser intertheoretic relationships and less formal ways in which scientists construct explanations in practice (for example, Cartwright 1983; Darden and Maull 1977; Kitcher 1981; van der Steen and Kamminga 1991). There is a growing recognition that there is more to science than logical deduction and that there is more to theories than universal statements. With this recognition, the domain of science that philosophers can usefully explore has been enlarged beyond those areas (mostly in theoretical physics) where theories are expressed in abstract formalisms. Far from involving a 'descent' into underdeveloped territory, studies of, say, biological cases can enrich philosophy of science by focussing attention on general (and respectable) scientific practices that had hitherto remained in the shadow. Such studies need not undermine fruitful philosophical notions established in other contexts: Post's general correspondence principle, for one, remains robust in the loosely structured, but coherent, theoretical domains discussed here.

## ACKNOWLEDGEMENTS

I am grateful to Steven French, Peter Lipton and Avinash Puri for very constructive comments on an earlier draft of this paper. If this account passes muster with Heinz Post, it will be thanks to their good efforts; if not, Heinz will no doubt know where to lay the blame.

*Wellcome Unit for the History of Medicine*
*Department of History and Philosophy of Science*
*University of Cambridge, Cambridge, U.K.*

## NOTES

1. The complexity of the situation in biological fields is well illustrated by Lindley Darden's study of strategies for theory change in genetics (Darden 1991). In her important and rich book, Darden discusses strategies for producing new ideas, strategies for theory assessment and strategies for the resolution of anomalies and for changing the scope of theories. Like Post, Darden emphasises that theory construction does not take place *de novo* and that theory evaluation and theory construction are intimately related: the nature of the failure of a theory can provide clues as to where and how that theory can most profitably be modified. Given this strong affinity between their starting positions, it is a pity that Darden does not refer to or comment on Post's paper, especially because Post's general correspondence principle might have served as a useful anchor for the many detailed heuristic guidelines that she offers. Darden's treatment goes significantly beyond Post's work in heuristics, but nothing in her book is in obvious conflict with Post's ground rules. The book is a beautiful illustration of the fruitfulness of a philosopher's concern with heuristics – it ought to have been written by a pupil of Post's!
2. Michael Scriven was one of the first philosophers to emphasise this point (see, for example, Scriven 1959). More recently, the role of antecedent conditions in historical explanation in biology has been discussed extensively by Meyer (1990). The status and role of non-universal statements in biology has been discussed more generally by van der Steen and Kamminga (1991).

3. Interestingly, Heinz Post also drew attention to what he called the "curious role" played by the concept of evolution "in converting biology into a historical science" (Post 1971, p. 245). My argument here is that the all-pervasiveness of the historicising consequence of Darwinian evolutionary theory for biology does not look quite so curious if the use of evolutionary theory as a device for avoiding teleology is recognised. (Especially as no other effective means for avoiding teleological reasoning in biology has been thought of thus far.) It is *by* converting biology into a historical science that teleology is avoided.
4. For explicit examples, see Haeckel (1866, Vol. I, p. 169), or Richter's statement, among others of this kind in the same article, that his own theory of the extraterrestrial origin of life on Earth "provides the cornerstone for Darwin's bold edifice" (Richter 1865, p. 249).
5. I owe the notion of explanatory obligations of theories to Peter Lipton (personal communication).
6. Ernst Haeckel was the most prolific writer on this subject, beginning with his *Generelle Morphologie der Organismen* (1866). See Huxley 1869, Nägeli 1884, Tyndall 1874 and Weismann 1882 for other typical examples.
7. Investigating the nitrogenous breakdown products of proteins, Pflüger had found that the metabolic decomposition of proteins in living organisms always yielded excretory substances containing cyanogen, while he never obtained such products upon the artificial cleavage of proteins in dead materials such as foodstuffs. He concluded that cyanogen is a fundamental characteristic of the living (see Pflüger 1875).
8. In fact, Haeckel later did incorporate Pflüger's proposals in his own writings; see Haeckel 1904, pp. 361–362.
9. The translation into English was published in 1938. Oparin published other works on the subject of the origin of life both before and after 1936. I have chosen to focus on this particular book because it presents the fully articulated theory that formed the foundation for subsequent developments in the field, including Oparin's own later writings. His earlier work lacks the rich chemical and biochemical content that gave the 1936 theory its concrete content and its heuristic value. In addition, the key condition of a reducing atmosphere was only introduced in 1936. More recently this notion has come under challenge, but that does not affect my argument here. Indeed the very nature of the charge that a reducing atmosphere is historically implausible reinforces the methodology adopted by Oparin, if not the detailed content of his theory (see also Meyer 1990, Ch. VI).
10. Steven French has suggested that the prediction and discovery of the $\Omega$ particle in the 1960s is another case in point, although here the background theory was not so well secured, so that the sense in which there was a flaw to be overcome is less clear. More generally, it is through changing antecedent conditions that new entities are introduced, with obvious implications for the predictive success of theories.

## BIBLIOGRAPHY

Cartwright, N. (1983), *How the Laws of Physics Lie*, Clarendon Press, Oxford.
Darden, L. and Maull, N. (1977), 'Interfield Theories', *Philosophy of Science*, 44, pp. 43–64.
Darden, L. (1991), *Theory Change in Science: Strategies from Mendelian Genetics*, Oxford University Press, Oxford.
Darwin, C. (1859), *On the Origin of Species by Means of Natural Selection*, John Murray, London.
Feyerabend, P. (1975), *Against Method*, New Left Books, London.
Haeckel, E. (1866), *Generelle Morphologie der Organismen*, Berlin.
Haeckel, E. (1904), *The Wonders of Life* (transl. by J. McCabe), London.
Huxley, T.H. (1869), *Protoplasm. The Physical Basis of Life*, Melbourne.
Kamminga, H. (1992), 'The Structure of Explanation in Biology and the Construction of Theories of the Origin 14 of Life on Earth', *Uroboros*, 2, pp. 47–64.
Kitcher, P. (1981), 'Explanatory Unification', *Philosophy of Science*, 48, pp. 507–531.

Kuhn, T.S. (1962), *The Structure of Scientific Revolutions*, University of Chicago Press, Chicago.
Lakatos, I. (1970), 'Falsification and the Methodology of Scientific Research Programmes', in I. Lakatos and A. Musgrave (eds.), *Criticism and the Growth of Knowledge*, Cambridge University Press, Cambridge, pp. 91–196.
Meyer, S.C. (1990), *Of Clues and Causes. A Methodological Interpretation of Origin of Life Studies*, Ph.D. thesis, University of Cambridge, Cambridge.
Mitchell, S. (1989), 'The Causal Background of Functional Explanation', *International Studies in the Philosophy of Science*, 3, pp. 213–229.
von Nägeli, C. (1884), *Mechanisch-physiologische Theorie der Abstammungslehre*, Munich-Leipzig.
Oparin, A.I. (1938), *Origins of Life* (transl. by S. Morgulis), MacMillan, New York.
Pflüger, E. (1875), 'Ueber die physiologische Verbrennung in den lebendigen Organismen', *Archiv für gesammte Physiologie*, 10, pp. 251–367.
Post, H.R. (1971), 'Correspondence, Invariance and Heuristics: in Praise of Conservative Induction', *Studies in History and Philosophy of Science*, 2, pp. 213–255.
Putnam, H. (1974), 'The 'Corroboration' of Theories', in P.A. Schilpp (ed.), *The Philosophy of Karl Popper*, Open Court, La Salle, Illinois, pp. 221–240.
Richter, H.E. (1865), 'Zur Darwin'schen Lehre', *Schmidt's Jahrbücher der in-und ausländischen gesammten Medicin*, 126, pp. 243–249.
Scriven, M. (1959), 'Explanation and Prediction in Evolutionary Theory', *Science*, 130, pp. 477–482.
van der Steen, W.J. and Kamminga, H. (1991), 'Laws and Natural History in Biology', *British Journal for the Philosophy of Science*, 42, pp. 445–467.
Tyndall, J. (1874), *Address Delivered Before the British Association Assembled at Belfast, 1874*, London.
Weismann, A. (1882), *Studies in the Theory of Descent* (transl. by L. Meldola), London.

ALLAN FRANKLIN

# The Rise of the 'Fifth Force'

## PROLOGUE

In academic life it is easy to acknowledge the work of others. You simply provide an appropriate reference or citation. But how do you acknowledge an influence and spirit that pervades an entire body of work? I don't know how to, but Heinz Post's critical spirit and his emphasis on experiment and on the importance of the actual practice of science to the philosophy of science has influenced everything I've done. The best I can do is to say thank you.

Perhaps a little autobiography may help to explain this. I arrived in London in January 1978 to spend my sabbatical leave as a visiting researcher in the Department of History and Philosophy of Science, Chelsea College, then headed by Heinz. When I first appeared at the college Heinz made it very clear that I was to consider myself a full member of the department and that one of my most important duties would be to attend and participate in the weekly departmental seminars. At my first seminar Heinz introduced the speaker and announced that the speaker had requested a straight run and that questions would be deferred until the speaker had finished. This was not true. At this talk, and at every subsequent talk I attended, chaired by Heinz, I watched as he became progressively more uncomfortable. Finally, after about an average of seven minutes, he could bear it no longer. He asked the first question. What followed can only be described as an intellectual free for all with the speaker, if they were lucky, getting to speak fifty percent of the time. What distinguished the questions and criticisms was both their severity and their good-heartedness. The intellectual questions were always the issue. It was never personal. You might be wrong, but you weren't stupid. The seminars are quieter now, and the speaker does get a straight run. I liked it better in the old days.

As a veteran who has given, at last count, thirteen of the seminars I have good empirical evidence to support my claims. Whenever I've given these talks I have always looked at Heinz's reaction. If he was nodding gravely and approvingly I felt I was doing well. I hope this essay merits that nod.



*S. French and H. Kamminga (eds.), Correspondence, Invariance and Heuristics*, 83–103.

## I. INTRODUCTION

One of the interesting questions in the history and philosophy of science is how a hypothesis is proposed and how it acquires sufficient plausibility to be considered worthy of further theoretical and experimental investigation by the scientific community. I am not denying that the suggestion of hypotheses or theories is the free creation of an individual scientist, but I doubt that such creative events occur in a vacuum. Thus, although Newton's thoughts on the universality of gravitation may have been triggered by the apple falling on his head, it seems unlikely that it would have had that effect had he not already been thinking about gravitation and the motion of the moon[1]. I suggest that when a scientist offers a hypothesis they, or the rest of the scientific community, may have been considering the problem for a time. In addition to solving the problem, the hypothesis is also likely to be supported by other empirical evidence, or has some theoretical plausibility because it resembles previous successful solutions of other problems[2]. It may also fit in with an existing research program or look like a fruitful or interesting line of research. Another factor may be that the theory has desirable mathematical properties. For example, the Weinberg-Salam unified theory of electroweak interactions did not receive much attention until 1971 when t'Hooft showed it was renormalizable[3]. These may also be reasons why a theorist may pursue an hypothesis.

An experimentalist planning to investigate an hypothesis may have similar reasons for their work. In addition, there may be what one might call experimental reasons for such pursuit. These may include the fact that the proposed measurement can be done with existing apparatus or with small modifications of it. The measurement may fit in with an existing series of measurements in which the experimenter(s) have expertise or the experimenter may think of a clever way to perform the measurement. If the hypothesis is sufficiently important the experimenter may even construct an entirely new apparatus. At this point the cost of the experiment, the availability of research funds, as well as the perceived interest and importance of the experiment and hypothesis will certainly enter into the decision to do the experiment, but that is left for future discussion.

As we shall see below, the suggestion of a 'Fifth Force' in gravitation occurred after the authors had been worrying about the problem for some time, did have some empirical support, and also resembled, at least in mathematical form, Yukawa's previous successful suggestion of the pion to explain the nuclear force. It also fit in with the previous work on modifications of gravitational theory by Fujii and others.

In this paper I will look at how and why Ephraim Fischbach, Sam Aronson, Carrick Talmadge and their collaborators came to suggest modifying gravitational theory by adding such a force[4]. I will also examine the evidential context at the time that led at least a segment of the physics community to investigate this hypothesis. At the present time, I believe it is fair to say that the majority of the physics community does not believe that such a force exists. The current experimental limits on the strength of such a force are approximately $10^{-4}$ that of the normal gravitational force (this depends on the choice of coupling, i.e. baryon, isospin, etc., and on the assumed range of the force). Although some experimental anomalies do remain they are not presently regarded as serious[5]. There are, in addition, other experimental results which contradict the anomalous results and the overwhelming preponderance of evidence is against the existence of the Fifth Force.

## II. K MESONS AND CP VIOLATION

The story of the Fifth Force begins with a seeming digression because it involves not a modification of gravitational theory, but rather new predictions and tests of that theory. In 1975 Colella, Overhauser and Werner had measured the quantum mechanical phase difference between two neutron beams caused by a gravitational field[6]. Although these experiments showed the effects of gravity at the quantum level, they did not, in fact, distinguish between General Relativity and its competitors, as Fischbach pointed out (1980, Fischbach and Freeman 1979). This was because these experiments were conducted at low speeds, and in the non-relativistic limit all existing gravitational theories, such as General Relativity and the Brans-Dicke theory, reduced to Newtonian gravitation. Fischbach went on to discuss how one might test general relativity at the quantum level by considering gravitational effects in hydrogen.

In this work, partly as a result of conversations with Overhauser, Fischbach went on to consider whether or not gravitational effects might explain the previously observed violation of CP symmetry (combined charge-conjugation or particle-antiparticle symmetry and parity or space reflection symmetry) in $K^o_L$ decays. He had shown that an external gravitational field resulted in an admixture of atomic states of opposite parity. For a two-fermion system, such as positronium or charmonium, this also leads to a change in the eigenvalue of CP. This made "it natural to attempt to connect $V^\sigma$ (the gravitational effect) with the known CP-violating $K_L$ decays" (Fischbach 1980, p. 371). Although, as Fischbach noted, there were both experimental and theoretical reasons against gravity as the source of CP violation, the relevance of the arguments to his case were not clear.

The arguments Fischbach was referring to concerned attempts to explain CP violation and will be relevant to the later history as well. Bell and Perring (1964) and Bernstein, Cabibbo and Lee (1964) had speculated that a long-range external field that coupled differently to the $K^o$ and $\overline{K}^o$, a hyperphoton, could explain the observation. Such a field predicted that the effect would be proportional to the square of the energy of the K mesons. Weinberg (1964) pointed out that because neither strangeness nor isotopic spin, the supposed origins of the field, were absolutely conserved, the hyperphoton must have a finite mass, related to the range of the interaction. Assuming that the range of the interaction was the size of our galaxy, he calculated the ratio ($K^o_2 \rightarrow 2\pi$ + hyperphoton) / $K^o_2 \rightarrow 2\pi$) as $10^{19}$. This implied that the K meson and all strange particles would be totally unstable, in obvious disagreement with experiment. He could explain the observations if he assumed that the range of the interaction was the size of the earth, which he regarded as implausible. The issue became moot when the experiments of Galbraith *et al.* (1965) and of DeBouard *et al.* (1965) at very different energies from both each other and from the original experiment of Christenson *et al.* (1964) failed to show the predicted energy-squared dependence. In fact, the experiments indicated that the CP violation was constant as a function of energy for the energy range 1 – 10 GeV.

Fischbach was also motivated by what he took to be a "remarkable numerical relation." Using his calculated energy scale for the gravitational effect, $\Delta m$, the known $K_L - K_S$ mass difference, and an enhancement factor of $m_K/\Delta m$, for which no justification was given, he found that his calculation of the gravitational effect for K mesons = $0.844 \times 10^{-3}$, whereas the CP violating parameter 1/2 $Re\ \varepsilon$ was equal to $(0.83 \pm 0.03) \times 10^{-3}$ or $(0.80 \pm 0.03) \times 10^{-3}$, for $K_L$ semileptonic decay or decay into two pions, respectively. This seems indeed to be a remarkable coincidence because there is no known connection between gravity and CP violation, or any accepted explanation of CP violation itself.

Fischbach continued to work on the question of how to observe gravitational effects at the quantum level. A relativistic version of the Colella, Overhauser and Werner experiment using neutrons did not seem feasible, so he turned his attention to K mesons, where such experiments did seem possible. He began a collaboration with Sam Aronson, an experimental physicist with considerable experience in K meson experiments. At this time Aronson and his collaborators had been investigating the regeneration of $K_S$ mesons (Roehrig *et al.* 1977 and Bock *et al.* 1979). Although the published papers stated that "the data are consistent with a constant phase (of the regeneration amplitude)" (Bock 1979, p. 351–2)[7], Aronson and Bock, two members of the group were troubled by what seemed to be an energy dependence of the phase. In fact, Bock had investigated whether changes in the acceptance could account for the effect. They couldn't. The data are shown in Figure 1, along with the constant phase prediction. Although the data are consistent with a constant phase, there is at least a suggestion of an energy dependence. The low energy points have a larger phase than the high energy points.

Aronson and Bock then asked Fischbach if there was a theoretical explanation of the effect. Fischbach had none to offer. This suggested energy dependence led them to examine the possible energy dependence of the parameters of the $K^o - \overline{K}^o$ system in some detail. They found suggestive evidence for such a dependence (Aronson 1982). They examined $\Delta m$, the $K_L - K_S$ mass difference, $\tau_S$, the lifetime of the short-lived K meson, $|\eta_{+-}|$, the magnitude of the CP-violating amplitude, and tan $\phi_{+-}$, the tangent of the phase

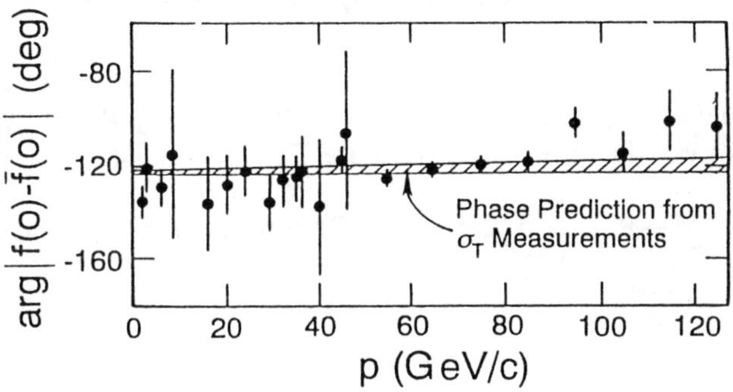


*Figure 1.* The phase of the regeneration amplitude as a function of momentum. From Bock et al. (1979).

of the CP-violating amplitude. They fitted these parameters to an energy dependence of the form $x = x_o\ (1+ b_x\gamma^N)$, $N = 1, 2$ and $\gamma = E_K/M_K$. They found that the coefficients differed from zero by 3, 2, 2, and 3 standard deviations, for the quantities noted above, respectively. One of their fits is shown in Figure 2.

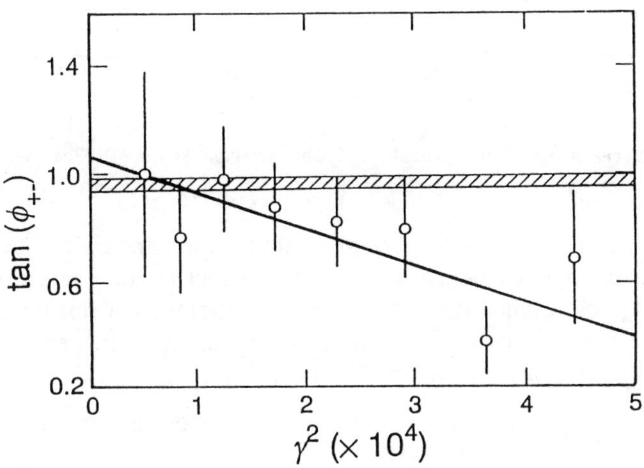


*Figure 2.* Tan $\phi_{+-}$ as a function of $\gamma^2$ (energy squared). From Aronson et al. (1982).

The fit to $\phi_{+-}$ depended on the value one assumed for $\phi_{21}$, the phase of the regeneration amplitude. (Recall that it was the suggestion of an energy dependence in this quantity that led to this investigation.) The measured quantity, in fact, depends on $\phi_{21} - \phi_{+-}$. One could attribute the energy dependence to either one of them separately, or to both of them. All theoretical models at the time (See Aronson 1983b for details) predicted that over this energy range, 30 – 110 GeV, the change in $\phi_{21}$ would be less than 2°. The observed change of approximately 20° was then attributed to an energy dependence in $\phi_{+-}$.

The group continued their study and presented a more detailed analysis that included data from other experiments. The most significant energy dependence, that in $\phi_{+-}$, is shown in Figure 3[8]. They concluded,

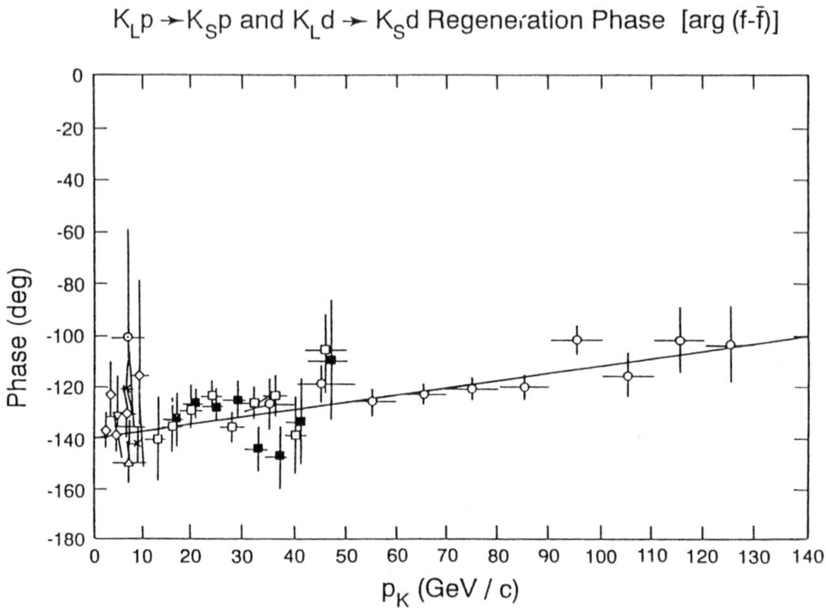


*Figure 3.* $\phi_{+-}$ as a function momentum. From Aronson *et al.* (1983a). See discussion in note 7.

> The experimental results quoted in this paper are of limited statistical significance. The evidence of a positive effect in the energy dependences of $\Delta m$, $\tau_S$, \$\eta_{+-}$\, and $\phi_{+-}$ is extremely tantalizing, but not conclusive. The evidence consists of $b_x^{(N)}$'s which are different from zero by at most 3 standard deviations (Aronson 1983a, p. 488).

A second paper (Aronson 1983b) examined possible theoretical explanations of the effects[9]. They found,

> Using this formalism we demonstrate that effects of the type suggested by the data (energy dependences) cannot be ascribed to an interaction with kaons with an electromagnetic, hypercharge, or gravitational field, or to the scattering of kaons from stray charges or cosmological neutrinos. (Aronson 1983b, p. 495)

They suggested that a tensor field mediated by a finite mass quantum might explain the effects and concluded,

> It is clear, however, that if the data . . . are correct, then the source of

> these effects will represent a new and hitherto unexplored realm of physics. (Aronson 1983b, p. 516)[10]

The subsequent history of measurements of these quantities seems to argue against any energy dependence[11]. Coupal *et al.* (1985) measured $\backslash\eta_{+-}\backslash$ at 65 GeV/c and found a value $\backslash\eta_{+-}\backslash = (2.28 \pm 0.06) \times 10^{-3}$ in good agreement with the low energy (5 GeV) of $(2.274 \pm 0.022) \times 10^{-3}$, and in disagreement with $(2.09 \pm 0.02) \times 10^{-3}$ obtained by Aronson (1982)[12]. Grossman *et al.* (1987) measured $\tau_S$, the $K_S$ lifetime, over a range 100–350 GeV/*c*. Their results, along with those of Aronson (1982), are given in Figure 4. The fits obtained by Aronson for possible energy dependence are also shown. They concluded,

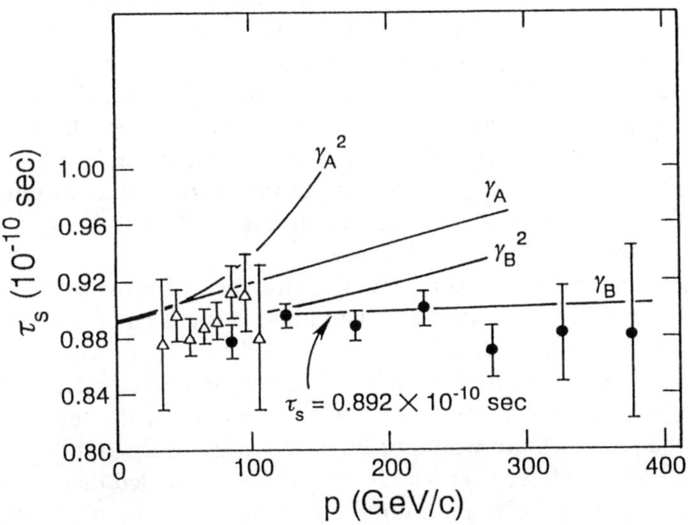


*Figure 4.* $\tau_S$, the $K_S$ lifetime, as a function of momentum. From Grossman *et al.* (1987).

> No evidence was found for the momentum dependence suggested by the intermediate-range 'fifth-force' hypothesis. (Grossman *et al.* 1987, p. 18)

Still, at the time of our story, the suggested energy dependence remained a "tantalizing" effect.

## III. MODIFICATIONS OF NEWTONIAN GRAVITY

A second strand of our story concerns the recent history of alternatives to, or modifications of, standard gravitational theory[13]. For a time, at least, this strand was independent of the K meson story. The Fifth Force story, per se, began when the two strands were joined. Newtonian gravitational theory and its successor, Einstein's general theory of relativity, although strongly supported by existing experimental evidence[14], have not been without competitors[15]. Thus, Brans and Dicke (1961) offered a scalar-tensor alternative to General Relativity. This theory contained a parameter $\omega$, whose value determines the relative importance of the scalar field compared to the curvature of spacetime. For small values of $\omega$ the scalar field dominates, while for large values the Brans-Dicke theory is indistinguishable from General Relativity. By the end of the 1970s experiment favored a large value of $\omega$, and thus, favored General Relativity. For example, the lunar laser-ranging experiments required $\omega > 29$, while the Viking time-delay results set a lower limit of $\omega > 500$ (Will 1981).

In the early 1970s, Fujii (1971, 1972, 1974) suggested a modification of the Brans-Dicke theory that included a massive scalar exchange particle in addition to the usual massless scalar and tensor particles. He found that including such a particle gave rise to a force that had a short range (of the order 10m – 30km) depending on details of the model. In Fujii's theory, the gravitational potential had the form $V = -GmM/r(1 + \alpha\, e^{(-r/\lambda)})$ where $\alpha$ was the strength of the interaction and $\lambda$ its range. The second term was Fujii's modification. O'Hanlon (1972) suggested the same potential. This model also predicted a gravitational constant $G$ that varied with distance[16]. Fujii calculated that the gravitational constant at large distances $G_\infty$ would be equal to 3/4 $G_{LAB}$, the value at short distances.

Fujii also looked for possible experimental tests of this theory. Most interestingly for our story, he discussed the famous experimental test of Einstein's equivalence principle that had been performed by Eötvös and his collaborators (1922). (This experiment, which will be very important later in our history, will be discussed below.) He noted that if his new field coupled equally to baryons and leptons there would be no effect, while if the field did not couple to leptons such an effect would be observed. He calculated that for an Eötvös-type experiment on gold and aluminum there would be a change in angle $\eta = 0.07 \times 10^{-11}$, for an assumed range of 40km. The best experimental limit at the time was that of Roll, Krotkov and Dicke (1964) of $3 \times 10^{-11}$, although that experiment which measured the equality of fall toward the sun cast no light on his short range force. A note added in proof remarked that he had learned that the best estimate of the range of such a force was considerably smaller. For an assumed 1km range he found that the change in angle was $0.5 \times 10^{-9}$. This was still smaller than the limit set by Eötvös, whose experiment did apply to his theory. He suggested redoing the Eötvös experiment and other possible geophysics experiments, although he noted that mass inhomogeneities would present difficulties.

Other modifications of gravitational theory were suggested by Wagoner (1970), Zee (1979, 1980), Scherk (1979), and others. Zee's modification had a much shorter range than that of Fujii, while Scherk suggested a repulsive force with a range of about 1 km. Long (1974) considered the question of whether or not Newtonian gravity was valid at laboratory dimensions. He

found that past $G$ (the gravitational constant) measurements in the labora-

> tory set only very loose limits on a possible variation in $G$ and that present technology would allow a considerable improvement. (p. 850)

He also made reference to the suggestions of Wagoner, of Fujii, and of O'Hanlon. Long (1976) proceeded to test his hypothesis experimentally, and found a small variation in $G$ of the form $G(R) = G_0(1 + 0.002 \ln R)$, where $R$ is measured in centimeters.

Long's work led Mikkelsen and Newman (1977) to investigate the status of $G$[17]. They used data from laboratory measurements, orbital precession, planetary mass determinations, geophysical experiments, and solar models. They concluded,

> Constraints on $G(r)$ in the intermediate distance range from $10\text{m} < r < 1\text{km}$ are so poor that one cannot rule out the possibility that $G_c$ ($G_\infty$) differs greatly from $G_0$ ($G_{LAB}$). (p. 919)

They pointed out that their analysis "does not even rule out Fujii's suggested value $G_c/G_0 = 0.75$" (p. 924).

The experimental study of possible violations of Newtonian gravity continued. Panov and Frontov (1979) found $G(0.3\text{m})/G(0.4\text{m}) = 1.003 \pm 0.006$ and $G(10\text{m})/G(0.4\text{m}) = 0.998 \pm 0.013$. They concluded that, despite the fact that their experimental uncertainty was larger than Long's measured effect, "These results do not confirm the data of D.R. Long, according to which spatial variations of $G$ do exist" (p. 852). Spero *et al.* (1980) agreed. Their measurements at distances of 2cm – 5cm had the required sensitivity to check Long's result and

> The results support an inverse-square law. Assuming a force deviating from inverse square by a factor $(1 + \varepsilon \ln r)$ (Long's suggested form) it is found that $\varepsilon = (1 \pm 7) \times 10^{-5}$. (p. 1645)

Long (1981) surveyed the literature and concluded that within the quoted uncertainties all the results, including that of Panov and Frontov, were consistent with his observation. He argued that Spero's result did not, in fact, contradict his. This was because his suggested cause of the deviation, a quantum gravity vacuum polarization effect, would be significant only for a non-zero gravitational field, whereas Spero's experiment was conducted in a zero gravitational field.

The most important summary of this work, from the point of view of the subsequent history of the Fifth Force, was that given by Gibbons and Whiting (1981). Their results are shown in Figure 5 for both attractive and repulsive forces. In both cases $\alpha$ is restricted to lie below each curve, except for curve *b*, which is Long's result, and in which $\alpha$ must lie between the two curves. They stated,

> However, conventional vacuum polarization effects in quantum gravity do not lead to the behavior required by Long: such effects are insignificant. (p. 636)

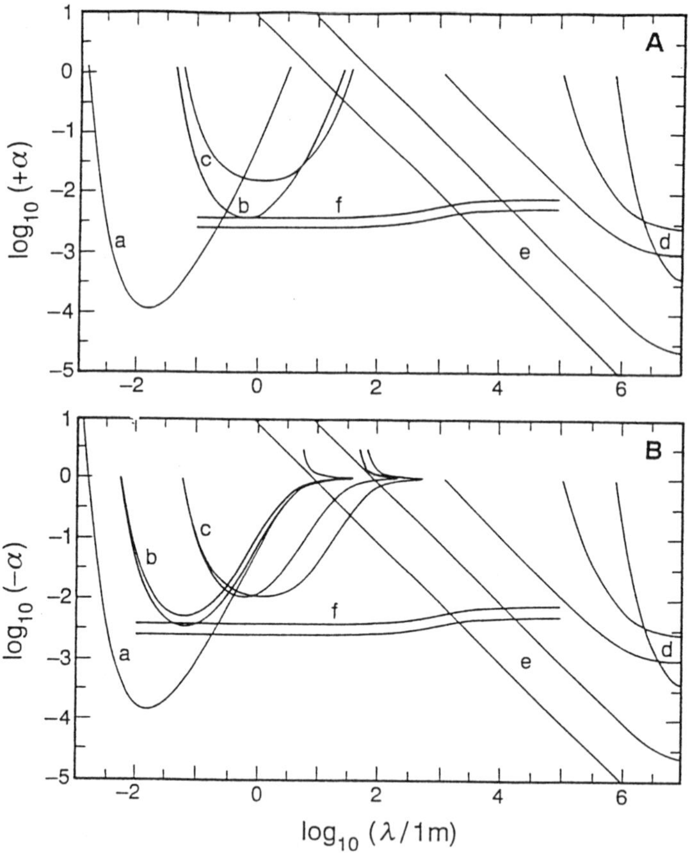


*Figure 5.* $\text{Log}_{10}\ \alpha$ versus $\log_{10}(\lambda/1\text{m})$. $\alpha$ is constrained to lie below the curves. From Gibbons and Whiting (1981).

Curve a was Spero's data, calculated at 1 standard deviation (s.d.); c was from Panov and Frontov; d from Mikkelsen and Newman using lunar surface gravity and Mercury and Venus flybys; e a comparison of satellite and geodesy data by Rapp (1974, 1977), the upper curve assumed agreement to 0.1 ppm (parts per million) and the lower curve 1 ppm. Rapp had reported an agreement to 2 ppm.

A different type of experiment, that of measuring gravity in either a mine (Stacey *et al.*, 1981) or in submarines (Stacey, 1978) was discussed in curve f. The curves were *calculated*, not measured, assuming a mine experiment with an accuracy of 1% (upper curve) and a submarine experiment with an accuracy of 0.1% (lower curve), both for a depth of 1 km. Stacey (1981) had measured $G$ and found it to be $G = (6.71 \pm 0.13) \times 10^{-11}\ \text{m}^3\text{kg}^{-1}\text{s}^{-2}$, in agreement with the laboratory value of $(6.672 \pm 0.004) \times 10^{-11}$. Their quoted uncertainty included an estimate of possible systematic effects, which increased the uncertainty by about a factor of three. They also surveyed other mine and borehole measurements of $G$ and found them to be, in general, systematically slightly high but "tantalizingly uncertain" because of possible mass anomalies. A somewhat later paper (Stacey and Tuck 1981) gave numerical details of that survey and reported values of $G$, calculated in two different ways, based on a comparison of sea floor and sea surface measurements of $G = (6.730 \pm 0.010) \times 10^{-11}$ and $(6.797 \pm 0.016) \times 10^{-11}$ $m^3kg^{-1}s^{-2}$, where the uncertainty is purely statistical and does not include possible systematic effects. Once again the results were higher than the laboratory value, but because of the uncertainty about possible systematic effects no firm conclusion could be drawn.

Gibbons and Whiting summarized the situation as follows.

> It has been argued that our experimental knowledge of gravitational forces between 1m and 10km is so poor it allows a considerable difference between the laboratory measured gravitational constant and its value on astronomical scales – an effect predicted in theories of the type alluded to above [these included Fujii and O'Hanlon, whose work was also cited in the experimental papers] . . . it can be seen that for $3m < l < 10^3km$ $\alpha$ is very poorly constrained. (p. 636)

> We conclude that there is very little scope for a theory which allows deviations > 1% from Newton's law of gravitational attraction on laboratory or larger length scales . . . Further large scale experiments are essential to improve bounds on $\alpha$ between 1m and 10km. (p. 638)

There was, however, a clear, although small, window of opportunity.

## IV. THE FIFTH FORCE

Until early 1983 the two strands, that of the energy dependence of the $K^o$–$\overline{K}^o$ system parameters and that of modifications of Newtonian gravity and their experimental tests, had proceeded independently. At about this time Fischbach became aware of the discrepancies between experiment and gravitational theory[18]. He made no connection, at this time, between the two problems because he was still thinking in terms of long-range forces, which produced an energy-squared dependence of the $K^o$– $\overline{K}^o$ parameters, and was ruled out experimentally. In early 1984, he realized that this would not be the case for a short-range force, and that the effect could be much smaller[19]. At this time he also became aware of the Gibbons and Whiting summary and realized that such a short-range force might be possible and that the two problems might have a common solution.

Fischbach, Aronson, and their collaborators looked for other places where such an effect might be seen with existing experimental sensitivity. They found only three: 1) the $K^o$–$\overline{K}^o$ system at high energy, which they had already studied; 2) the comparison of satellite and terrestrial determinations of $g$, the local gravitational acceleration[20]; and 3) the original Eötvös experiment, which measured the difference between the gravitational and inertial masses of different substances. If a short-range composition-dependent force existed it might show up in this experiment. They noted that the very precise modern experiments of Roll, Krotkov and Dicke (1964) and of Braginskii and Panov (1972) would not have been sensitive to such a force because they had compared the gravitational accelerations of pairs of materials toward the sun, and thus looked at much larger distances.

The apparent energy dependence of the $K^o$–$\overline{K}^o$ parameters along with the discrepancy between gravitational theory and experiment led Fischbach, Aronson and their colleagues (Fischbach 1986) to reexamine the original data of Eötvös *et al.* (1922) to see if there was any evidence for a short-range, composition-dependent force[21]. By this time they knew of Holding and Tuck's 1984 result which gave $G$ measured in a mine as $G = (6.730 \pm 0.003) \times 10^{-11}\ m^3kg^{-1}s^{-2}$ in disagreement with the best laboratory value of $(6.6726 \pm 0.0005) \times 10^{-11}$. This result was still uncertain because of possible regional gravity anomalies. Fischbach used the modified gravitational potential $V(r) = -G_\infty\, m_1 m_2/r(1 + \alpha\, e^{(-r/\lambda)})$. They remarked that such a potential could explain the geophysical data quantitatively if $\alpha = (-7.2 \pm 3.6) \times 10^{-3}$, with $\lambda = 200 \pm 50$ m. (This was from a private communication from Stacey. Details appeared later in Holding 1986). This potential had the same mathematical form as that suggested much earlier by Fujii. Recall also that Fujii had suggested redoing the Eötvös experiment. Fujii's work does not seem to have exerted any direct influence on Fischbach. No citations of it are given in this paper[22].

The apparatus for the Eötvös experiment is shown schematically in Figure 6[23]. The gravitational force is not parallel to the fiber due to the rotation of the earth. If the gravitational force on one mass differs from that on the other the fiber will rotate. Reversing the masses should give a rotation in the opposite direction.

Fischbach attempted to combine the gravitational discrepancy with the energy dependence of the $K^o$–$\overline{K}^o$ parameters. They found that if they considered a hypercharge field with a small, finite mass hyperphoton (the $K^o$ and $\overline{K}^o$ have different hypercharges) they obtained a potential of the same form as shown above[24]. They also found that $\Delta k = \Delta a/g$, the fractional change in gravitational acceleration for two substances, would be proportional to $\Delta(B/\mu)$ for the two substances, where $B$ was the baryon number of the substance (equal, in this case, to the hypercharge) and $\mu$ was the mass of the substance in units of the mass of atomic hydrogen.

They plotted the data reported by Eötvös as a function of $\Delta(B/\mu)$, a quantity unknown to Eötvös, and found the results shown in Figure 7. The linear dependence visible is supported by a least-squares fit to the equation $\Delta k = a\, \Delta(B/\mu) + b$. They found $a = (5.65 \pm 0.71) \times 10^{-6}$ and $b = (4.83 \pm 6.44) \times 10^{-10}$. They concluded,

We find that the Eötvös-Pekar-Fekete data are sensitive to the composi-

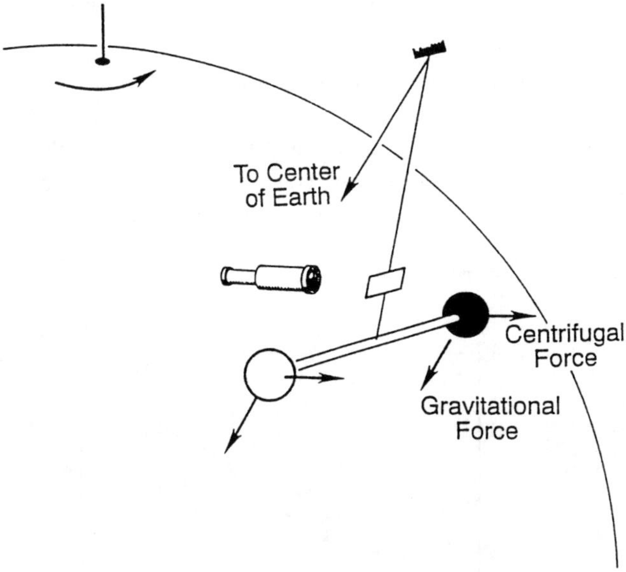


*Figure 6.* A schematic view of the Eötvös experiment. From Will (1984).

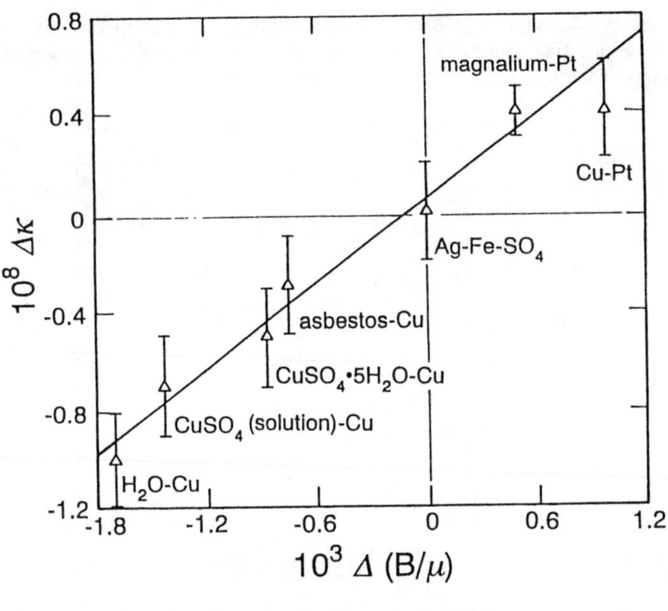


*Figure 7.* $\Delta k$ as a function of $\Delta(B/\mu)$. From Fischbach et al. (1986a).

> tion of the materials used, and that their results support the existence of an intermediate-range coupling to baryon number or hypercharge. (Fischbach 1986, p. 3)[25]

They calculated the coupling constant for their new interaction for both the Eötvös data and for the geophysical data and found that they disagreed by a factor of 15, which they found "surprisingly good" in view of the simple model of the earth they had assumed. Not everyone was so sanguine about this[26]. (See Figures 8–10 and Table I in conjunction with note 26.)

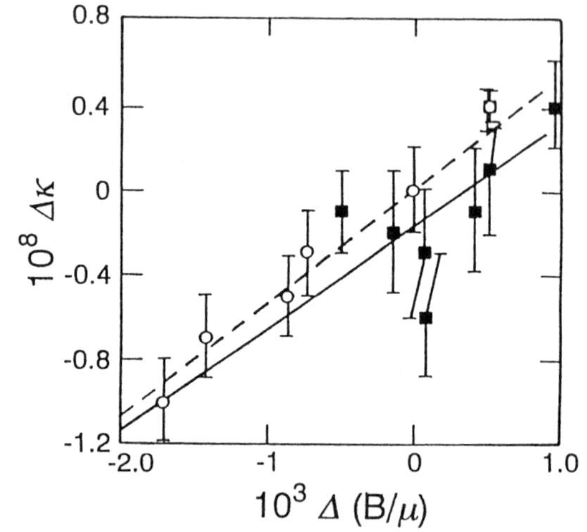


*Figure 8.* $\Delta k$ as a function of $\Delta(B/\mu)$. The circles are the data from Fischbach et al. (1986)a). The squares are the final summary from Eötvös et al. (1922). The best fit straight lines to both sets of data are shown.

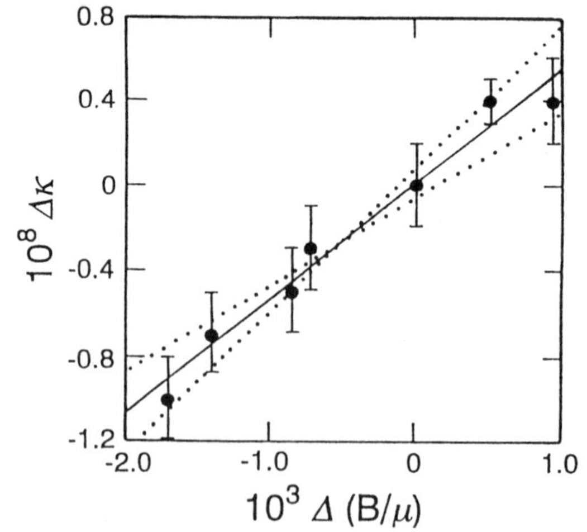


*Figure 9.* $\Delta k$ as a function of $\Delta(B/\mu)$ from Fischbach et al. (1986). The best fit line along with the 95% confidence level fits are shown.

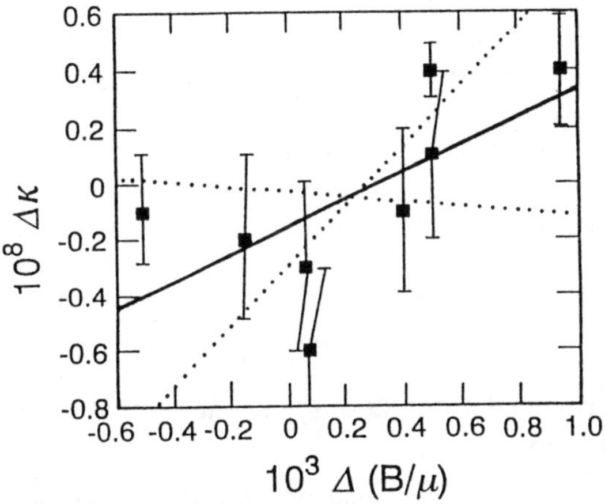


*Figure 10.* $\Delta k$ as a function of $\Delta(B/\mu)$. The data are from Eötvös et al. (1922). The best fit straight line along with the 95% confidence level fits are shown.

*Table I.* The value of $\varkappa - \varkappa_{Pt}$ ($\Delta k$ relative to platinum) for various substances. From Eötvös et al. 1922, p. 164.

| Substance | $(\varkappa - \varkappa_{Pt}) \times 10^{-6}$ |
|---|---|
| Magnalium | +0.004 ± 0.001 |
| Snake-wood | –0.001 ± 0.002 |
| Copper | +0.004 ± 0.002 |
| Water | – 0.006 ± 0.003 |
| Crystalline cupric sulfate | – 0.001 ± 0.003 |
| Solution of cupric sulfate | – 0.003 ± 0.003 |
| Asbestos | +0.001 ± 0.003 |
| Tallow | – 0.002 ± 0.003 |

It seems fair to summarize the Fischbach paper as follows. A reanalysis of the original Eötvös paper presented suggestive evidence for an intermediate-range, composition-dependent force, which was proportional to baryon number or hypercharge. With a suitable choice of parameters, one could relate this force to anomalies in mine measurements of gravity and to a suggested energy dependence of the parameters of the $K^o–\overline{K}^o$ system. It was these three pieces of suggestive, although certainly not conclusive, evidence that led others to extend the investigation of the Fifth Force, both experimentally and theoretically[27]. The fact that it also fit in with the tradition of modifying the gravitational theory of Fujii and others no doubt strengthened that decision. The context of pursuit is not devoid of evidence.

The fact that the subsequent history has argued against the existence of the Fifth Force should not cause us to overlook the fact that the suggestion of that force was the result of a sequence of reasonable and plausible steps.

This started with the Colella, Overhauser and Werner measurement, Fischbach's attempt to connect CP violation and gravity, and the subsequent observation of suggestive energy dependence of the $K^o$–$\overline{K}^o$ parameters. At the same time the work on the modifications of Newtonian gravity and the tantalizing results on the measurement of gravity in mines were proceeding. When these two strands were joined together it led the collaborators to reanalyze the original Eötvös experiment, where again a suggestive effect appeared. The suggestion of the Fifth Force then followed. There may not be a logic of discovery, but, at least in this case, it is not a totally mysterious process.

As long ago as 1971, Post discussed heuristics, how successive theories are related (Post 1971). Although he concentrated on theoretical constraints, he noted that experimental observations may enter into the construction or discovery of new theories.

> Experiment by itself does not lead to a theory so new as to break out of the theoretic framework in which the experiment is conducted, although general observational experience provides some basis which limits the choice of theories. (1971, p. 217)

Post cites the observation of line spectra as calling for a new theory. The episode recounted above would seem to illustrate this. The gravitational measurements, as summarized, for example, by Gibbons and Whiting, certainly set limits on both the range and the strength of any proposed modification of Newtonian gravitation, including the Fifth Force.

As Post suggested, we can certainly say more about the process of discovery than that it is merely trial and error. The theoretical work of Fujii and others were attempts to make plausible extensions of existing gravitational theories, and to look at their consequences. This work certainly fit Post's criterion of "retaining the old theory, in a certain sense." Because both Fujii's work and the Fifth Force hypothesis included only a small correction to Newton's Law of Gravity and had significant experimental consequences in a region in which that law had not, in fact, been tested, all of the evidence supporting Newton's Law was retained.

That work along with subsequent theoretical work on the Fifth Force also fit Post's criterion of simplicity. Post offered only a general sense of this simplicity criterion, i.e. preferring a straight line to a more complex curve. In practice, scientists do make such rough and ready judgments. Thus, Fujii added only a massive scalar particle to the already present massless scalar and tensor exchange particles of the Brans-Dicke theory. This led, as did the Fifth Force hypothesis to a simple, single exponential modification of the gravitational potential. When this simple extension failed to agree with experimental results, more complex scenarios were offered (see Franklin 1993 for details). It was true, however, that the theoretical work only had a major impact on the work of scientists when it was combined with the three apparent empirical anomalies for existing gravitational theories; the geophysical anomaly, the energy dependence of the $K^o$–$\overline{K}^o$ parameters, and the reanalyzed Eötvös experiment. It seems clear that philosophers of science do have something to say about both the context of discovery and about the context of pursuit. Post has, for a long time, urged us to do so.

*Dept. of Physics,*
*University of Colorado,*
*Boulder, Colorado*

NOTES

1. Professor Sam Westfall, the noted Newton scholar and biographer, believes the story of Newton and the apple is true. He reports that Newton repeated it on at least four occasions during his lifetime (private communication).
2. I do not suggest that these are necessary or sufficient conditions for such hypothesis creation, after all an individual might come up with a solution at the first try, but my speculation is that these are the usual circumstances.
3. Pickering (1984, p. 106) noted that the citation history of Weinberg's paper clearly shows this. t'Hooft showed the theory was renormalizable in 1971. The citations were 1967, 0; 1968, 0; 1969, 0; 1970, 1; 1971, 4; 1972, 64; 1973, 162.
4. These three physicists played the leading roles in the formulation of the Fifth Force hypothesis. I will refer to papers written by them and their collaborators by the first author listed. This may give the impression that only a single author was involved. This is definitely not the case. Virtually all of these papers had multiple authors.
5. More informally, during a discussion held at the 1990 Moriond Workshop, Orrin Fackler, a member of the Livermore Tower group, remarked, "The Fifth Force is dead". No dissenting voice was heard. The group included many of those who have been active in working on the Fifth Force including Ephraim Fischbach and Sam Aronson, who originally suggested it. (I was present at the discussion.) Perhaps a bit more formally, the Fifth Force did not appear on the program for the 1991 Moriond Workshop. There are some who might question whether or not the force was ever alive.
6. Overhauser and Colella (1974) contains a discussion of the theoretical aspects of this experiment.
7. A stronger statement had appeared earlier. "The results are clearly consistent with constant phase . . . " (Roehrig *et al.* 1977, p. 1118)
8. The graph actually shows the energy dependence of $\phi_{21}$, assuming $\phi_{+-}$ was constant. If $\phi_{21}$ is considered to be a constant, the graph shows the energy dependence of $\phi_{+-}$.
9. An earlier paper (Fischbach 1982) had examined the same question, although in less detail, and reached the same conclusion.
10. These results were not greeted with enthusiasm or regarded as reliable by everyone within the physics community. Commenting on the need for new interactions to explain the effects, an anonymous referee remarked, "This latter statement also applies to spoon bending." (A copy of the referee's report was given to me by Fischbach.) The paper was, however, published.
11. This possible energy dependence played an important role in the genesis of the Fifth Force hypothesis, as discussed in detail below. It may very well have been a statistical fluctuation.
12. There is a further oddity in this history. The six measurements of $\backslash\eta_{+-}\backslash$ made prior to 1973 had a mean value of $(1.95 \pm 0.03) \times 10^{-3}$. The value cited by Coupal *et al.* was the mean of post-1973 measurements. For details of this episode see Franklin (1986, ch. 8).
13. I will be discussing here modifications of the Newtonian inverse square law, and not the well-established relativistic post-Newtonian corrections, which are of order $GM/c^2r$.

14. For an excellent and accessible discussion of this see Will (1984). For more technical details see Will (1981).
15. The history of gravitational theory is not a string of unbroken successes. Newton himself could not explain the motion of the moon in the *Principia* and his later work on the problem, in 1694–5, also ended in failure. (Westfall 1980, pp. 442–3, 540–8). The law was also questioned during the 19th century when irregularities were observed in the motion of Uranus. The suggestion of a new planet by Adams and LeVerrier and the subsequent discovery of the planet Neptune turned the problem into a triumph. During the 19th century it was also found that the observed advance of the perihelion of Mercury did not match the predictions of Newtonian theory. This remained an anomaly for years until the advent of Einstein's General Theory of Relativity, the successor to Newtonian gravitation.
16. Some readers might worry that a variable constant is an oxymoron, but it does seem to be a useful shorthand.
17. The influence of Long's work is apparent in the first sentence of the abstract:

> D.R. Long and others have speculated that the gravitational force between point masses in the Newtonian regime might not be exactly proportional to $1/r^2$. (Mikkelsen and Newman 1977, p. 919)

18. Fischbach (private communication) attributes this to a conversation with Wick Haxton, who told him about it.
19. Fischbach's first calculation was for a δ-function force.
20. Rapp (1974, 1977) had already found $\Delta\ g/g \approx (6 \pm 10) \times 10^{-7}$. For the proposed Fifth Force parameters (see below) the predicted effect would be approximately $2 \times 10^{-7}$.
21. Because the energy dependence of the $K^o - \overline{K}^o$ parameters might have indicated a violation of Lorentz invariance, Fischbach (1985) had looked at the consequences of such a violation for the Eötvös experiment.
22. Fischbach keeps detailed chronological notes of papers read and calculations done. He reports that he has notes on Fujii's work at this time, but does not recall it having any influence on his work.
23. Eötvös was originally interested in measuring gravity gradients so the weights were suspended at different heights. This introduced a source of error into his tests of the equivalence principle, the equality of gravitational and inertial mass.
24. Fischbach noted that in the limit of infinite range their suggested force agreed with that proposed by earlier by Lee and Yang on the basis of gauge invariance.
25. In a later paper (Aronson 1986) the group suggested other experiments, particularly on K meson decay, that might show the existence of such a hyperphoton.
26. An interesting sidelight to this reanalysis is reported in a footnote to the Fischbach paper. Instead of reporting the observed values of $\Delta k$ for the different substances directly, Eötvös and his colleagues presented their results relative to platinum as a standard. The effect of this combining say $\Delta k(H_2O\text{-}Cu)$ and $\Delta k(Cu\text{-}Pt)$ to infer $\Delta k(H_2O\text{-}Pt)$ is to reduce the magnitude of the observed nonzero effect (for water and platinum) from 5σ to 2σ (Fischbach 1986, p. 6). $\Delta k(H_2O\text{-}Cu) = (-10 \pm 2) \times 10^{-9}$ and $\Delta k(Cu\text{-}Pt) = (+4 \pm 2) \times 10^{-9}$, respectively. Adding them to obtain $\Delta k(H^2O\text{-}Pt)$ gives $(-6 \pm 3) \times 10^{-9}$.

Figure 8 shows both the final summary reported by Eötvös as well as Fischbach's reanalysis, along with best-fit straight lines for both sets of data separately (This is my own analysis.) Although several of the experimental uncertainties have increased, due to the calculation process, the lines have similar slopes. The major difference is in the uncertainty of the slopes. If one looks at the 95% confidence level, as shown separately for the Fischbach and Eötvös data, respectively, in Figures 9 and 10, one finds that at this level the original Eötvös data are, in fact, consistent with no effect, or a horizontal straight line. This is certainly not true for the Fischbach reanalysis. A skeptic might remark that the effect is seen only when the data are plotted as a function of $\Delta(B/\mu)$, a theoretically suggested parameter. As De Rujula remarked:

> In that case, Eötvös and collaborators would have carried their secret to their graves: how to gather ponderous evidence for something like baryon number decades before the neutron was discovered. (1986, p. 761)

It is true that theory may suggest where one might look for an effect, but it cannot guarantee that the effect will be seen. Although one may be somewhat surprised, along with De Rujula, that data taken for one purpose take on new significance in the light of later experimental and theoretical work, it is not unheard of. There is a possibility that Eötvös and his collaborators might actually have seen something of this effect, but discounted it. They report:

> The probability of a value different from zero for the quantity $\underline{x}$ ($\Delta k$) even in these cases is vanishingly little, as a review of the according observational data *shows quite long sequences with uniform departure from the average* (emphasis added), the influence of which on the average could only be annulled by much longer series of observations. (Eötvös 1922, p. 164)

The original summary, given in Table 1, gives an average value for $\underline{x} = (-0.002 \pm 0.001) \times 10^{-6}$, which seems to justify Eötvös's original conclusion:

> We believe we have the right to state that $\underline{x}$ relating to the Earth's attraction does not reach the value of $0.005 \times 10^{-6}$ for any of these bodies. (Eötvös 1922, p. 164)

27. Not everyone in the physics community was impressed by the evidence presented. Glashow, a Nobel Prize-winning particle theorist, was quite negative:

> Unconvincing and unconfirmed kaon data, a reanalysis of the Eötvös experiment depending on the contents of the Baron's wine cellar (a humorous allusion to the importance of local mass inhomogeneities), and a two-standard-deviation geophysical anomaly. Fischbach and his friends offer a silk purse made out of three sows' ears, and I'll not buy it. (quoted in Schwarzschild 1986, p. 20)

For further details see Franklin (1993).

## BIBLIOGRAPHY


Aronson, S.H. *et al.* (1982), 'Determination of the Fundamental Parameters of the $K^o$–$\overline{K}^o$ System in the Energy Range 30–110 GeV', *Physical Review Letters*, 48, pp. 1306–9.

Aronson, S.H. *et al.* (1983a), 'Energy Dependence of the Fundamental Parameters of the $K^o$–$\overline{K}^o$ System I. Experimental Analysis', *Physical Review D*, 28, pp. 476–94.

Aronson, S.H. *et al.* (1983b), 'Energy Dependence of the Fundamental Parameters of the $K^o$–$\overline{K}^o$ System II Theoretical Formalism', *Physical Review D*, 28, pp. 495–523.

Aronson, S.H. *et al.* (1986), 'Experimental Signals for Hyperphotons', *Physical Review Letters*, 56, pp. 1342–5.

Bell, J.S. and Perring, J. (1964), '$2\pi$ Decay of the $K^o_2$ Meson', *Physical Review Letters*, 13, pp. 348–9.

Bernstein, J., Cabibbo, N., and Lee, T.D. (1964), 'CP Invariance and the $2\pi$ Decay of the $K^o_2$', *Physics Letters*, 12, pp. 146–8.

Bock, G.J. *et al.* (1979), 'Coherent $K_S$ Regeneration on Protons from 30 to 130 GeV/c', *Physical Review Letters*, 42, pp. 350–3.

Braginskii, V.B. and Panov, V.I. (1972), 'Verification of the Equivalence of Inertial and Gravitational Mass', *Zhurnal Eksperimental'noi i Teoreticheskoi Fiziki* (JETP), 34, pp. 463–6.

Brans, C. and Dicke, R.H. (1961), 'Mach's Principle and a Relativistic Theory of Gravitation', *Physical Review*, 124, pp. 925–35.

Christenson, J.H. *et al.* (1964), 'Evidence for the $2\pi$ Decay of the $K^o_2$ Meson', *Physical Review Letters*, 13, pp. 138–40.

Colella, R., Overhauser, A.W., and Werner, S.A. (1975), 'Observation of Gravitationally Induced Quantum Interference', *Physical Review Letters*, 34, pp. 1472–4.
Coupal, D.P. *et al.* (1985), 'Measurement of the Ratio ($K_L \rightarrow \pi^+\pi^-$) ($K_L \rightarrow \pi/\nu$) for $K_L$ with 65 GeV/c Laboratory Momentum', *Physical Review Letters*, 55, pp. 566–9.
DeBouard, X. *et al.* (1965), 'Two Pion Decay of $K^o2$ at 10 GeV/c', *Physics Letters*, 15, pp. 58–61.
De Rujula, A. (1986), 'Are There More Than Four?' *Nature*, 323, pp. 0660–1.
Eötvös, R., Pekar, D., and Fekete, E. (1922), 'Beitrage zum Gesetze der Proportionalität von Tragheit und Gravität', *Annalen der Physik (Leipzig)*, 68, pp. 11–66. I have used the translation that appeared in *Annales Universitatis Scientarum Budapestinensis de Rolando Eötvös nominate, Sectio geologice* 7, 1963, pp. 111–65.
Fischbach, E. (1980), 'Tests of General Relativity at the Quantum Level', in P. Bergmann and V. De Sabbata (eds.), *Cosmology and Gravitation*, Plenum, New York, pp. 359–73.
Fischbach, E. and Freeman, B. (1979), 'Testing General Relativity at the Quantum Level', *General Relativity and Gravitation*, 11, pp. 377–81.
Fischbach, E. *et al.* (1982), 'Interaction of the $K^o$–$\overline{K}^o$ System with External Fields', *Physics Letters*, 116B, pp. 73–6.
Fischbach, E. *et al.* (1985), 'Lorentz Noninvariance and the Eötvös Experiments', *Physical Review D*, 32, pp. 154–62.
Fischbach, E. *et al.* 'Reanalysis of the Eötvös Experiment', *Physical Review Letters*, 56, 1986, pp. 3–6.
Franklin, A. (1986), *The Neglect of Experiment*, Cambridge University Press, Cambridge.
Franklin, A. (1993), 'The Rise and Fall of the Fifth Force', forthcoming.
Fujii, Y. (1971), 'Dilatonal Possible Non-Newtonian Gravity', *Nature (Physical Science)*, 234, pp. 5–7.
Fujii, Y. (1972), 'Scale Invariance and Gravity of Hadrons', *Annals of Physics (N.Y.)*, 69, pp. 494–521.
Fujii, Y. (1974), 'Scalar-Tensor Theory of Gravitation and Spontaneous Breakdown of Scale Invariance', *Physical Review D*, 9, pp. 874–6.
Galbraith, W. *et al.* (1965), 'Two-pion Decay of the $K^o{}_2$ Meson', *Physical Review Letters*, 14, pp. 383–6.
Gibbons, G.W. and Whiting, B.F. (1981), 'Newtonian Gravity Measurements Impose Constraints on Unification Theories', *Nature*, 291, pp. 636–8.
Grossman, N. *et al.* (1987), 'Measurement of the Lifetime of $K^o{}_S$ Mesons in the Momentum Range 100–350 GeV/c', *Physical Review Letters*, 59, pp. 18–21.
Holding, S.C. and Tuck, G.J. (1984), 'A New Mine Determination of the Newtonian Gravitational Constant', *Nature*, 307, pp. 714–16.
Holding, S.C., Stacey, F.D., and Tuck, G.J. (1986), 'Gravity in Mines – an Investigation of Newton's Law', *Physical Review D*, 33, pp. 3487–94.
Long, D.R. (1974), 'Why Do We Believe Newtonian Gravitation at Laboratory Dimensions?', *Physical Review D*, 9, pp. 850–2.
Long, D.R. (1976), 'Experimental Examination of the Gravitational Inverse Square Law', *Nature*, 260, pp. 417–18.
Long, D.R. (1981), 'Current Measurements of the Gravitational 'Constant' as a Function of Mass Separation', *Nuovo Cimento*, 62B, pp. 130–8.
Mikkelsen, D.R. and Newman, M.J. (1977), 'Constraints on the Gravitational Constant at Large Distances', *Physical Review D*, 16, pp. 919–26.
O'Hanlon, J. (1972), 'Intermediate-Range Gravity: A Generally Covariant Model', *Physical Review Letters*, 29, pp. 137–8.
Overhauser, A.W. and Colella, R. (1974), 'Experimental Test of Gravitationally Induced Quantum Interference', *Physical Review Letters*, 33, pp. 1237–9.
Panov, V.I. and Frontov, V.N. (1979), 'The Cavendish Experiment at Large Distances', *Zhurnal Eksperimental'noi i Teoreticheskoi Fiziki* (JETP), 50, pp. 852–6.
Pickering, A. (1984), 'Against Putting the Phenomena First: The Discovery of the Weak Neutral Current', *Studies in History and Philosophy of Science*, 15, pp. 85–117.

Post, H.R. (1971), 'Correspondence, Invariance and Heuristics: In Praise of Conservative Induction', *Studies in History and Philosophy of Science*, 2, pp. 213–255.
Rapp, R.H. (1974), 'Current Estimate of Mean Earth Ellipsoid Parameters, 1, pp. 35–8.
Rapp, R.H. (1977), 'Determination of Potential Coefficients to Degree 52 by 5° Mean Gravity Anomalies', *Bulletin Geodesique*, 51, pp. 301–23.
Roehrig, J. *et al.* (1977), 'Coherent Regeneration of $K_S$'s by Carbon as a Test of Regge-Pole-Exchange Theory', *Physical Review Letters*, 38, pp. 1116–19.
Roll, P.G., Krotkov, R., and Dicke, R.H. (1964), 'The Equivalence of Inertial and Passive Gravitational Mass', *Annals of Physics (N.Y.)*, 26, pp. 442–517.
Scherk, J. (1979), 'Antigravity: A Crazy Idea', *Physics Letters*, 88B, pp. 265–7.
Schwarzschild, B. (1986), 'Reanalysis of Old Eötvös Data Suggests 5th Force . . . to Some', *Physics Today*, XX, pp. 17–20.
Spero, R. *et al.* (1980), 'Tests of the Gravitational Inverse-Square Law at Laboratory Distances', *Physical Review Letters*, 44, pp. 1645–8.
Stacey, F.D. (1978), 'Possibility of a Geophysical Determination of the Newtonian Gravitational Constant', *Geophysical Research Letters*, 5, pp. 377–8.
Stacey F.D. and Tuck, G.J. (1981), 'Geophysical Evidence for Non-Newtonian Gravity', *Nature*, 292, pp. 230–2.
Stacey, F.D. *et al.* (1981), 'Constraint on the Planetary Scale Value of the Newtonian Gravitational Constant from the Gravity Profile Within a Mine', *Physical Review D*, 23, pp. 1683–92.
Wagoner, R.V. (1970), 'Scalar-Tensor Theory and Gravitational Waves', *Physical Review D*, 1, pp. 3209–16.
Weinberg, S. (1964), 'Do Hyperphotons Exist?', *Physical Review Letters*, 13, pp. 495–7.
Westfall R.S. (1980), *Never at Rest*, Cambridge University Press, Cambridge.
Will, C. (1981), *Theory and Experiment in Gravitational Physics*, Cambridge University Press, Cambridge.
Will, C. (1984), *Was Einstein Right?*, Basic Books, New York.
Zee, A. (1979), 'Broken-Symmetric Theory of Gravity', *Physical Review Letters*, 42, pp. 417–21.
Zee, A. (1980), 'Horizon Problem and the Broken-Symmetric Theory of Gravity', *Physical Review Letters*, 44, pp. 703–6.

ELSPETH CRAWFORD

# Michael Faraday's Thought: Discovery or Revelation?

## I. PSYCHOLOGICAL FACTORS, HEURISTICS AND DEMARCATIONS

During my time of study with Heinz Post, among many things which I will always appreciate, I learnt the value of making demarcations, not arbitrarily, but by searching for the relevant criteria by means of which lateral impressionistic thinking could be given communicable structure and identifiable form.

In this paper, I use insights from psychoanalysis to examine the possible process by means of which Michael Faraday made some of his discoveries of electromagnetic phenomena. In this first section I identify the kind of demarcation I am making, its relevance to heuristics and some aspects of value in it, so that readers unfamiliar with this approach can hopefully appreciate that relevance and value with respect to my conclusion.

I begin by noting Post's demarcation, as given in the introduction to his paper 'Correspondence, Invariance and Heuristics':

> This paper is concerned with the better part of philosophy of science, namely heuristics. It is intended to combat two common views:
> (1) That it is impossible or improper to define any procedure for the constructing of new theories.
> (2) That the most one can say of the heuristic process is that it is a trial and error procedure.
>
> A variant of the first view relegates an enquiry into heuristics to psychology, as if that disposed of it.
>
> To deal with this last point immediately: I do believe that there is a demarcation of the legitimate field of the philosopher of science, but that this line runs through the field of psychology. This demarcation is provided by a criterion of relevance: the line is to be drawn between what is intersubjectively relevant, and what is at best of private relevance . . . The cut occurs between the private circumstances leading the scientist to acquire a prejudice on the one hand, and the rational connections between the prejudice in question and further developments in science on the other hand. (Post 1971, p. 215.)



*S. French and H. Kamminga (eds.), Correspondence, Invariance and Heuristics*, 105–124.

Post went on to state that in his paper the field of heuristics which he wished to discuss was even firmer than intersubjectively relevant psychology. He was interested in the objective features of theories which provide a spur for theory change (their formal flaws) and with the objective features (explanatory power, etc.) in terms of which it can be decided whether a theory change is progressive or not. He successively outlines heuristic procedures which are objectively identifiable in relation to these various features. In using the phrase "the structure of the heuristic *objects*" (my italics; *ibid.*, p. 216), I think he was stating that his central concern was with understanding those concepts of heuristic thinking which could be objectively delineated. Thus although he saw intersubjectively relevant psychology as part of the field of heuristics, he himself did not address it in his paper. I am therefore thinking of 'heuristic objects' as all those concepts which Post does address.

I hold to Post's demarcation of relevance as identified above. I am not however attending to heuristic objects, except indirectly, but to the *subjective* thinking processes by means of which they are arrived at, or recognized as objects for thought. I consider a heuristic object to be a result of these processes. Such an object can be named, whether by definition or by conceptual understanding. Once named, it can be located within the general structure of objective understandings and discussed to articulate its meaning, its validity within theory, etc. Discussion, as in Post's paper, produces consensual agreement, or disagreement, or clarification of these issues.

In considering thinking processes, there is a conceptual difficulty, in that one is at the same time using some kind of thinking process. It is essential that it is the *operation* of the process which is considered and not its name. This is analogous to the appreciation of art, or music, where one does not appreciate something by knowing what it is said to be, but by engaging with it, as it is, in a personal experience. In this paper, I am trying to *create engagement* with the heuristic process, using Faraday's 'electrotonic state' as an example and applying a psychoanalytic view of thinking. I believe this is a rational and communicable activity, but it is not purely objective, nor accessible by reasoned thinking alone. Thinking about thinking is a subjectively reflective enquiry, in which the emotional qualities are first identified, then lifted into the realm of 'objects' for rational thought.

I am exploring thinking *processes*, some of which involve prejudice and block scientific development and some of which may enable escape from prejudice and allow development to proceed. The historical evidence that discovery has occurred and been located within an explanatory system indicates a change or development in the processes of thought and a capacity to escape from prejudice. In the personal sphere of undergoing psychoanalysis, such a change would be considered fundamental to psychoanalytic process, not its theory. I am not, however, looking at the processes by means of which new formulations in psychoanalytic theory have been reached. I am taking some of those formulations, particularly the insights into unconscious thinking process made by Wilfred Bion[1] and applying them to the sphere of scientific discovery, to see whether successful heuristic processes can be inferred from Michael Faraday's written record of discoveries. In other words, I am exploring subjective mental change in a non-private sphere by considering the historical evidence of a fundamental change in scientific knowledge and theory.

## II. PROCESSES OF THINKING

### II.1. *Prejudice, Lie and Truth*

Bion and Faraday were both inspiring thinkers. Faraday in the nineteenth century made the discoveries which laid the foundations for modern electromagnetic field theory; Bion in this century moved forward from the thought of Freud and Melanie Klein (see Klein 1952 and 1975 and Meltzer 1978) to open up new developments in psychoanalytic thinking with regard to the nature of thought itself (see below).

Central to the thinking of both Bion and Faraday was the concept of *truth* and their struggle was the difficulty of finding truth. Both reflected closely on the reasons for the difficulty. Faraday tried all his life to identify and defeat that kind of unthinking, apparently scientific assumption which he called *prejudice*. He often referred to the difficulties of establishing facts and described the failure to do so as a consequence of prejudice. For example, he said,

> . . . men are so often bowed down and carried forward from fallacy to fallacy, their eyes not being opened to see what that fallacy is . . . (Faraday 1833)

He also perceptively realized that the problem of failure was not a matter of intelligence or hard work as is sometimes thought, because he followed the above with this comment:

> . . . the more acute a man is, the more he is bound by the chains of error; for he only uses his ingenuity to falsify the truth which lies before him. . . . (*ibid.*)

Faraday's most formal attempt to discuss the nature of prejudice and the means by which he believed it could be overcome or at least mitigated was made towards the end of his long career at the forefront of scientific discovery in his lecture 'Observations on Mental Education' (Faraday 1855). I refer to this further in section III, Faraday's Thinking.

In considering Bion's ideas, I shall focus on his paper 'A Theory of Thinking' (Bion 1962a) and subsequent books which developed the ideas put forward in this seminal paper (Bion 1962b, 1963, 1965, 1970). Some comments which he made concerning science were influential in my thinking concerning the development of science and the nature of prejudice. For example, he said,

> The difficulties of the patient suffering from a disorder of thought are similar to those that beset scientists, and others concerned with the establishment of facts, in that they arise through failure to ascertain facts and so involve the investigation of the nature of failure. (Bion 1962b, p. 66)

He went on to say that this failure was impossible to psychoanalyse

> . . . without understanding the problem of the philosopher of science and conversely that his problem is incompletely stated without the aid of psychoanalytic experience of disorders of thought. (*ibid.*)

It is this mutual relevance between philosophy of science and psychoanalytic experience as illuminated by Bion that I want to explore here.

II.1.1. *Bion's Lie*

Bion identified a thinking process which produced what he called a *lie*. I will try to build a realization of the meaning of this concept by commenting on an example[2].

Consider the following statement about the effect of pre-conceived ideas on genuine understanding, made by John Tyndall[3]. (It is recorded that Faraday overheard and agreed with the remark (Tyndall 1868, p. 63).) Tyndall said:

> In our conceptions and reasonings regarding the forces of nature, we perpetually make use of symbols which, when they possess a high representative value, we dignify with the name of theories. Thus, prompted by certain analogies, we ascribe electrical phenomena to the action of a particular fluid, sometimes flowing, sometimes at rest. Such conceptions have their advantages and their disadvantages; they afford peaceful lodging to the intellect for a time, but they also circumscribe it, and by and by, when the mind has grown too large for its lodging, it often finds difficulty in breaking down the walls of what has become its prison instead of its home.

This quotation describes a well-known picture of science; the construction of theories, development of models, testing, etc., which is often supposed to be the nature of scientific method. It is undoubtedly a realistic description of much of the activity we call science. Elsewhere (Crawford 1985 b, p. 222) I have called this process 'benevolent prejudice'.

Bion would have called it a 'lie', regardless of whether the particular theory concerned was at its "peaceful lodging to the intellect" stage, or at its "prison" stage. *He* was not concerned with its explanatory power, but was concerned instead with the intentions or unconscious phantasies of the thinker.

This concern needs explication. It could be said that a 'thought' such as Tyndall describes implies a mental activity *characterized* by reasoning, theorizing, model-making, speculating and so on. But, is 'characterized' an accurately used word when most, if not all, of the description refers only to conscious activity? Bion asks us to look at the totality of mental activity, and the psychoanalytic insight that unconscious phantasy pervades our mental activity[4]. He suggests that, unconsciously, most thoughts will be produced by the thinker to cope with the emotional circumstances in which he/she finds him/herself.

This is Bion's lie: a thought which requires a thinker for its construction because of the thinker's need, and thus exists because of the thinker. The lie cannot exist without the thinker who has thought it. In relation to narcissistic states of mind, Bion said,

> The lie depends on the thinker and gains significance through him. (Bion 1970, p. 103)

Another way of putting this is that the lie is a thought constructed by the thinker for some purpose relevant to his internal needs, whether that need is a peaceful lodging for the intellect in the face of unknowns, or the significance craved by the narcissist, or a need of some other kind. This thinking is not necessarily in opposition to the establishment of facts, because internal need is not necessarily in opposition to reality. Neither is it necessarily a self-deception, unless the intention or need of the thinker is to deceive the self. However, *it is a thinking process in which the notion of truth is irrelevant at the time the process is in operation.*

I agree with Bion's view of the 'lie' process. He said,

> . . . the weakness [of scientific method] may be closer to the weakness of psychotic thinking than superficial scrutiny would admit. (Bion 1962 b, p. 14)

II.1.2. *The Problem*

At this stage in my presentation of the ideas, it would appear that whether or not the product of thought is a true fact can only be decided by hindsight; during a thinking process, as described, there will be no way to judge the truth or otherwise of the result. The problem has been put to me as follows:

> . . . if the lie (which does require a thinking person) is produced because of the distorting influence of the thinkers emotional needs, how can a thinker ever be, or know to be, in a state where such distorting influences are absent (since any thought process in a thinking being will be embedded in some emotions)? (H. Kamminga and S. French, personal communication)

It would thus seem that the only thing which can be done would be to apply testing by hindsight as frequently as possible. However, this would not answer Faraday's perceptive understanding of the nature of prejudice: that it gets worse, because ingenuity can be used in falsification.

The questions can be answered: a) by noting more carefully the nature of the lie process, b) by accepting as possible Bion's idea of 'truth' and c) by considering this idea in relation to discovery (Faraday's account in particular) and seeing this as the achievement of an emotional being (which is how Faraday himself saw it).

II.1.3. *The Nature of the Lie Process*

If it is accepted that all thinkers are emotional beings, it does not necessarily follow that objective unbiased thought cannot exist. A thinker interested in objectivity is usually aware of the sorts of problems inherent in supposedly objective concepts. John Tyndall's comment which I have previously used to illustrate Bion's notion of lie is a good example of such awareness. I will now use that same comment to illustrate the nature of the lie process.

First, the comment itself may not be a lie; it seems to be a truthful complaint; it is the description of scientific method contained in it which is the lie. Tyndall describes the use of 'conceptions' and 'reasonings', etc. and how they "afford peaceful lodging to the intellect" or "become a prison". However, when he and many others are engaged in that scientific method, actually reasoning, there is little or no acknowledgement whatever of this emotional quality, even when, as in this case, at least some of it appears to be quite within the range of consciousness. However often scientists or philosophers acknowledge that thinking may be biassed by subjectivity of some sort, they often continue to engage in reasoning *as if* the conceptions and symbols they use in the process were indeed objective.

As Bion said, their problem is incompletely stated. It is as though knowing how difficult it is to reach a certainty that something is objective, knowing how much objectivity and its clarity is desired, that they say to themselves "I need peace, I'll treat this as if it is certain and clear and I'll eventually find out when I see how it fits with the other things I'm also treating as if they were certain and clear too". The lie is located in the *as if*: in order to find a peaceful lodging in the face of unknowns, the 'liar' turns away from 'unknown', and *for the present*, treats it as 'known'.

II.1.4. *Bion's Idea of Truth*

The words 'truth' and 'lie' are frequently used as opposites. Neither Faraday's 'prejudice', nor Bion's 'lie', is the opposite of 'truth' as though both the *true* and the *not true* were propositions of equal status in a logical system. *Truth* is in a different category of thought, independent of any particular system; it appears the same from every frame of reference, if every frame could be encompassed in thought[5], and exists regardless of systems of lies, prejudices or local consistencies which might exist within a particular frame.

Bion distinguished 'truth' from 'knowledge' as follows: He said that an 'ultimate reality' could be conjectured, but that it was formless, unknown and unknowable (Bion 1970, p. 26). It is also infinite and evolving. His belief or faith in the conjecture of ultimate reality is not religious, although he is aware that many religions would use a name or 'godhead' for such ineffable conjecture. He says that it is scientific, and

> That it exists is an essential postulate of science but it cannot be scientifically discovered. (*ibid.*, p. 30)

That is, some of the phenomena from which the existence of 'ultimate reality' is conjectured are the observable behaviours of scientists, and the observed successes of the scientific enterprise.

Bion illustrated his views with reference to the phenomena of psychoanalytic events, with only incidental comments on science, such as that quoted. The following is therefore my version of how his work applies to science and scientific knowledge.

I think he was saying that knowledge (not truth) of the world was gained from phenomena by virtue of the senses, and that mental events, such as the construction of scientific concepts, could be regarded as either having irrelevant defects which obstructed an intersection with the experience of ultimate reality, or having pointers which initiated such an intersection. In the latter case, the evolution of awareness could lead to discovery and the reinforcement of the conjecture, or faith, that ultimate reality exists.

Thus, in a thinking being, the 'truth' or that part of truth which is a 'true thought' is a representation of existent reality; knowledge of it may be imprecise or partial or experienced as changing and can be recognized as not whole, but it is a representation which is good enough in that it does not distort or falsely suggest certainty. It creates further possibilities of intersection with the evolving whole which is ultimate reality.

Therefore Bion said that the 'truth' or a 'true thought' could exist whether or not it was thought of by any person at all. It is a possible intersection with the formless ultimate reality and thus exists regardless of any systems of lies or prejudice or emotional states which might be part of someone's thinking. It exists whether or not any person is aware of the intersection.

The task of the scientist is to attempt to initiate intersection, to be aware of it, and to *interpret* the conjunction between personal experience and the unknowable reality, thus creating some form of knowledge. The scientific task is misconceived if it is an attempt to *construct* knowledge.

II.1.5. *Truth in Relation to Discovery*

The problem faced by the thinker is that of getting in touch with truth and the further problem of a scientist is that of knowing whether or not this has been done.

To express the problem in another way, the question *Is this true?* is a question asked when the thought is already framed (and a philosopher may say that it is not an answerable question, and pose a different one about validity or consistency). Bion might have said "Has intersection been initiated?" However, it is too late. If intersection has not taken place, it cannot now be engendered by looking at the lie. If intersection has taken place, there will be an awareness, however inarticulate[6]. It is this awareness of truth which excites and stimulates the discoverer or searcher. A scientist's problem is therefore not that of knowing whether *this* is true, as *this* is not yet there to be looked at; his/her worry is: *Am I thinking truly?* (Am I becoming aware?) The question is not whether or not one has thought truly in the past, but whether or not one is doing so at this very moment.

Bion expressed his understanding of how this kind of thinking was distinguished from the lie only (so far as I know) in terms of the mode of enquiry of a psychoanalyst, describing how 'truth' and 'lie' could be distinguished in psychoanalytic investigation (Bion 1970, p. 33). The two kinds of thinking are emotionally distinguished by *emotional need for certainty or clarity*, and *emotional capacity to tolerate deep uncertainty and ignorance*.

In transposing Bion's distinction between kinds of thinking to the context of scientific thought, there is no change in (a) recognition that all kinds of thought have both emotional and cognitive elements and (b) the classes of thinking differ because the mode of selection of objects for thought differs.

### II.2. *Different Kinds of Thinking in Scientific Investigation*

Fortunately for science, the study of Faraday's work, particularly his Diary (see below, Section III), shows two classes of thinking process. The first is that 'scientific method' already described as a 'lie' because of its unconscious intention. In the scientific context, I personally prefer my own term, 'benevolent prejudice', because I have experienced its power to explain, sort out confusions, and generally construct mental representations which enable memory or cognition of numerous complex events. However, this kind of thinking *cannot discover something which is unknown because it does not operate on ideas which have no prior mental representation.*

Henri Poincaré described the process of creation of a mathematical formulation as follows:

> If a new result is to have any value, it must unite elements long since known, but until then scattered and seemingly foreign to each other, and suddenly introduce order where the appearance of disorder reigned. Then it enables us to see at a glance each of these elements in the place it occupies in the whole. Not only is the new fact valuable on its own account, but it alone gives a value to the old facts it unites. Our mind is as frail as our senses are; it would lose itself in the complexity of the world if that complexity were not harmonious; like the shortsighted, it would only see the details, and would be obliged to forget each of these details before examining the next, because it would be incapable of taking in the whole. The only facts worthy of our attention are those which introduce order unto this complexity and so make it accessible to us. (Poincaré 1913)

This can help us to understand the reasons why prejudice can be benevolent, but it also contains the clue to a different mode of selection. "To lose our mind in the complexity of the world" is a frightening idea, especially if our emotional strength is frail, and we only have reasoning or cognitive strength. There is a different mode of selection of phenomena, which involves knowingly facing loss of apparent knowledge. This second kind of thinking is not an opposing process, but one of a different category, where we deliberately acknowledge and face our need, seeking clarity only in terms of distinction between that which is certain and that which is not. Mental representation of something previously unknown begins with an unfulfilled indefinite state in which the complexity of phenomena is part of awareness but not part of knowledge.

Faraday tried to inform us of this process:

> This education has for its first and last step *humility* . . . founded . . . on the increase of that internal knowledge which alone can make us aware of our internal wants. . . . And though such self-schooling must continue to the end of life to supply an experience of deficiency rather than of attainment, still there is abundant stimulus to excite any man to perseverance. What he has lost are things imaginary, not real; what he gains are riches before unknown to him, yet invaluable. . . (Faraday 1855, p. 491)

Of course this is what many scientists and philosophers do, and it is the reason the scientific enterprise succeeds. However, the process is not made explicit: this kind of thinking may or may not happen in any individual, or may happen only occasionally and without understanding of the emotional reflectiveness which allows it. Most importantly, it tends to happen only because of good fortune in education or training and can disappear as arbitrarily as it has appeared when social or institutional conditions are conducive to the 'lie'.

### II.3. *Thinking Truly; The Ideas of Negative Capability and Revelation*

The understanding of the meaning of 'truth' and 'lie' and of the processes in thinking by which the first is brought into the domain of knowledge, and by which the second is produced from emotional need, have implications concerning the nature of thoughts. It seems to me that both kinds of thinking can take place. The conscious result of either process is the *experience* of an idea in the mind. Therefore when we have an idea, *the real scientific task is to know from which category of thinking it has come*, (although it would appear that conscious thinking very rarely discriminates in this respect). As I shall show, Faraday thought in both ways (see below, sections III.2 and III.3[7]), and, in his lecture on 'Mental Education', struggled to describe the meaning of the latter task in what he called "exercise of judgment".

Bion tried to describe what would happen in the experience of a person thinking truly. As I have said above, Bion's idea was that the true thought, an "evolution of ultimate reality", could exist independently of the thinker. Hence this thought would be experienced as *received* by the thinker, not constructed. Again quoting Bion,

> Nobody need think the true thought: it awaits the advent of the thinker . . . his significance depends on whether or not he will entertain the thought, but the thought remains unaltered. (1970, p. 103)

When, or if, such a truth is received by someone, it may or may not be possible to articulate it. 'Entertainment' will depend on whether or not the state of mind of the thinker can create space allowing its truth to evolve in some kind of communicable form[8]. The feelings could vary from exalted awareness, the sense of intuition, the tantalizing solution just out of reach, to the abrasive stress of contradiction, but no feeling is a reliable guide – that needs analysis of the genesis of the thought and the work which places it in the context of knowledge.

Bion encapsulated the complexity of his understanding of the state of mind in which a person would think truly by reference to a quotation from John Keats,

> I had not a dispute but a disquisition with Dilke on various subjects; several things dovetailed in my mind, and it at once struck me what quality went to form a Man of Achievement . . . I mean Negative Capability, that is when a man is capable of being in uncertainties, mysteries, doubts, without any irritable reaching after fact and reason. (Keats 1952)

Faraday stated the same idea in reference to conscious exercise of judgment. He said,

> Occasionally and frequently the exercise of the judgment ought to end in *absolute reservation*. It may be very distasteful, and great fatigue, to suspend a conclusion . . . [but] we should suspend our conclusions. Our knowledge of the fact itself, and the many varieties of it, is not the less abundant or sure; and when the truth shall hereafter emerge from the mist, we ought to have no opposing prejudice, but be prepared to receive it. (Faraday 1855, pp. 483–484)

*Revelation* was a concept of great importance to Faraday, the notion of receiving truth being part of his Sandemanian faith (See Cantor 1991). The ultimate reality which was God would be known only in future life, but in the present, some of God's truth, the truth of nature, could be revealed to the enquirer by means of the 'spirit'. 'Negative capability' might well have expressed for him the manner in which one had to work while waiting for revelation.

## III. FARADAY'S THINKING

In Faraday's writing, one of his ideas, that of the *electrotonic state*[9], illustrates very clearly that both kinds of thought can co-exist, or appear to develop one after the other. The history of this idea also illustrates that the crucial problem in thinking is how to judge when one should be using reasoning and when it is the totality and complexity of the phenomena which needs attention.

I shall describe some of the story of the electrotonic state, which I think was a revelation (or 'intersection with truth'). Faraday used and even misused the idea in the process of constructing theories over a period from 1831 to 1853 when he finally seemed satisfied that he understood it in relation to other concepts. In other words, it was a thought, which existed, but which needed a very long time before it could be articulated with conceptual understanding in relation to other knowledge. Many of his earlier conceptions had to change in order that this could happen.

### III.1. *The Electrotonic State*

In September 1831, Faraday discovered the phenomenon of electromagnetic induction (Faraday 1932–1936, Diary, September 1831, para. I-2–I-8). His discovery of the effect was made as follows: Two wires were wound, one on each side of a large iron ring. Electrical energy from a chemical battery was supplied to one of the wires, the primary wire, and electric current flowed in the other wire, the secondary wire. That is, the secondary wire, with no electrical connection to the primary, had had current induced in it. As Faraday realized, the only link between the wires was a *magnetic* link, through the iron of the ring. Faraday immediately suggested an explanation: he said that 'influence' from the primary coil created a peculiar or 'tonic' state in the wire of the secondary coil, and that this 'electrotonic state' caused current to flow. Over forty years later, James Clerk Maxwell (see Crawford 1985a, Chapter 2) said that Faraday's 'electrotonic state' was a recognition of 'potential difference', our present term for the condition which gives rise to current. However he also said that,

> . . . without a much greater degree of development than Faraday gave it, this conception does not easily lend itself to the explanation of phenomena.

This sums up an important point. The electrotonic state as Faraday first expressed it does not actually give an explanation of electromagnetic induction. In 1831, Faraday was faced with this problem: he had observed a new phenomenon, he had had what I think could be considered a revelation regarding its cause, as Maxwell later confirmed, but he could not, in spite of valiant attempts at articulation, actually show that the idea of the state did explain the phenomenon. There was no independent evidence that any state existed and thus no plausible link between the supposed cause and its effect.

What in fact happened at this time (Faraday 1932–1936, January 1832) was that Faraday went on to find and state the laws of electromagnetic induction without further reference to an explanation. He found that he could describe what happened in terms of an easily observed and well-known phenomenon, the directions of magnetic curves, or lines of force, as shown by iron filings. He gave up trying to say why it happened and wrote regretfully that the electrotonic state was not needed:

> Thus the reasons which induce me to suppose a particular state in the wire have disappeared; and though it still seems to me unlikely that a wire at

> rest in the neighbourhood of another carrying a powerful electric current is entirely indifferent to it, yet I am not aware of any distinct *facts* which authorize the conclusion that it is in a peculiar state. (Faraday 1839–1855, para. 242)

In spite of saying that it was no longer needed, he was again making use of the electrotonic state idea in the following year, 1832, to attempt electrochemical explanations (Faraday *ibid.*, para. 76 and Crawford 1985a, Chapter 2). However, again, his final formulation of the laws and behaviour of electrochemical phenomena made the electrotonic state idea redundant. The behaviour was better explained by simple measurements of masses of substance and duration of current flow (*ibid.*; also Faraday 1932–1936, Vol. 2, especially pp. 684–701).

Two years later, in 1834, Faraday again referred to the electrotonic state when exploring the phenomenon of self-induction (Faraday 1839–1855, Series 9 and 1932–1936, para. 2092 ff). Then, in the following year, 1835, when William Whewell wrote to him expressing support for the electrotonic state concept, Faraday replied that he had given the idea up. He said,

> I have given up this electrotonic state for the time as an experimental result because I could find no fact to prove it but I cling to it in fancy or hypothesis from general impressions produced by the whole series of results . . . I feel that my safety consists in facts; and even these I am but too anxious to pervert through the influence of pre-conceived notions . . . (letter from Faraday in Williams *et al.* 1971, pp. 293–297)

### III.2. *Constructed Thought* (*Benevolent Prejudice, or Need For Safety Produces a Lie*)

At this time, the articulation of the idea of the electrotonic state idea was difficult, if not impossible, in terms of other concepts with which Faraday was accustomed to think. These other concepts, of *matter, space, power*, were his mental home. They were not queried; the query, *prejudice*?, was addressed to the notion of the electrotonic state.

Nevertheless, the notion persisted. By 1838, many advances in his knowledge and explanation of electrical phenomena had been made, but the underlying cause of electromagnetic phenomena was still a mystery. Because electromagnetic effects occurred in the space around a wire carrying current, they were known as the 'lateral' or 'transverse' actions of electricity. Faraday determined to explain their nature. His Experimental Researches describe in detail how he set about this task (Faraday 1839–1855, Series 12, 13, 14). This is a classic example of 'scientific method' or 'benevolent prejudice' as described in Section II.1.1., above.

Faraday outlined two hypotheses: one, that action occurred by means of action-at-a-distance, the other that the electrotonic state, still unsupported by independent evidence of its existence, was a pre-condition of electromagnetic action. These hypotheses gave different predictions for a particular type of experiment in which material was interposed in the space between a primary and a secondary coil, so that the coils were 'screened' from each other. Faraday expected the current induced to vary depending on whether the screening material was a conductor or an insulator, and that the kind of variation found would then decide between the two hypotheses.

Unfortunately for him, at least in the short-term, all his results were negative. No screening effects were found and thus his prior reasoning had to be reconsidered. Following this re-examination, which led nowhere, he said,

> I cannot conclude these general remarks upon the relation of the electric and magnetic forces without expressing my surprise at the results obtained with the copper plate . . . Almost the only way of reconciling this effect with generally received notions is, as it appears to me, to admit that magnetic action is communicated by the action of the intervening particles . . . (Faraday 1839–1855, para. 1735)

This is equivalent to saying that action was not at-a-distance but was via the electrotonic state; *but*, the predictions bring no evidential results, it simply appears to Faraday to be the better supposition. It has not occurred to him to question the "generally received notions"; it is his hypotheses of action which are being questioned and tested.

Faraday's published papers and his Diary at this period display evidence of classic reasoning, always leading nowhere, or coming back full circle to the same "generally received notions", proposed hypotheses and missing or contradictory evidence. This is Tyndall's prison; to use Faraday's own words, previously quoted, *the more acute he was, the more tightly was he bound by the chains of error*. The difficulties of reasoning evident in Faraday's writing at this time demonstrate that he was in the grip of the *lie*, that construction type thought processes were in control and that he felt the chains of prejudice most acutely, although, prejudice being blind, he could not see or identify them[10].

### III.3. *Faraday Thinking with Negative Capability*

The situation began to change in 1839 and 1840. Unknown to Faraday, while he had been performing his screening experiments, in America, Joseph Henry had been performing very similar experiments. Because Henry's previous work had been concerned with magnitudes of voltage and current, he had constructed very powerful coils, and these were what he used in his screening experiments. His results, for various materials, were all positive. He found himself unable to explain all positive results any better than Faraday could explain all negative ones. Henry's results are in actuality the correct set as there is a screening effect for all materials, *but*, it depends on *magnetic permeability*, a property then unknown. It is ironic that the effect is minute for the materials copper, sulphur and shellac which Faraday had carefully chosen because they have very different values in respect of *electrical conductivity*, a property we now know to be irrelevant to the screening effects.

However we now understand it, in 1839 screening effects had been found. Henry had even found effects when the materials were not placed between the coils. He discovered that anywhere nearby would do. He could think of no reasons for this, made certain that it was so, and faithfully published his discoveries, saying that,

> . . . this part of the investigation was, for a time given up almost in despair . . . (1839, p. 319)

Faraday read Henry's paper in 1839, accepted his results, but was aware that they did not suit his reasoning any better than his own results had done (Faraday 1932–1936, para. 5234).

In 1840, he began another series of explorations into the topic. The only record of this work is the account given in his Diary (Faraday 1932–1936, August 1840, para. 6086–6187). This account is quite different in character from the reasoned arguments of 1838. There is little developed reasoning or ordered description except for some experimental ideas and a convoluted theory of 'partial current'. It is very difficult to follow, some is chaotic. Ideas appear, are jotted down, are briefly developed, are abandoned. A new track appears, seemingly without connection, an old one re-appears. Reminders about phenomena are interspersed with thought experiments, references to other work, experimental proposals, and apparently random comments.

The whole tone of these Diary entries makes one realize that in these pages Faraday is not writing down *what* he thought, but is instead exhibiting *how* he was thinking when a major change in his conceptual framework was taking place. A mistake[11] shows that he was getting lost in the complexity, forgetting thc details which he already knew.

It is tempting to speculate that this happened because of Henry's generous admission of inadequacy. Faraday was no longer alone with bewilderment and despair. One of the insights of psychoanalysis is that change does not occur via description or explanation of phantasy, but via shared experience of its *feeling*.

Bion has said that to reach a new formulation one must inhibit 'memory and desire' (Bion 1970, p. 41). In these pages, Faraday sets aside his desire for order and his memory of "generally accepted notions", to think with negative capability. The result was a new formulation: a conviction, without the need for evidence, that the supposed electrotonic state was indeed the mechanism of all electric and magnetic action, allied to a realization that supposedly certain concepts about matter and space had to be re-examined. The true thought had finally been accommodated. The cost was very high; Faraday first had to doubt and then change his deeply-rooted concepts about matter, space and electrical properties. Having done so, he had to learn how to articulate these changes. The meaning of the electrotonic state could not be expressed until the re-organization of other concepts was completed. There is not space to follow this part of the story here (Crawford 1985a, Chapter 5), but it should be noted that it was more than three years before his first presentation of these developments took place, when he gave a Friday Evening Discourse at the Royal Institution entitled 'A Speculation Touching Electric Conduction and the Nature of Matter'.

### III.4. *Physical Reality and Personal Experience*

At the beginning of this section, I said that the real scientific task was to know from which category of thinking an idea had come. The suggestion is that this would be a genuinely successful heuristic *process*, even though the truth of resulting theory could not be certainly stated and even though the knowledge acquired would remain partial, dependent on the current state of other conceptions. By taking up the story of the electrotonic state much later, and attending to Faraday's own comments concerning it, I think that, however difficult it may be, there is a recognizable possibility that scientists and others can distinguish ideas founded in truth from those whose sources are prejudice.

In 1852, Faraday wrote a paper, 'On the Physical Character of the Lines of Magnetic Force' (Faraday 1839–1855, Vol. 3, para. 3243–3299; see also Series 23–29 and other shorter papers in Volume 3, which were written during 1852–1854). This paper distinguishes the idea by which a theory represents a physical phenomenon and the assertion that the idea stated by the theory or representation is in itself the real physical phenomenon. Much of later nineteenth century discussion of the new concept of *the field* addressed this question of physical reality versus the alternative view that field theory was a mental or mathematical representation of something else. Is a field line real, or an enabling description? Although very conscious of both sides of the argument, Faraday declared that the lines of force were physically real on the grounds *of his own experience in their discovery*. He had to account for his personal awareness in the discovery process, and could not limit his argument to objective considerations attending only to explanatory power. He said,

> Indeed what we really want, is not a variety of different methods of representing the forces, but the one true physical signification of that which is rendered apparent to us by the phenomena . . . I do not perceive that the mathematician, even though he may think that each [representation] contains a higher principle than any that I have advanced, can tell the true from the false, or say that either is true . . .
>
> . . . Neither of these views could have led the mind to the phenomena of diamagnetism, and I think not to the magnetic rotations of light; and I suppose that if the question of the possibility of diamagnetic phenomena could have been asked beforehand, a mathematician, guided by either hypothesis, must have denied that possibility . . . (Faraday 1839–1855, Vol. 3, para. 3303)

In 1845, Faraday had discovered the new effects, referred to in this quotation: an interaction between light and magnetism and the phenomenon of diamagnetism in all materials. He had then developed magnetic field theory (Faraday 1839–1855, Vol. 2, papers from 1845–1850). An explanation of these phenomena can indeed be given without asserting the reality of magnetic field lines. However, like Faraday, I cannot see how the initial discoveries could have been made if Faraday had not 'known', however inarticulate the knowing, that magnetic lines of force were real in the same sense as a light ray is real. I do not see that the idea for the experiments could have come into mind, let alone have been performed, if Faraday had not thought that he was directing a real magnetic line through his piece of glass.

Therefore,when in 1852 and again in 1854, Faraday directly addressed the question of physical existence, and the distinction which had to be made between theory and physical reality, his chief line of argument was his subjective recollection of the making of discovery.

The heuristic question is, where had this inarticulated conception of reality come from, which was available to Faraday in 1845? There is evidence that at earlier times, for example in 1831 (Faraday *ibid.*, para. 232, 238), he had treated the magnetic lines as an expedient representation which enabled the expression of the magnetic laws. At that time he had hoped that the electrotonic state idea was the statement of the deeper, more fundamental reality. In 1853, he was saying that the magnetic lines had a real physical existence analogous to the physical existence of a light ray, and, further, that these lines were one and the same thing as the old idea of the electrotonic state.

He said,

> . . . again and again the idea of the electrotonic state has been forced on my mind, such a state would coincide and become identified with that which would then constitute the physical lines of magnetic force . . . (*ibid.*, para. 3269)

The word "forced" is meaningful. It implies that Faraday experienced the idea of the electrotonic state arising outside his own mental framework, that he felt it as a pressure from the perception of external phenomena. I believe it was accommodated internally along with changes in other concepts via the chaotic but accommodating thought processes of 1840. As the electrotonic state became real, so too did the lines by which it had been represented. Then, after his work on its meaning in relation to other ideas, new discovery was possible.

## IV. CONCLUSION

Faraday had both faith and experience of the idea of revelation. He had also endured over a long period the kind of thinking processes which had enabled him to entertain the revelatory thought.

He thus felt able to declare that the process of discovery justified his assertion that the lines were physically real, in spite of his continued concern and fear of prejudice. He did not distinguish, as far as I know, any difference in categories of thinking. However, he did try very hard and very consciously to distinguish between internal and external sources of ideas. This is one reason why I am confident that his use of the word "forced" described above was no accident.

In 1854, in his lecture on 'Mental Education', he said,

> I believe that a very large proportion of the errors we make in judgement is a simple and direct result of our perfectly unconscious state. (Faraday 1855, p. 465)

Then, having given examples of his own and others errors, he said,

> Among those points of self-education which take up the form of mental discipline, there is one which is of great importance, and moreover, difficult to deal with, because it involves an internal conflict and equally touches our vanity and our ease. It consists in the tendency to deceive ourselves regarding all that we wish for, and the necessity of resistance to these desires . . . (p. 475)

Faraday continues with references to feelings of deficiency, humility and need. These words, like "vanity", "ease" and "internal conflict" are all words describing emotional experience. Faraday was saying that to be objective one must know oneself and that this was a matter of emotions, not of reason.

Bion would have agreed. To think truly requires an emotional state of negative capability, in which thoughts are not produced to satisfy internal need, but are instead an accommodation in feeling to the thought which has been perceived.

It is easy, following the work of Bion, to write about two different categories of thought. The problem is that when the conscious thinking apparatus experiences a thought it very rarely discriminates carefully between internal and external sources. There is in fact much to indicate that it will often favour the internal source and resist external pressure quite strenuously. However, by acknowledging emotional levels in thought, and attending to the emotional meaning of *truth* and *lie* which Bion has articulated, I think that one can hope for the possibility that scientists and others can distinguish ideas founded in truth from those whose sources are a prejudice.

Post began his paper with the sentence,

> Philosophers of science should be concerned with Science; that is with the activity of scientists, whether the concern be descriptive, prescriptive or both. (Post 1971, p. 215.)

His thesis was 'In Praise of Conservative Induction'; throughout the paper he made various remarks about the meaning of this, such as,

> Quite generally, the thesis may be put this way: no theory that ever 'worked' adequately turned out to be a blind alley. Once a theory has proved itself useful in some respects, has shown its semantic simplicity or explanatory power, it will never be scrapped entirely. . . .
>
> My claim is not merely that there is an element of continuity in change, which is necessarily the case in all human activity, but that in science part of the content of the old theory is preserved, as far as one can tell, for ever; not just in the next theory, but throughout all future theories. . . .
>
> I do, however, see links and commensurability in all theories so far investigated that ever had any success at all. By success we mean here, of course, not just popular acclaim, but measurable pragmatic success (e.g. correct predictive power), though these usually go hand in hand. (pp. 237, 238, 255)

My own thesis is that knowledge results from intersection with the 'ultimate reality' as described by Bion and that a particular kind of thinking process enables this to happen. Post's thesis provides evidence for Bion's view of truth: that is, the observed commensurability of successful theories is a phenomenon encouraging the conjecture that 'truth', however unknown it remains in its totality, can be brought partially into the domain of knowledge.

In stating my thesis, I have attended to another part of the activity of scientists, the emotional part of their thinking, alongside the objects of their thought. Emotionality is another class of observable phenomena.

A 'lie' could for a time achieve popular acclaim, as human beings have some common emotional needs, and thus might welcome a 'theory', or variety of 'theories', which gave particular sets of needs satisfaction (this might explain Feyerabend's thesis also, and the persistence of sterile theories such as witchcraft or scientology which function when accorded emotional dependence).

However, as Post claims, there are also theories which are successful in predictive power and fruitfulness. My analysis of the processes of thinking indicates a possible explanation of why both kinds of theories should exist.

*Moray House, Institute of Education,*
*Heriot - Watt University, Edinburgh*

## NOTES

1. Neither a definitive biography nor a full critique of Bion's work has yet been published, although several authors have attempted partial accounts. See Meltzer 1978 and Bleandonu forthcoming.
2. I am solely responsible for the attempt to interpret and explain Bion's idea in this way. I hope that I have not introduced a distortion of his thinking and am grateful to Bion's widow, Francesca Bion, who some years ago encouraged me in my attempt to explore Faraday's thinking via the thought of Bion.
3. John Tyndall, Faraday's successor as professor at the Royal Institution, also his colleague and biographer. See Tyndall 1868.

4. Consider for example the following quote:

> All a child's activities, and those of adults too, even those most geared to external reality, also express and contain their phantasies. External reality affects our phantasies, and our phantasies influence our perceptions of external reality. (Daniel 1992)

5. Karl Popper's recognition that "science can never claim to have attained truth" is not contradicted by Bion's understanding of 'truth'.
6. The exploration of children indicates awareness linked with inadequacy of knowledge.
7. Faraday's Diary contains many entries besides those referred to in this paper, which also illustrate how he was thinking at the time of writing.
8. For example, the same truth might be expressed in art or poetry or scientific concept; few of us are able to articulate in separate forms, although we can respond to each sometimes.
9. By this term Faraday meant a *state* or *condition* which would exist in matter, caused by electrical action. Maxwell stated that,

> The scientific value of Faraday's conception of an electrotonic state consists in directing the mind to lay hold of a certain quantity, on the changes of which the actual phenomena depend. (Maxwell 1873, Article 540)

The modern equivalent of the "certain quantity" is the electric field potential. Faraday used his term in a more concrete sense.

10. There is also evidence for this from beyond the boundary of that which is scientifically relevant. I do not think it is coincidence that the most severe onset of Faraday's recurrent illness occurred at this time, and lasted until 1845, when, as I describe below, he had demolished the prison walls. Concerning Faraday's illness, Cantor, Gooding and James say:

> Faraday's massive research output in the 1820s and especially the 1830s contrasts strikingly with the early 1840s when, for a time, research almost ceased. Moreover, he lectured less frequently at the Royal Institution. The reason for these changes was illness. Although he had earlier complained of various problems, he now suffered from extreme exhaustion, dizziness, headaches and, worst of all, loss of memory . . . He had recovered significantly by 1844 but thereafter often complained of memory problems. It has proved impossible to identify a single cause of this (and other) illnesses. (Cantor, Gooding and James 1991)

11. There is at least one major mistake concerning the phenomenon of electromagnetic induction which Faraday of all people ought not to have made. In considering 'partial' currents, which he thought might build tertiary or higher order currents from primary and secondary ones, he declared that, at these orders, current was induced from *continuing, not transient*, current in the order below. This is of course not so.

## BIBLIOGRAPHY

Bion, W.R. (1962a), 'A Theory of Thinking', *International Journal of Psychoanalysis*, 43, pp. 306–310; also in *Second Thoughts*, London (1967), reprinted 1984, 1987, 1990; and in E. Bott Spillius (ed.), *Melanie Klein Today, Vol. 1, Mainly Theory*, London, (1988).

Bion, W.R. (1962b) *Learning From Experience*, London, reprints 1984, 1988, 1991.

Bion, W.R. (1963) *Elements of Psychoanalysis*, London, reprints 1984, 1989.

Bion, W.R. (1965) *Transformations*, London, reprint 1984.

Bion, W.R. (1970) *Attention and Interpretation*, London, reprints 1984, 1988.

Bleandonu, G. (forthcoming) *Wilfred Ruprecht Bion, 1897–1979; His Life and Work*, Free Association Books.

Cantor, G. (1991) *Michael Faraday, Sandemanian and Scientist, A Study of Science and Religion in the Nineteenth Century*, Macmillan, London.

Cantor, G., Gooding, D. and James, F.A.J.L. *Faraday* (London, 1991).

Crawford, E. (1985a) *The Ideas of Particle and Field in Michael Faraday's Work, 1831–1845*, Ph.D. thesis, University of London.

Crawford, E. (1985b) 'Learning from Experience' in D. Gooding and F.A.J.L. James (eds.), *Faraday Rediscovered, Essays on the Life and Work of Michael Faraday, 1791–1867*, London.

Daniel, P. (1992) 'Child Analysis and Unconscious Phantasy' in R. Anderson (ed.), *Clinical Lectures on Klein and Bion*, London.

Faraday, M. (1833) 'Address delivered at the Commemoration of the Centenary of the Birth of Rev. Joseph Priestley', *Philosophical Magazine*, 2, pp. 390–391.

Faraday, M. (1839–1855) *Experimental Researches in Electricity, Reprinted from the Philosophical Transactions of 1831–1852, with other electrical papers from the Quarterly Journal of Science, Philosophical Magazine, the Proceedings of the Royal Institution*, 3 volumes, London.

Faraday, M. (1855) 'Observations on Mental Education, a lecture delivered before His Royal Highness The Prince Consort and the members of the Royal Institution on 6th May, 1854' in M. Faraday *Experimental Researches in Chemistry and Physics*, Taylor and Francis, 1859, reprinted 1991.

Faraday, M. (1932–1936) *Faraday's Diary, Being the Various Philosophical Notes of Experimental Investigation made by Michael Faraday, D.C.L., F.R.S., during the years 1820–1862 and bequeathed by him to the Royal Institution of Great Britain* (ed. T. Martin), 7 volumes, London.

Henry, J. (1839) 'Contributions to Electricity and Magnetism, number 3: On Electrodynamic Induction', *American Philosophical Society Transactions*, 6, pp. 303–338.

Keats, J. (1952) 'Letter to G. and T. Keats, 21 December 1817' in H.B. Forman (ed.) *J.Keats, Letters*, 4th edition, London.

Klein, M. (1952) *Developments in Psychoanalysis*, London.

Klein, M. (1975) *Writings of Melanie Klein*, London.

Maxwell, J.C. (1873) *A Treatise on Electricity and Magnetism*, Oxford.

Meltzer, D. (1978) *The Kleinian Development*, Clunie Press, Perthshire.

Poincaré, H. (1913) 'Mathematical Creation' in *The Foundations of Science*, translated by G. Bruce Halstead, The Science Press, New York.

Post, H.R. (1971) 'Correspondence, Invariance and Heuristics: In Praise of Conservative Induction', *Studies in History and Philosophy of Science*, 2, pp. 213–255.

Tyndall, J. (1868) *Faraday as a Discoverer*, London, reprinted 1870, 1894, 1898.

Williams, L.P., Fitzgerald, R. and Stallybrass, O. (eds.) (1971) *The Selected Correspondence of Michael Faraday*, 2 volumes, Cambridge.

NORETTA KOERTGE

# Ideology, Heuristics and Rationality in the Context of Discovery

## I. INTRODUCTION

Philosophy often progresses by introducing sharp distinctions and then dismantling them. A prime example for those interested in heuristics is Reichenbach's demarcation between the context of discovery and the context of justification. Reichenbach himself thought the distinction an important one and referred to it on many occasions[1], but he provides little detail about exactly what the distinction is or its proper function in the history and philosophy of science. Whatever Reichenbach's own intentions may have been[2], historians of science, philosophers of discovery, and, more recently, cognitive scientists have vehemently objected to his claim that the context of discovery " . . . can be analyzed only psychologically, not logically . . . and cannot be portrayed by a rational procedure . . . " (Reichenbach 1949, p. 434). And in his ground-breaking 1971 paper 'Correspondence, Invariance, and Heuristics', Heinz Post was actually able to articulate a variety of what he called " 'theoretic' guide lines to new theories" (1971, p. 217) and to provide a rationale for them. Even if the *Entdeckungszusammenhang* turned out to be a private, subjective affair, the *Entwicklungszusammenhang* could profitably be studied by the philosopher of science. So Reichenbach's demarcation, if interpreted as an attempt to define a water-tight boundary between the proper domains of psychology and epistemology[3], has proved to be a very leaky one indeed.

Today it is a commonplace amongst philosophers and cognitive scientists to say that Reichenbach just got it wrong when he said that "epistemology is *only* occupied in constructing the context of justification" (1938, p. 7; my italics) – there is also the possibility of providing a philosophical analysis of heuristics and the logic of pursuit. However when we look back at Reichenbach's original discussion, it seems clear that his major purpose in drawing the distinction was not to keep philosophers *out* of the domains of psychology and the social sciences[4]. Rather the primary job of his distinction was to *protect* the context of justification *from* extraneous elements. Perhaps if pressed, he would have said that the distinction should act like a



*S. French and H. Kamminga (eds.), Correspondence, Invariance and Heuristics*, 125–136.

semi-permeable membrane. If philosophers found grist for their mill outside of the *Rechtsfertigungszusammenhang*, if it turned out that there *were* logical aspects of discovery, I doubt that Reichenbach would have been too surprised[5]; it certainly would not have made much difference to his core views. His main concern was to forestall criticisms of his theory of induction on the grounds that it was incomplete:

> [Those] who believe that induction could become a sort of philosopher's stone, supplying methods that automatically transform facts into theories, misunderstand the task of logical analysis and burden the theory of induction with an unsolvable problem. (Reichenbach 1949, p. 434)

In addition he wished to defuse any attempts to fault his rational reconstructions of mature science on the grounds that they were inaccurate descriptions of actual history (1965, p. 67). Thus the most important function of his demarcation was not to keep logic out of discovery but to keep epistemology free of historical contingencies.

Although semi-permeable membranes can easily survive in nature, they are difficult to maintain in the conceptual world. So perhaps it was inevitable that as philosophers attacked Reichenbach's demarcation from one side, showing how logic and the tools of rational analysis played crucial roles in the context of discovery, on the other side were social scientists, especially advocates of the sociology of scientific knowledge, arguing that socio-psychological factors were inextricably implicated in the context of justification.

Many friends of discovery would be appalled by the latter development – certainly Heinz Post would. His exciting inaugural lecture, given on the occasion of becoming the Professor of Philosophy of Science at Chelsea, is entitled *Against Ideologies* and it contains a two-pronged message: "By all means try any ideology you like for heuristic purposes *but keep your critical powder dry*" (1974, p. 6; italics in original). He also says his talk might be sub-titled 'In Praise of Water-Tight Compartments' (*ibid.*, p. 1) and he gives historical examples of how science is distorted when ideology enters into the process of theory evaluation (*ibid.*, p. 14).

The problem I wish to address in this paper is the following: If it is rationally permissible to use a motley array of ideas, beliefs, prejudices, interests and ideologies for heuristic purposes but illegitimate to let them enter into the context of justification or evaluation and if, as Gigerenzer (1991) and others argue, the two sorts of processes are not temporally disjoint (scientists may alternate between hypothesis formation and error elimination over long periods of time), then is it realistic to expect them to keep their critical powder dry? Or to change the metaphor: Once we invite the nose of the ideological camel into the discovery side of the tent of rational scientific procedure, can we keep it out of the justification side, especially when scientists are constantly moving back and forth across the interface?

I think Post's philosophy provides us with clues about how to alleviate this problem, although, speaking from my past experience as his student, I doubt if he will go along entirely with my solution! But first let us see how the alleged infiltration of ideology into the output of science is supposed to occur.

## II. THE SEEPAGE SCENARIO[6]

Once one abandons positivism, so the story goes, it is impossible to keep ideology out of science, where 'ideology' is construed broadly to include everything from a metaphysical commitment to determinism to the hope that women and men do not differ in brain lateralization. Traditionally it was deemed a fallacy to let one's evaluation of the desirability that *p* be true influence one's evaluation of how likely[7] it is that *p* is true. But, the claim is, the influence of ideology is ubiquitous. The choice of scientific problems is influenced by pressing medical or political problems and the availability of funding or equipment. And even in the most isolated of ivory towers, problems are structured by metaphysical research programs and what Popper called 'background knowledge', a mishmash of uncritically examined common-sense beliefs and prejudices. The generation of solutions to these problems and the selection of the most promising paths to pursue are likewise influenced by pragmatic and doctrinal considerations.

As one looks at the process of testing and appraising hypotheses, so the scenario continues, the influences of ideology are less obvious, but nevertheless important. The design of a severe test depends on judgments about which variables are likely to be relevant – these also draw heavily on 'background knowledge' and the availability of scientific apparatus (which in turn depends on earlier pragmatic decisions about research directions). Some philosophers have argued that theory appraisal is better viewed as a comparative affair, but this hardly helps eliminate ideological tendencies which are shared by the entire scientific community.

Even when a theory fails a test, the scenario continues, it will be replaced with another variant of the same old metaphysical research program. And if it passes the test, our decision as to how far and long to push critical scrutiny will depend on a judgment as to how much harm would result from accepting it even if it were false and our assessment of the consequences of error will typically include more than just epistemic damage.

So, the conclusion goes, science is fallible, but not just because of the inevitability of more or less random missteps. Rather it is infiltrated in a systematic way with its own ideological history.

## III. HAIR OF THE DOG 'CURE'

No one is pleased with the conclusion that science is inevitably infected with the dominant ideology. Old-fashioned objectivists would like to curtail or eliminate non-epistemic elements. Political activists would like to replace what they see as the reactionary, oppressive ideological bent of mainstream science with a progressive, liberating ideology which would produce a science-for – all-the-people.

Curiously, there are parties in both camps who agree that the cure for the problem of ideology in science is, in the short run at least, more ideology! The hope of the objectivists is that if a plurality of ideologies figures in the context of discovery and pursuit, there will be more competition and a better chance of coming up with the truth in the context of justification and error elimination. Those who are striving for an ideologically correct science, on the other hand, argue for a sort of affirmative action effort on behalf of their under-represented ideologies. Thus Longino (1990) says that scientists should be held accountable for the political commitments of their scientific research programs:

> . . . if faced with a conflict between [political] commitments and a particular model of brain-behavior we allow the political commitments to guide the choice. (Longino 1990, p. 191)

> . . . I am suggesting that a feminist scientific practice admits political considerations as relevant constraints on reasoning . . . (*ibid.*, p. 193)

Both the objectivist and those who favor the-correct-ideology hope that the pluralistic struggle will be of finite duration and will produce a single victor. But for the objectivist the outcome is intended to be superior because it is free of non-epistemic values, while the other camp hopes for knowledge which is somehow shaped and constrained by progressive social values.

I have several objections to both variants of the idea that the best cure for the ideological hangover is "the hair of the dog that bit you." To the pluralists it suffices to point out that all they are really asking for are set-breaking radically different alternative approaches, which can be generated by scientists who are at least provisionally able to suspend belief in the concepts of absolute space and conservation of matter, or the idea that altruism could not be adaptive. It is not required that these scientists be committed to the radical alternative or that they be bolstered by its ideological attractiveness. One doesn't have to be a "true believer" to give a crazy idea a run for its money! As Post put it, try any ideology you like, "preferably several ideologies . . . [but] *do not take ideology seriously*" (1974, p. 1; italics in original).

To the affirmative action advocates, I would voice general agreement with the proposition that scientists take responsibility for the political aspects of their actions. Thus I think scientists *are* accountable for their decisions about whether to do research on nerve gas, 'Star Wars', or racial differences. I do not think it is possible to give general recipes (such as those provided by ideologies) about what one should do in such cases, but individual scientists, like all citizens, should think about these issues and it is entirely appropriate that groups of scientists, like other professional groups, should draw up ethical guidelines. However, I would consider it extremely unwise to advise scientists to try to constrain the set of hypotheses to be tested by considerations of political desirability or ideological correctness. One can invoke the specters of Nazi science or Stalinist science as reminders of how knowledge deteriorates when the personal characteristics of the investigator or the political correctness of the research program are allowed to influence scientific appraisals. In reply, Longino might emphasize that the additional values which she is advocating be incorporated into scientific practice should themselves to be subject to careful scrutiny. To argue that science should be shaped by political considerations is not to endorse the use of bad politics. Furthermore, she is not advocating that we cease other scientific activities such as experimentation and critical discussion. Rather her proposal is this: Why not try to generate knowledge which is good science from both an epistemic and political perspective?

However, I would oppose the deliberate incorporation of even a benevolent ideology into science for many reasons. First, we should remember the extreme fallibility and liability of our judgments about the political desirability of scientific ideas. To cite a couple of contemporary examples: In telling the story of the infamous Tuskegee Syphilis experiment in which scientists deliberately withheld treatment from over 400 black men in order to chart the course of untreated syphilis, historians claim that there were racist reasons for even setting up the experiment in the first place, since by 1932 a longitudinal study of white syphilitics had already been carried out in Sweden[8]. Yet in 1992 American medical scientists are being faulted for *not* studying the difference between hypertension in blacks and whites. The ideological correctness of studying racial differences has changed. And after years of campaigning against 'raging hormone' theories about women, feminists are now calling for research on how the effect of medication varies with hormone levels.

It is hard enough to make one's hypothesis consistent with the ever-increasing volume of empirical data. To try to impose the further constraint of political progressiveness is even more difficult because there is so much variation in the social contexts in which the ideas will be utilized. For example, when Bell and Weinberg's massive study of homosexuality appeared in 1978, many gay activists in Germany were appalled at the speculation in the concluding chapters that there might be a congenital predisposition towards homosexuality. They remembered all too well the Nazi 'solution' to unwanted genetic factors and feared that such a hypothesis might lead to compulsory sterilization laws.

In America, on the other hand, many activists welcomed the same hypothesis, feeling that it might alleviate discrimination and even provide protection under the U.S. Constitution – after all, how can one be blamed for a sexual preference that one is born with? Thus Bell and Weinberg discovered that the political attractiveness of their tentative conclusion depended on which side of the Atlantic they were on. In this case, it would have been impossible for them to put consistent political constraints on their research program. Often the result of too much pressure to be politically accountable has the effect of driving responsible scientists out of such areas of research entirely, leading either to the neglect or to the politicization of entire fields of inquiry.

## IV. RE-EXAMINING THE ROLE OF NON-EPISTEMIC FACTORS IN THE CONTEXT OF DISCOVERY

The claim that the content of mature science is inevitably influenced by human hopes, fears and prejudices can be sub-divided. First, there is the relatively non-controversial statement that the object or domain of study depends in part on non-epistemic factors, ranging from the availability of funds to the aesthetic properties of the phenomena under investigation. The follow-up statement posits that not only *what we study* is under external influence, but also that *what we come to believe about it* depends in part on our hopes, fears and social interests.

The second part of the claim is the most troubling (although I shall also have some things to say about the first). I have already argued against the solution of more ideology – neither the pluralist nor the progressivist variant seems likely to improve science. The only remaining move seems to be to call for less ideology. But isn't this tantamount to a positivistic rejection of all of the important philosophical analyses of the heuristic and programmatic aspects of scientific discovery and development? I think not, but to show this we need to look more carefully at exactly what choices arise in this context and what judgments of value must be made.

For purposes of analysis, I will use a decision-theoretic approach. This framework has the advantage of forcing us to separate the desirability of an option from the probability of its success. As do all users of rational choice models, I enter the usual caveats about not assuming that actual decision-making always involves the sort of explicit deliberation portrayed in these analyses.

Although most writers on scientific discovery begin their account with an analysis of the moves people make in generating hypotheses or selecting which ones to pursue, I wish to push the discussion back a step further into the context of problem identification and problem choice. What factors enter into a rational choice amongst research options? There are two basic types of consideration: how the problems rank as to *importance* (e.g., which is the most desirable to have solved) and how they rank as to *solvability* (e.g., which is the most likely to be amenable to solution)[9].

Let us look at the various factors which enter into the assessment of importance. First of all there are purely cognitive criteria: If two comprehensive, well-tested theories are mutually inconsistent, that poses a major epistemological problem for our present knowledge system. In such a case the importance of the problem increases with the scope of the two inconsistent fragments because we realize that a lot of our present knowledge will need to be revised.

Another source of cognitive problems are unexpected 'gaps' in our knowledge. When chemists undertook the problem of searching for the missing elements in Mendeleev's Periodic Table, they were not just looking for new elements (itself a task fundamental to chemical theory); they were also checking out the predictive power of his Periodic Law. Again the cognitive stakes were high.

Post's 1971 discussion of formal flaws or unexplained degeneracies in theoretical systems provides further examples of objective determinants of cognitive importance. There are also cases, especially in mathematics, where the search for a particular result becomes important, not because of its own intrinsic epistemic interest, but because we believe it will serve as a lemma or bridge leading to other more important results.

This is not an exhaustive list of the objective determinants of importance, but it perhaps suffices to remind us that in appraising the *desirability* of pursuing one problem rather than another we rely on a variety of traditional epistemic values such as consistency, completeness and content.

We may, if we wish, add to our decision matrix the social desirability of giving certain problems priority. This complicates the picture in a couple of interesting ways. First of all, there is the apples-and-oranges problem of how to compare the virtues of pure vs. applied research problems, a difficulty which already arises within the purely epistemic context – which is more serious, a gap or a flaw? There is a second complication which is more troublesome. In the case of most epistemically pressing problems, both positive and negative outcomes are of at least some intellectual interest. In fact some of the most important results in theoretical science are 'impossibility proofs', such as Bell's Theorem which rules out a wide variety of hidden variable theories as solutions to puzzles raised by Quantum Mechanics. However, in research directed toward the solution of a social problem, negative results are typically viewed as failures. No applied research funding agency wants to hear about the interesting theoretical reasons why it is so difficult to find an AIDS vaccine or produce energy through controlled fusion. This means that even if one manages to finesse the apples/oranges difficulty by choosing problems which are both epistemically and socially important, one's research could still be shaped by the 'success bias' inherent in applied research[10].

Bacon pointed out that we value science because it gives us both light and fruit and we have sketched how the importance of problems can be rated on both an epistemic and a social scale. Science policy experts may disagree about the appropriate mix of the two types of desirability and it may well be that too much stress on immediate practical results would stunt the epistemic growth of science, but a mere prioritizing of politically relevant problems would not by itself lead to ideological distortion in workable solutions to those problems unless the stress on success led to fraud.

Let us now turn to the appraisal of solvability. Before one actually solves a problem, can one really judge how difficult it is? It is at this point that the various items generally discussed under the loose rubric of heuristics become relevant to problem choice. Sometimes there are formal measures of complexity. Have we previously solved problems of a similar structure so that we have at hand a useful analogy? Are there available invariance principles or correspondence conditions which put constraints on the search space? Judgments of how likely we are to be able to solve a problem can also be influenced by the past success of metaphysical research programs. Is this the kind of problem where a reductionist, or environmentalist, or social constructionist approach looks promising? Note that all of these appraisals, though tenuous, are in principle objective. Even though one's allegiance to a research program may be based on ideological factors, its actual degree of success to date is, as Lakatos pointed out, a matter of public record. (Of course, partisans of an unsuccessful approach may reply by presenting evidence that it has never been given a fair run for its money).

There are also purely pragmatic factors which properly enter into our appraisal of solvability. The absence of funding, trained personnel or appropriate equipment may prevent us from solving a problem (but their presence in no way guarantees success).

Our analysis of the choice of research problems could be extended to include factors such as personal ambition and the reward system in science but we have already gone far enough to answer questions about the roles open to ideology. To the extent that our ideology affects our judgments about the social importance of problems, it may well influence problem choice. Our ideology may also make us unrealistically sanguine about how tractable certain problems are. But neither of these influences are especially worrisome. If the problem can in fact easily be solved, no epistemic harm is done. If it cannot, then just because work on social problems is so strongly success-oriented, research in a non-productive direction should eventually taper off. If a dominant ideology devalues certain social problems or is unduly pessimistic about our ability to solve them, society may suffer but the objectivity of science need not be damaged.

## V. FROM PURSUIT TO TENTATIVE ACCEPTANCE

So far, the camel's nose is under control. But what happens when we look at the process of actually pursuing a solution to a research problem? Does the desirability that the solution be of a certain type influence our judgment about what the correct answer is?

Again what is called for is a careful decision-theoretic analysis of the options open to the scientist at each step in the process of inquiry. As in the choice of research problem two quite different sorts of appraisal will be involved – the desirability of a certain result and the probability of success. Thus it is quite legitimate for scientists to begin by pursuing solution candidates which are linear or mathematically tractable, easy to test or technologically attractive. By the same token scientists may well choose to pursue research programs which they find ideologically attractive. David Hull has recently (1988) argued that non-epistemic factors such as personal ambition and wanting to prove your opponents are wrong may provide valuable motivation for putting in 80 hour weeks in the laboratory! Likewise the ideological attractiveness of a certain type of solution may spur us to greater industry, but no harm is done to the objectivity of science as long as the scientific community does not confuse the *desirability* of a certain sort of solution with its *plausibility*.

As the testing and critical evaluation of a favored hypothesis continues, the scientific community is often able to generate evidence which renders the earlier plausibility judgments irrelevant. The rationality of the process may be illustrated with a simple example.

Suppose one is trying to decide whether a given book is reliable (R). One may form a preliminary impression of the text by just looking at its cover. Is it published by a vanity press (V)? Most establishment types, reasoning that prob(R, V) < prob(R, ~V), would form a negative preliminary opinion of the book. (Counter-cultural types might assign an opposite appraisal). However, if one is quite interested in the topic of the book, one may read part of it and even check some of the footnotes. If the book holds up well under critical scrutiny (C), even the person who believes most vanity press publications are unreliable may nevertheless admit that prob(R, C.V) = prob(R, C.~V). In other words, once one has actually read and critically evaluated a book, assessments based on its cover become irrelevant.

There are many interesting parallels to science which might be extracted from this example. The primary morals which I wish to draw are these: Scientists may not agree on the appraisals of plausibility and desirability which inform their decisions in the context of pursuit. Seeing this variability, the naive observer may conclude that scientific research is guided by subjective or ideological concerns. However, as long as scientists remain genuinely interested in understanding nature and maintain a tradition of critical evaluation, then the myriad factors which influence pursuit are eventually overridden by the on-going procedures of experimental testing and skeptical appraisal. The more confident we are that the process of criticism will weed out empirically inadequate scientific claims, the less tempted we will be to try to impose ideological constraints on the ideas we are willing to entertain as hypotheses. On the other hand, the more social harm which would result from mistakenly accepting an empirically inadequate hypothesis, the higher the level of critical scrutiny to which it should be subjected.

## VI. CONCLUSION

The original demarcation between discovery and justification coincides with neither the line between psychology and logic nor the distinction between the external and internal aspects of science. Instead what we find is a complicated process of inquiry which can be modeled as a tree of choice points and decisions. At each step of the process two distinct sorts of appraisal must be made – one requires us to assess the *desirability* of an outcome. This evaluation may be based on traditional epistemic values such as simplicity and explanatory power or it may include social and political values. The important thing is that this appraisal of the desirability of a possible outcome be made completely independently from the assessment of *how likely* it is that the outcome will be realized. It is interesting to note that on my analysis what have commonly been called *heuristic factors* often straddle this demarcation. Theoretical simplicity (however that is understood) is typically a *desired* attribute, but in most situations we have no reason to believe that the simplest theory is *likely* to be correct. On the other hand, the fact that a new theory stands in a correspondence relation to a refuted but largely successful older theory may in some circumstances confer a degree of prior *plausibility* on that new theory. In heuristic decision making, i.e., decisions about pursuit worthiness, we combine these considerations of attractiveness and plausibility and so it is easy to overlook the separability of the components.

When we turn to epistemic decision making, however, it is crucial to keep the distinction clearly in mind. The assessment of the probability that the O-ring on the Challenger will fail must be independent of how personally, politically or financially undesirable the result of that assessment may be.

In order for science to aspire to objectivity, it is not necessary to try to seal it off from its political and social context. The desirability of pursuing certain problems, the attractiveness of certain types of solutions, the costs or benefits of various technological applications – all of these may properly be influenced by what is often called ideology. What we must insist on is that evaluations of any kind, whether they be of social desirability or epistemic desirability, *not* contaminate our appraisals of our chances of success. It is the demarcation between what we wish were the case and what is in fact likely to be possible which is crucial for both the rationality and the objectivity of science.

Conversely, those critics of science who feel that the products of the scientific establishment too often serve as tools of a repressive political establishment may well find that what they are upset about is not well-described by saying that scientific objectivity is a pretense, that science is, to use Sandra Harding's adaptation of Clausewitz' aphorism, merely "politics by other means" (Harding 1991). Perhaps what the critics should stress instead is the *selectivity* of the scientific enterprise. Not every phenomenon can be studied so inevitably scientists must direct their inquiry. In this paper I have briefly described the variety of factors which enter into these heuristic decisions. It may turn out that the paths *not* taken are the ones which political activists or ordinary citizens wish that scientists had pursued. But that does not entail that the travelers' tales brought back from the path they did in fact explore are not close approximations to the truth. And sometimes, when we are lucky and the perennial scientific quest for a grand theory is successful, the highly selective path of scientific inquiry actually ends up providing us with an objective overview of a vast terrain.

*Dept. of History and Philosophy of Science,*
*Indiana University,*
*Bloomington, Indiana*

## NOTES

1. The canonical references are to Reichenbach's *Experience and Prediction* (1938) and *The Rise of the Scientific Philosophy* (1951), but he also invokes the distinction in his *Philosophic Foundation of Quantum Mechanics* (1944), *Elements of Symbolic Logic* (1947), and *The Theory of Probability*, 2nd ed. (1949). (The original German edition of 1935 does not contain the relevant section).
2. See Curd (1980) on this point.
3. For Post's own views on this boundary see his (1971, p. 216).
4. As examples of Reichenbach's own philosophical excursions into these areas, see his 1932 interview, 'Is The Human Mind Capable of Change?' (*Selected Writings*, Vol. 1, pp. 275–279) or his 1959 essay, 'The Freedom of the Will' (*op.cit.*, pp. 431–473).
5. Although Reichenbach quotes with approval Schrödinger's remark about "the darkness of instinctive guessing" (1965, p. 67), he notes in his 1929 'The Aims and Methods of Physical Knowledge' that intuitive models can be employed by the physicist as "a valuable aid in both the process of research and the process of comprehension (1978, II, p. 208). He also gives examples in which he says " . . . the criterion of simplicity has played a major role in the construction of physical theory." (*op. cit.* p. 162), but then goes on to distinguish descriptive simplicity, which is merely of pragmatic value, from inductive simplicity.
6. For a fuller account see (Koertge 1991).
7. This formulation is not intended to commit one to any particular view of evidential support and could be rephrased so as to satisfy Popperians.
8. For a good discussion of 'racial medicine', see Jones (1981). Obviously the manner in which the experiment was carried out was grotesquely unethical but it would also appear that if one had put enlightened political constraints on permissible hypotheses the question of whether there are racial differences in the progression of syphilis could not have been raised.
9. This distinction is used to explain a puzzle about Mendeleev's behavior in (Koertge 1978).
10. I conjecture that it is this aspect of medical research which makes it especially susceptible to fraud. Of course, negative results are probably always somewhat under-valued. It would be interesting to have a philosophical account of why some negative results are more interesting than others.

## BIBLIOGRAPHY

Bell, A. and Weinberg, M. (1978), *Homosexualities, A Study of Diversity Among Men and Women*, Simon and Schuster, New York.

Curd, M. (1980), 'The Logic of Discovery: An Analysis of Three Approaches', in T. Nickles (ed.), *Scientific Discovery, Logic and Rationality*, Reidel, Dordrecht, pp. 201–219.
Gigerenzer, G. (1991), 'From Tools to Theories: A Heuristic of Discovery in Cognitive Psychology', *Psychological Review*, 98, pp. 254–267
Harding, S. (1991), *Whose Science? Whose Knowledge? Thinking from Women's Lives*, Cornell University Press, Ithaca, New York.
Hull, D. (1988), *Science as a Process*, University of Chicago Press, Chicago.
Jones, J. (1981), *Bad Blood, The Tuskegee Syphilis Experiment*, The Free Press, New York.
Koertge, N. (1978), 'Towards a New Theory of Scientific Discovery', in G. Radnitzky and G. Andersson (eds.), *Progress and Rationality in Science*, Reidel, Dordrecht, pp. 253–278.
Koertge, N. (1991), 'Ideology, Science and a Free Society', in G. Munevar (ed.), *Beyond Reason*, Kluwer, Dordrecht, pp. 225–241.
Longino, H. (1990), *Science As Social Knowledge*, Princeton University Press, Princeton.
Post, H. (1971), 'Correspondence, Invariance and Heuristics', *Studies in History and Philosophy of Science*, 2, pp. 215–255.
Post, H. (1974), *Against Ideologies*, Inaugural Lecture, Chelsea College, University of London.
Reichenbach, H. (1938), *Experience and Prediction*, The University of Chicago Press, Chicago.
Reichenbach, H. (1949), *The Theory of Probability*, University of California Press, Berkeley and Los Angeles.
Reichenbach, H. (1964), *The Rise of Scientific Philosophy*, University of California Press, Berkeley and Los Angeles.
Reichenbach, H. (1965), *Philosophic Foundations of Quantum Mechanics*, University of California Press, Berkeley and Los Angeles.
Reichenbach, H. (1966), *Elements of Symbolic Logic*, The Free Press, New York.
Reichenbach, H. (1977), *Der Aufstieg der wissenschaftlichen Philosophie*, University of California Press, Berkeley and Los Angeles.
Reichenbach, H. (1978), *Selected Writings, 1909–1953*, Volume I, Reidel, Dordrecht.
Reichenbach, H. (1978), *Selected Writings, 1909–1953*, Volume II, Reidel, Dordrecht.

NEWTON C.A. DA COSTA AND STEVEN FRENCH

# Towards an Acceptable Theory of Acceptance: Partial Structures, Inconsistency and Correspondence

## I. INTRODUCTION

In the second appendix to his 'Correspondence, Invariance and Heuristics' (Post 1971), Heinz Post notes that

> Acceptance, rather than rejection, is the more remarkable exercise of scientific judgment. (p. 253).

It is, perhaps, just as remarkable that, within the philosophy of science, the nature of acceptance and, in particular, its relationship to belief, have not received the analysis and consideration they deserve. Our suspicion is that it is for this reason that, in the case of many issues within the field, the participants often seem to be talking past one another. This is nowhere more evident than in recent discussions of the much debated distinction between justification and discovery. In particular, many appear to subscribe to a holdover from positivism according to which, within the domain of justification, theories are accepted as true, whereas within the domain (or domains) of discovery and pursuit they are, at best, merely 'entertained', to use a recent suggestion. Abandoning unconditional assent in the latter domain(s) is then often taken to licence the importation of sociological or psychological criteria of theory pursuit and discovery.

Like many before us, including Heinz Post, we shall effectively be questioning this distinction but in a somewhat roundabout manner which allows for the possibility that discovery and pursuit may possess sufficient structure so as to rule out as unnecessary non-epistemic or 'cultural' factors. If, as we claim, we are justified in accepting theories, not as true in the simple correspondence sense, but as provisionally or partially true only, then the need for an alternate to denote the attitude adopted in discovery or pursuit simply evaporates. Thus, our aim here is to lay down the framework for a view of acceptance which is broad enough to accommodate the various senses in which Quantum Field Theory is currently accepted, Newtonian mechanics is accepted within a limited domain and Bohr's theory of the atom was accepted, albeit only provisionally.



S. French and H. Kamminga (eds.), Correspondence, Invariance and Heuristics, 137–158.

The latter two examples are particularly interesting, given that Newtonian mechanics is empirically false and Bohr's theory was, famously, inconsistent. We shall begin by attempting to articulate the sense in which they can be said to be 'accepted' in terms of an attitude of commitment which distinguishes acceptance from mere 'factual' belief. This will provide the philosophical underpinning of our framework. The technical counterpart is supplied by recent work which introduces the notion of 'partial structures' into the semantic, or model-theoretic, approach to scientific theories. It is this which allows us to conveniently formalise the notion of 'partial' or 'pragmatic' truth which is central to our account.

An understanding of the sense in which Newtonian mechanics may be regarded as 'partially' or 'pragmatically' true, and accepted as such, involves Post's 'General Correspondence Principle'. Put simply, this states that we never lose the best of what we have, where empirical success ultimately determines what is retained. It is this success, albeit over a limited domain, which supports the claim that Newtonian mechanics is pragmatically true. The Correspondence Principle also plays a heuristic role in delimiting the field of possible successors to a theory. It is in this capacity, in particular, that it may be used in moving from an inconsistent theory, such as Bohr's theory of the atom, to a consistent successor. From this point of view, inconsistencies are to be regarded as, at best, heuristically fruitful but not as permanent features of our representations of reality. Such representations which are inconsistent should therefore not be regarded as true, as certain neo-Hegelians might suggest, but neither should they be immediately dismissed out of hand, as a more conservative approach might dictate. Rather, we claim, they should be regarded as partially true and accepted as such.

These seem to be perfectly straightforward points (at least they seem so to us!). Nevertheless, in following them through a number of difficult issues must be addressed, such as the imposition of deductive closure, the role of non-classical formalisms in accommodating inconsistency and the nature of a logic of acceptance. We do not pretend to have resolved all these issues here; however, we hope that we have at least shed some light on them.

## II. ACCEPTANCE, COMMITMENT AND BELIEF

The fundamental question with which we shall begin is

> What is the nature of the attitude of acceptance that we adopt in relation to scientific theories?' (de Oliveira 1985, p. 129[1]).

Various answers have been given to this question. The most straightforward, and also most naive, is that to accept a theory is to believe it to be true, in the correspondence sense of truth. However, simple reflection upon the historical record suggests that even the best scientific theories currently accepted will some day come to be rejected as false and this answer is now generally rejected (*ibid.*, p. 130)[2].

A well known alternative view retains the linkage between acceptance and belief but restricts the latter to belief in the empirical adequacy of the theory only (van Fraassen 1980). The deficiencies in this 'constructive' empiricism are well known and it is particularly worth noting that the only justification that is given for confining belief to the truth of the empirical sub-structures of a theory, rather than to the truth of the theory as a whole, is essentially sceptical in nature and should be looked upon with a jaundiced eye for that reason alone (French 1989).

In response to such an approach, Horwich has recently argued that there is, in fact, no real difference between believing a theory and accepting it in the sense of being disposed to use it to make predictions, design experiments, etc. (Horwich 1991). Since the latter is precisely the sense in which the constructive empiricists understand acceptance, the distinction on which their philosophy is based is one, he claims, that is without a difference. According to Horwich, to believe a theory is the same as being disposed to act *as if* it were true. Clearly, then, he adopts a behaviouristic construction of '*x* believes that *p*'. But, it might be asked, why not adopt the disposition-to-explicit-assent or the commitment-to-assent constructions[3]? The former approach, in particular, can accommodate situations where a scientist might use a theory to make predictions, design experiments etc., and therefore accept it, but would withhold her complete assent, for various reasons, including the very good one given above in response to the first answer. Indeed, such situations seem to be the rule rather than the exception in science.

A more radical answer to our question than any of the above, cuts the connection between belief and acceptance entirely. Thus, Cohen has recently drawn a sharp distinction between the two: with regard to belief, he agrees with Hume that it is essentially passive and involuntary in nature. Concerning acceptance, however, he adopts a Kantian, or rather, Cartesian, approach in arguing that it involves a conscious and voluntary choice. Beliefs come over us 'willy-nilly', whereas acceptance involves an element of commitment (Cohen 1989; this latter emphasis is not uncommon in the philosophy of science, at least; see, for example, van Fraassen *op. cit.*, p. 12, Kaplan 1981). More specifically, when we *accept* a proposition, hypothesis, or whatever, we commit ourselves to

> . . . going along with that proposition (either for the long term or for immediate purposes only) as a premiss in some or all contexts for one's own and others' proofs, argumentations, inferences, deliberations, etc. (Cohen 1989, p. 386).

With regard to the acceptance of theories, Cohen counsels the scientist to dispense with belief in the truth of hypotheses and rest content with accepting them:

> Scientific enquiry, whether in pursuit of empirical uniformities or of theoretical explanations, is not to be regarded as a procedure that is consummated when justifiable beliefs, with appropriate content, arise in or come over those engaged, who have meanwhile been waiting patiently for this to happen to them. Guesses and hunches, welling up from the subconscious, may make a very considerable contribution to the progress of an enquiry. In some cases an early conviction that *p* may even usefully fortify a scientist's resolution to seek those research facilities that are necessary in order to test whether *p*. But the culmination envisaged – the culmination that adds to our resources for explanation, prediction, technology, or further research – is a conscious and voluntary act of appropriately reasoned acceptance that is echoed throughout the relevant scientific community. (Cohen 1989, p. 385)

However, if acceptance were completely divorced from belief in this manner, theory choice would become wholly a matter of convention. Cohen's response is that this is not so, since the 'observational element' of scientific knowledge does, in fact, require belief (*ibid.*, p. 386). But if this observational element is cited in support of accepting a theory, as Cohen admits it is, then isn't the connection between belief and acceptance restored, at least at the level of the empirical sub-structures? Given the above analysis, what sense are we to make of the claim that scientists " . . . accept that *p* in the light of the evidence that *p*." (p. 385)?

These questions suggest that the link between belief and acceptance cannot be completely severed in the way that Cohen wants. But now we face something of a dilemma: at one extreme we have the view that identifies acceptance and belief, at the other there is the view that regards them as entirely distinct. Both views have their faults, yet both seem to capture a part of what is involved in accepting a theory: belief, of some kind, on the one hand and commitment on the other. Is there an alternative, some sort of 'middle way', which embodies what is right about each of these two extremes?

We think that there is and in order to understand the essential idea which lies behind it, it is important to note that both of the above views suffer from a defect common to almost all accounts of these notions, namely that of identifying 'belief that *p*' with 'belief that *p* is true', where truth is understood in the standard correspondence sense. It is not just that, as Cohen suggests, the conditions for rational belief are commonly regarded as indistinguishable from those of rational acceptance, but also that the nature of belief itself is treated in too coarse-grained a manner. In by far the vast majority of philosophical discussions, 'belief', as such, is analysed in terms of holding a proposition to be true in the simple correspondence sense. Indeed, it is often *identified* in this manner. Now this might be appropriate if the objects of knowledge are taken to be propositions of the form 'the grass is green', but, we suggest, it is hopelessly inadequate for accommodating those forms of knowledge which have as their objects theories, world-views, or complex representations in general. Thus, whereas beliefs are typically defined to be propositions accepted as true, scientists use 'belief' with a certain vagueness, a vagueness which reflects their epistemic fallibility (cf. Sperber 1982, pp. 174–175). Elsewhere we have suggested that this 'vagueness' which is characteristic of a scientist's doxastic attitudes, can be captured by means of the notion of the 'partial' or 'pragmatic' truth of a theory. We now wish to suggest that this alternative view can be expanded to accommodate the "more remarkable exercise" of scientific acceptance.

## III. THE TRUTH, THE WHOLE TRUTH AND NOTHING BUT THE TRUTH

In a series of recent works (da Costa 1989; da Costa and French 1989a; 1990a; 1990b) we have argued that theories should be, and, at least within the scientific community, generally are, regarded as 'partially' or 'pragmatically' true only. This, of course, is not a particularly original idea and, indeed, it seems the most obvious line to take in the present context. However, it has failed to receive the respect it deserves because, at least in good part, of its perceived lack of formalisation. Truth in the correspondence sense achieved respectability following Tarski's renowned model-theoretic formulation but 'partial' truth has languished in the doldrums of vagueness and ambiguity and has been dismissed because of it. This situation has been rectified through the work of Mikenberg, da Costa and Chuaqui (1986) which, moving along Tarskian lines, sets down a model-theoretic formalisation of the notion of 'pragmatic truth'. Briefly and informally put, the idea is as follows:

When we say that some sentence $\mathfrak{S}$ is 'true', we may interpret it as strictly saying that $\mathfrak{S}$ is true in a certain structure or model which represents a portion of reality. In fact, $\mathfrak{S}$ can be said to 'point' to the world by means of a model, which normally is not made explicit. Let us suppose that we are investigating a given domain of knowledge $\mathfrak{D}$ ; if $\mathfrak{S}$ refers to $\mathfrak{D}$, this is accomplished with the help of a structure $\mathfrak{A}$ which effectively substitutes for $\mathfrak{D}$ in our thought. Sometimes $\mathfrak{A}$ models $\mathfrak{D}$ so well that we have no doubts in accepting that there is an isomorphic correspondence between the two, as in the case, perhaps, of the empirical substructures of a theory.

One of the main points of Tarski's formalisation of the concept of correspondence truth is that in order to talk rigourously of truth we require a language $L$ and an interpretation $\mathfrak{I}$ of $L$ in a structure $\mathfrak{A}$. A sentence of $L$, then, is true or false only with reference to $\mathfrak{I}$. In other words, truth and falsity are properties of sentences of a particular language $L$, in accordance with an interpretation $\mathfrak{I}$ of $L$ in some structure $\mathfrak{A}$. Informally, of course, we may assert $\mathfrak{S}$ of $L$ as being true, without explicit mention of the interpretation and may even forget that $\mathfrak{S}$ is part of $L$. If we are to be rigourous, however, we must consider the interpretation.

Analogously, when we talk of 'pragmatic' or 'partial' truth, we must also introduce the twin notions of interpretation and structure. In this case, however, the models underlying the interpretations are not viewed as completely or exactly mapping the given domain; rather they are regarded as only partially modeling it[4]. A 'partial structure' is a structure of the form,

$$\mathfrak{A} = \langle \mathbb{A}, \mathbb{R}_k, \mathbb{P} \rangle_{k \in \mathcal{K}}$$

where $\mathbb{A}$ is a non-empty set, $\mathbb{R}_k$, $k \in \mathcal{K}$, is a partial relation defined on $\mathbb{A}$ for every $k \in \mathcal{K}$, and $\mathbb{P}$ is a set of sentences of a language of the same similarity type as that of $\mathfrak{A}$ and which is interpreted in $\mathfrak{A}$. For some $k$, $\mathbb{R}_k$ may be empty; $\mathbb{P}$ may also be empty (the $\mathbb{R}_k$ are described as 'partial' because any relation $\mathbb{R}_k$, $k \in \mathcal{K}$, of arity $n_k$ is not necessarily defined for all $n_k$-tuples of elements of $\mathbb{A}$; also, in the general case partial functions could also be included in a partial structure). In general, $\mathbb{A}$ denotes the set of individuals of the domain of knowledge modeled in the particular case considered (human beings in the case of genetic theory, for example), and the family of partial relations $\mathbb{R}_k$ models the various relationships which hold among these individuals. The $\mathbb{P}$ are then regarded as a set of distinguished sentences which may include, for example, observation statements regarding the domain.

Loosely speaking, a usual, total structure $\mathfrak{B}$ is called $\mathfrak{A}$-normal if it has the same similarity type as $\mathfrak{A}$, its relations extend the corresponding partial relations of $\mathfrak{A}$, and the sentences of $\mathbb{P}$ are true, in the Tarskian sense, in $\mathfrak{B}$. Then $\mathfrak{S}$ is said to be partially or pragmatically true in $\mathfrak{A}$, or in the domain $\mathfrak{D}$ that $\mathfrak{A}$ partially reflects, if there is an interpretation $\mathfrak{I}$ of $L$ (and consequently of $\mathfrak{S}$) in an $\mathfrak{A}$-normal structure B and $\mathfrak{S}$ is true in the Tarskian sense in $\mathfrak{B}$. Clearly, $\mathfrak{A}$ is not conceived of as reflecting the (total) structure of $\mathfrak{D}$, but as only partially mirroring $\mathfrak{D}$; in particular, the sentences of $\mathbb{P}$ should be true, in the usual sense in $\mathfrak{D}$. Thus the partial model $\mathfrak{A}$ has to capture some fundamental aspects of $\mathfrak{D}$, or some 'elements of truth', although it does not mirror $\mathfrak{D}$ perfectly. Then, we can say that $\mathfrak{S}$ is pragmatically, or partially, true in $\mathfrak{A}$, if all logical consequences of $\mathfrak{S}$ or of $\mathfrak{S}$ plus the true primary statements $\mathbb{P}$ are compatible with any true primary statement. This definition of pragmatic, or partial, truth captures the gist of the idea of a proposition being such that everything occurs in a given domain *as if* it were true (in the correspondence sense of truth).

The introduction of partial structures into the model-theoretic approach generates a number of interesting consequences. Such structures offer the possibility of a formal framework within which the role and nature of models and approximations in science can be investigated (da Costa and French 1990a). It has been suggested that they may be employed in the treatment of the twin problems of the individuation of theories and theory evolution within the model-theoretic approach in general (Suppe 1989, p. 427). Furthermore, as we shall show in detail in future works, the above theory of pragmatic truth, based on classical logic (every $\mathfrak{A}$-normal model is classical) can be extended to non-classical logics. For example, instead of invoking *classical* partial structures, we can consider *paraconsistent* partial structures and develop a theory of paraconsistent partial (or pragmatic) truth (for more on paraconsistent logics see Arruda 1980 and 1989; da Costa 1974, 1982; da Costa and

Marconi 1988; Priest, Routley and Norman 1989; da Costa and French 1989b and 1990b). Moreover, the logic appropriate to handle partial truth, as delineated above, is so-called 'Jaskowski's logic' (see Jaskowski 1969; da Costa 1989; da Costa and Dubikajtis 1977; da Costa and Chuaqui, forthcoming); analogously, one can show that the logic which is convenient for handling non-classical partial truth, in particular paraconsistent partial truth, is a kind of generalized Jaskowski's logic.

Finally, the formalisation of pragmatic or partial truth in this manner removes any last inhibitions on the use of these notions to resolve a wide range of issues, from those underlying the realism-empiricism debate and the problem of induction (da Costa and French 1989b) to the question of theory acceptance considered here.

## IV. PARTIAL TRUTH AND REPRESENTATIONAL BELIEF

Having given a formal framework for partial truth what, then, is the connection between this notion and belief? More specifically, what is the attitude evinced by the statement that a theory is believed to be partially true only?

Let us begin by drawing a distinction between 'factual' and 'representational' beliefs: the former can be taken to be beliefs in propositions held to be true in the simple correspondence sense, whereas the latter have as their objects, not propositions per se, but rather conceptual representations whose propositional contents are not completely fixed (The distinction is taken from Sperber *op. cit.*, p. 168). Such representations may be called 'semi-propositional' and are distinguished from 'propositional representations' in that, unlike the latter, they fail to identify one and only one proposition (*ibid.*). Each semi-propositional representation thus determines a range of propositional interpretations. If the goal is complete understanding in the sense of arriving at a unique propositional representation for a given domain then each semi-propositional representation may be regarded as a step on the road to full comprehension in this sense. Representational beliefs, then, are beliefs in semi-propositional representations and cover a range of doxastic attitudes up to, but obviously not including, belief in the (correspondence) truth of the representation concerned.

Elsewhere we have argued that scientific beliefs should be regarded as representational in this manner, noting how smoothly this account meshes with the kind of model-theoretic approach outlined above (da Costa and French 1990b; French 1991; da Costa and French forthcoming). In particular, the empirical sub-structures of a theory can be considered the objects of factual beliefs, whereas a scientist's belief in the theoretical model itself, taken to be pragmatically true, is more properly characterised as representational. This combination, of representational beliefs in theories considered to be pragmatically true only, fully captures, we claim, the vagueness, uncertainty and fallibility of scientists' doxastic attitudes.

## V. REPRESENTATIONAL BELIEFS AND 'PARTIAL' ACCEPTANCE

Our emphasis in the present work, however, is somewhat different. Factual and representational beliefs also differ in that with regard to the former there is awareness only of a 'fact', whereas in the case of the latter there is an awareness of a commitment to a representation (Sperber *op. cit.*, p. 171). Factual beliefs simply 'come over' one, whereas their representational cousins involve a conscious and voluntary act of 'throwing in one's lot', as it were, with the representation concerned. It is precisely this attitude of *commitment* which, following Cohen, we wish to focus on here as the hallmark of acceptance. To accept a theory is to be committed, not to believing it to be true, but to holding it *as if* it were true, for the purposes of further elaboration, development and investigation. Thus, Stalnaker writes,

> To accept a proposition is to act, in certain respects, as if one believed it . . . (1987, p. 80).

However, regarding a theory as if it were true, within the limited domain of knowledge which it models, is precisely what is captured by the notion of pragmatic truth. Thus, the statement that 'scientist *X* accepts theory *T*' can be unpacked as 'scientist *X* holds a representational belief in *T* as pragmatically true only and, therefore, is committed to 'going along' with *T* for her epistemic purposes'.

This view of acceptance clearly distinguishes it from *factual* belief, which can be described in terms of a disposition to hold a certain proposition as true, irrespective of whether it is used as a premiss for other purposes. (In *this* respect Cohen's analysis is correct.) Factual beliefs, being firmly and tightly linked to the evidence, are 'sensitive' beliefs in the Humean sense, forced upon one by the evidential circumstances at hand. Attempts to block or dull the impact of the evidence in such cases very quickly leads one down the road to self-delusion (da Costa and French 1990b). Representational beliefs, however, are very different in that, as we have said, there is an awareness, not just of a set of facts, but also of the semi-propositional representation used to model those facts. For any such set, or empirical sub-structure, there will generally be many such representations and the decision to accept one, based on its pragmatic truth, involves a commitment to that representation or model. Acceptance is not, therefore, to be divorced from belief in *this* sense. (It is in *this* respect that Cohen's account is faulty.)

## VI. ACCEPTANCE AND THE GENERAL CORRESPONDENCE PRINCIPLE

This conception can then accommodate two interesting scenarios: acceptance of theories that have been (strictly) refuted and acceptance of theories that are (formally) inconsistent.

The empirical inadequacy of a given model may be revealed through either an extension of the model's domain to cover new phenomena or further exploration of the domain the model was intended to cover. Nevertheless, within the limited domain in which the model remains empirically adequate it may continue to be regarded as pragmatically true and, thus, accepted, *within that limited domain* . An obvious example of this is classical mechanics, known to be 'strictly false' yet still employed, and for good reason, in the construction of bridges, buildings etc. and as the theory of scientific instrumentation in many situations. This theory can therefore be regarded as pragmatically true within a certain delimited domain and within this domain the theory may be accepted as if it were true for the purposes indicated above.

Indeed, it was a concern to capture the attitudes adopted towards such theories that provided one of the principal motivations for the development of the formalisation of pragmatic truth (Mikenberg, da Costa and Chuaqui, *op. cit.*). Clearly we have here a situation in which acceptance is appropriate but belief (in the truth of the theory) would be quite out of place. Thus Cohen writes,

> . . . we should be wrong to believe the theory, in the sense of feeling it to be true. But we can certainly accept it, in the sense of going along with it as a premiss, for all the purposes to which it is applicable. Similarly, even when we regard a physical or chemical law as a simplification or idealization we can use it as a premiss for calculations about the actual world, if we make relevant allowances and corrections. So in this sense we can accept the law even when we do not believe it to be true of the actual world. (Cohen 1989, p. 386).

In other words, Newtonian mechanics may be regarded as accepted in the sense that it is believed to be pragmatically or partially true only and this, in turn, is to be understood as meaning that it is not refuted within the domain which it models and from which it draws its empirical support. To be more explicit, it is accepted for certain purposes and subject to certain provisions regarding the limits of its applicability. Although a more comprehensive theory is available, it would clearly be unreasonable to expect engineers and scientists to use this within the domain of applicability of Newtonian theory. There is, in these cases, some form of economy-accuracy trade-off (cf. Cherniak 1984) and the reasoning of scientists in these matters is not different in kind from that of laypersons in making decisions in general (da Costa and French forthcoming). The point is a perfectly general one and applies, not just to strictly refuted theories such as classical mechanics, but also to approximations and idealizations in general. (Some form of model-theoretic approach would appear to be admirably suited for dealing with such approximations and idealizations; see Laymon 1988; da Costa and French 1990a).

In similar vein, de Oliveira argues that Newtonian mechanics contains an "element of truth" , where this notion is explicated in terms of the "established sub-theory of a theory" (1985, pp. 133–134). However, he claims that to accept a theory,

> . . . means to believe that it has not been refuted and that all predictions are true which can be deduced from its established sub-theory. (*ibid.*, p. 134)

and that Newtonian mechanics is "no longer accepted" (p. 131) since it has been refuted. There are, of course, well known dangers in tying acceptance to refutability in this manner. Furthermore, the scope of acceptance in this sense is clearly narrower than that which we envisage here. In particular, it rules out the possibility of theories being accepted in a provisional or pragmatic sense, which is precisely the possibility we find interesting.

Nevertheless, de Oliveira does seem to be on the right track in pushing his "thesis of the unrefutedness of established sub-theories of accepted theories" which asserts that, " . . . no established sub-theory of an accepted theory has ever been refuted." (*ibid.*, p. 134). In developing this claim de Oliveira draws on Post's discussion of the 'General Correspondence Principle' (Post 1971). Presented as the most important of the heuristic constraints on the construction of theories, this principle states that,

> . . . any acceptable new theory *L* should account for the success of its predecessor *S* by 'degenerating' into that theory under those conditions under which *S* has been well confirmed by tests. (Post 1971, p. 228)

More prosaically, it claims that,

> . . . no theory that ever 'worked' adequately turned out to be a blind alley. Once a theory has proved itself useful in some respects . . . it will never be scrapped entirely. (*ibid.*, p. 237)

Thus the Principle makes the strong claim that a new theory must reduce to its predecessor in the domain successfully modeled by the latter. It implies, of course, that there are no 'Kuhn losses' in science and it is important to realise that Post's work, with its wealth of historical examples, constitutes a concerted attack upon the Kuhnian viewpoint. Even during a scientific revolution, the whole of the lower-level structure of the theory being overthrown is not destroyed, although the top levels may be lopped off and the lower levels may be given a different interpretation (*ibid.*). Bluntly put, we never lose the best of what we have and this can be mirrored within our account by what might be called the 'Principle of the Absolute Nature of Pragmatic Truth': once a theory has been shown to be pragmatically true in a certain domain, it remains pragmatically true, within that domain, for all time. It is this, of course, which lies behind the justification for continuing to use Newtonian mechanics within certain limits.

But what of other theories, such as phlogiston, which are now abandoned and have no such pragmatic value? The difference in this case can perhaps be most clearly identified by noting the narrower domain in which phlogiston theory can be regarded as pragmatically true as compared to Newtonian theory:

> . . . the phlogiston theory 'worked' in that it assigned consistent levels of phlogistication (explaining many features) to chemical substances related in more than one way by reactions. On the other hand it tried to establish a connection between colour and phlogistication, and this part of the theory was not successful even at the time. (Post 1971, p. 228; see also Koertge 1968).

Heuristically, the Correspondence Principle may be used to eliminate certain candidates as serious possibilities for the successor theory (Post *op. cit.*, p. 235). As Post explains, this use of the Principle is generally *post hoc* in nature, in the sense that it is employed to eliminate those candidates for the successor theory which fail to explain the well – confirmed part of the predecessor. Of course, the true extent of the latter is only conjecture at the time these developments are taking place (*ibid.*, p. 231) and considerable skill may be required in identifying those aspects of the predecessor theory which are to be retained by the successor.

Rueger has recently given an account of this process in the development of Quantum Field Theory (Rueger 1990). Thus he claims that the central question in the development of S-matrix theory was,

> Which traits of a quantum theoretical formalism, valid at low and moderately high energies, would remain unchanged if the high energy behavior of the formalism were, more or less radically, to be modified? Heisenberg tried to separate physical concepts which could not be applied in the future theory from those concepts which (probably) remained unaffected by the high energy difficulties. (Rueger 1990, p. 209).

Any theory which failed to incorporate these latter concepts would be regarded as an unfit candidate for a successor. An earlier example is represented by Einstein's 1916 derivation of Planck's radiation law, where

> . . . there was reason to believe that the parts of classical theory used by Einstein would be retained in any more comprehensive theory that would ultimately resolve the conflict between classical electrodynamics and the quantum postulates. (Smith 1988b, p. 437)

A similar attempt to make the results independent of the details of any future theory can be seen in Bohr's use of his correspondence principle which allowed him to obtain empirical results from a theory that was strictly speaking inconsistent (Rueger *op. cit.*; Smith *op. cit.*). In this case the problem of delineating the well-confirmed parts of the theory was particularly acute and Bohr used the principle both as a 'defence-mechanism', which allowed for the temporary accommodation of the inconsistent foundations, and as a heuristic guide which effectively helped to generate empirical support for the theory (Lakatos 1970, p. 144).

## VII. INCONSISTENCY AND THE LOGIC OF ACCEPTANCE

It was this support, of course, which justified the belief that Bohr's theory, although inconsistent (*ibid.*, p. 140), was at least partially true and which also justified accepting the theory as such. Thus, even inconsistent theories may be accepted, in our sense. The prevalence of inconsistency in both our scientific and 'everyday' belief structures is something which is being increasingly recognised. Bohr's theory, of course, is one of the more well known examples; others include the old quantum theory of black-body radiation (Norton 1987), the (early) theory of infinitesimals in calculus (Priest and Routley 1984, p. 14), the Dirac δ-function (*ibid.*), Stokes' analysis of pendulum motion (Laymon 1985) and Michelson's 'single-ray' analysis of the Michelson-Morley interferometer arrangement (Laymon 1988).

The existence of inconsistency poses obvious problems for any view which construes theories as sets of sentences expressed in terms of the classical predicate calculus. Two approaches for accommodating inconsistent theories can be broadly delineated. The first retains the syntactic view of theories in general but abandons the underlying classical logic for some non-classical counterpart. The second way of dealing with inconsistency drops the syntactic view entirely and favours the alternative model-theoretic, or semantic, approach[5].

As a representative example of the former approach we take Priest, who includes the occurrence of inconsistency in the history of science among his pragmatic grounds for adopting the Hegelian position that there exist true contradictions, explicitly citing Bohr's theory in this context (Priest 1987). The classical generation of all possible propositions from an inconsistent theory is then blocked through the adoption of a form of paraconsistent logic in which certain contradictions are tolerated. Thus, the 'logic of science', for Priest, is paraconsistent (for a discussion of the differences between Priest's and da Costa's approaches to such logics, see da Costa and French 1989c). However, the doxastic attitude of scientists towards Bohr's theory was clearly much weaker than that of belief in its *truth* and in general this approach runs afoul of the considerations expressed at the beginning of this essay.

Brown has also recently espoused the use of a form of paraconsistent logic in this context, while rejecting the position that inconsistent theories should be regarded as true *per se* (Brown 1990). Acceptance, for Brown, is essentially 'contextual' in character, in the sense that a theory is accepted when,

> . . . we choose to treat [it] as if true, observing contextual limits on that commitment so as to avoid bringing incompatible claims into play at any point, . . . (Brown 1990, p. 285).

The central idea here is that the context of application of an inconsistent theory may be broken down into subcontexts in which the conflicting principles laying behind the inconsistency may be isolated. Thus the conjunctive application of contradictory claims is effectively blocked (*ibid.*, p. 284). Since there is no context in which all the principles are collectively true, in the sense that the evidence supporting scientific theories is essentially local in nature, Brown adopts a paraconsistent 'non-adjunctive' logic of acceptance according to which the closure of a set of claims under conjunction is not implied by the set (*ibid.*, pp. 289–292; the particular form of non-adjunctive logic cited by Brown is that constructed by Schotch and Jennings 1980).

Similarly, Smith has recently argued that, with regard to the drawing of inferences from inconsistent sets of beliefs, these sets can be broken up into (self-) consistent sub-sets from each of which implications can be derived in classical fashion (Smith 1988a[6]). Concerning the adjunction of these sub-sets, however, he rejects the use of paraconsistent logics, preferring to abandon the requirement of deductive closure in such cases and follow Harman in claiming that acceptance of such closure should not be confused with inference (Harman 1986, Ch. 2[7]).

Against these attempts, whether classical or 'non-adjunctive', two criticisms may be raised. The first is that if inconsistent theories are to be regarded as fragmented in this way, it is difficult to see how they can still be regarded as coherent theories (Priest and Routley *op. cit.*, p. 8). In what sense is such a collection of 'contexts' or 'sub-sets' an integrated unit? Certainly scientists themselves seem to regard inconsistent theories as such. Thus, in considering the suggestion that the old quantum theory of black-body radiation may be accommodated within a non-adjunctive framework, Norton has argued that scientists took Wien's Law, the classical expression for the energy density and the quantised formula for the average energy of a resonator and conjoined them to obtain the Planck distribution law (Norton 1987). From the distributed truth of the former principles, he claims, they inferred the collective truth of the conjunction expressed in the latter. Adjunction was therefore not abandoned[8]. (As should now be clear, our point of attack is this emphasis on 'truth' per se, rather than adjunction.) Likewise, if we are not committed to the deductive closure of a set of statements, in what sense does such a set constitute a coherent, unified theory? Smith avoids the problem on this point by referring to inconsistent proposals as 'prototheories' (1988a, p. 246); relatedly he argues that we need something weaker than acceptance to refer to the attitude adopted towards such proposals. We shall question these distinctions.

Secondly, one may wonder if it is even possible to effect the clear-cut division between different 'contexts' or 'sub-sets' within a theory that the above accounts require. Brown, at least, acknowledges the difficulty, noting that,

> When we accept claims in this way, we are committed to giving systematic division of the contexts in which the incompatible claims will have their roles. This division must allow for closure of the contexts of acceptance under combination. And it must have some connection with the evidence we are able to give for the claims in question: the division of contexts should

> allow us to use particular claims in the contexts where we find evidence confirming those claims. (Brown 1990, p. 288)

Smith would certainly agree with the last requirement, but it might be asked whether, without the imposition of closure, this is sufficient to prevent the division into sub-sets from becoming an unsystematic affair, with 'anything goes' being the order of the day with regard to the application of claims from these sub-sets[9]. Certainly abandoning closure is a radical move to make under any circumstances and one might wonder whether the price of doing so is lower than that incurred by introducing non-classical logics! Harman's reasons for rejecting what he calls the 'Logical Closure Principle' hinge on the fact that we are not logically omniscient: we are capable of holding only a finite number of 'explicit' beliefs and many implications of our beliefs are not immediately obvious (Harman 1986, p. 14). This is used to support his overall position that " . . . there is no clearly significant way in which *logic* is specially relevant to reasoning." (*ibid.*, p. 20). It is worth noting, however, that a lack of logical omniscience can in fact be accommodated within a coherent formal system, by introducing a notion of truth related to that of 'pragmatic truth' (da Costa and French 1988). It does not follow, therefore, that because we are not omniscient in this respect, that we must give up deductive closure.

What the failure of logical omniscience says is that, given a set of beliefs which are held, there are many consequences of these beliefs which are not, and cannot be, consciously held, or of which we are not explicitly aware, due to the finite nature of our minds. However, this does not imply a lack of commitment to such consequences. If it did, the whole notion of revising our sets of beliefs in the light of their consequences would make little sense (this is particularly crucial in the scientific context where such consequences are not always immediately drawn). Since it is this aspect of commitment which distinguishes acceptance from (factual) belief, it would seem that our lack of logical omniscience could be acknowledged, while at the same time retaining the view that the set of propositions which we accept is closed under (some form of) logical implication[10]. Thus, as Cohen remarks:

> While the direct act of acceptance involves a conscious and voluntary choice of a premiss or premisses, a person may be said to accept indirectly all the logical or mathematical consequences of any conjunction of the propositions that he accepts directly, whether or not he is himself conscious of those consequences or disposed to work them out. That is because, when $p$ entails $q$ and he decides to adopt the policy of taking $p$ as a premiss, he stays effectively on course by taking $q$ as a premiss, because anything that follows from $q$ will also follow from $p$. So, if he announces to other people that he has accepted certain propositions, he implies acceptance also of their logically or mathematically necessary consequences or equivalents, and thus creates a presumption about how he will think and act. And, even if he does not announce to others his decision to accept that $p$, this decision

certainly reaches indefinitely far . . . beyond any consequences or equivalents that he may perceive at the moment. (Cohen 1989, pp. 370–371; cf. Brown *op. cit.*, p. 290, fn. 14).

We have noted above that the failure of logical omniscience can be accommodated within formal systems which, contrary to Harman, are deductively closed, *in some sense* . The systems we are referring to are, of course, non-classical and the closure is with respect to connectives, defined within the system, which behave in a non-classical fashion. Let us be more specific. As we have already suggested, the logic of 'partial' truth, as defined above, is paraconsistent and, to be more specific still, the logic of acceptance, in the sense in which we are outlining here, is, we claim, a form of Jaskowski's discussive logic (da Costa and Chuaqui *op. cit.*). When we reason with consistent premises, Jaskowski's logic reduces to the classical form; with inconsistent premises, of course, it does not. Acceptance, in our terms, is closed under the Jaskowski system but not in relation to classical logic (unless the theory concerned is consistent). More precisely, there is no closure under classical conjunction and material implication but we can define 'discussive' forms of conjunction and implication with respect to which acceptance can be regarded as closed. It is worth noting that there may be inconsistent theories whose 'degree of inconsistency', as it were (and it makes sense to talk in these terms when one is considering paraconsistent logics) is so great that they are worthless; one example might be a theory possessing an axiom of the form $p \leftrightarrow \sim p$. Hence the acceptance of inconsistent theories or of theories which are (mutually) inconsistent does not imply that we accept everything.

With these points in mind we are better able to evaluate Ullian's argument that truth is different from acceptability since truth is deductively closed, in the sense that what you deduce from a set of truths is also true, whereas "[a]cceptability is unlikely to have any such closure property" (Ullian 1990, p. 345). The point of making the distinction between truth and acceptability, for Ullian, is to suggest that perhaps talk of truth should be dropped altogether at the higher theoretical levels which are " . . . far from the soil of observation sentences" (*ibid.*, p. 344).

The conclusion, then, is that perhaps truth may not offer the best vantage point from which to view the scientific process (*ibid.*, p. 345) and that what is needed is " . . . a reasonable notion of acceptability to step into the breach" (*ibid.*, p. 344). However, to disassociate acceptance entirely from truth, even at the level of the empirical basis, is to open the door to conventionalism once more. Although the theoretical parts of the models employed in science appear to reside in the treetops, they are still connected to their observational roots (even trellised vines must touch ground somewhere!). Thus, while we agree that talk of truth in the strict correspondence sense must be dropped, *some* form of truth must be retained and what we are trying to do here can be seen as part of a programme to delineate just exactly what that form is. As for a "reasonable notion of acceptability", if one persists in regarding this in classical terms only, then the slide into Harman's position would seem to be inevitable. To our minds the consequences of such a view are sufficiently disagreeable as to make the alternative, non-classical, approach even more attractive! (Of course, as Ullian correctly notes, the form of acceptability we are invoking here does not possess the *classical* property of closure but it does possess a non-classical or paraconsistent 'analogue'.)

In essence then, we adopt the semantic alternative indicated above, according to which theories are regarded as classes of extra-linguistic structures, or models. This offers a number of advantages over the syntactic approach, one of which is that it

> . . . allows for easy and revealing consideration of computational concerns and . . . allows for the coherent use within a calculation of inconsistent theories. (Laymon 1988, pp. 262–263)

It is precisely the empirical consequences of a theory which in general are of principal interest to scientists themselves and which override concerns about consistency when viewed from a particular logical framework. However, the existence of inconsistencies can only be outweighed in this manner if the theory concerned is regarded, not as true, in the literal or correspondence sense, but as partially or approximately so. It is by acknowledging that there is a certain 'looseness of fit' between a representation and that which it represents, that inconsistency can be comfortably accommodated. By introducing 'partial structures' into the semantic approach this notion of a 'looseness of fit' can be formally captured and in a way which reflects the doxastic attitudes of scientists themselves. These attitudes can, in turn, be set down in terms of a 'logic of acceptance' which is fundamentally non-classical. Indeed, as we have just argued, once the partial nature of theories is acknowledged, the choice is either to accept such a view or expel logic from the field of reasoning altogether.

The 'looseness of fit' we are referring to here reveals itself through scientists' doxastic attitudes towards theories in general and inconsistent theories in particular: the belief involved is representational in character, being belief in a representation of semi-propositional form. It is the conceptual incompleteness of such representations which underpins this 'looseness of fit' and which allows for their acceptance, whether they be internally or mutually inconsistent[11]. Furthermore, they may be regarded, in general, not merely as steps on the road towards some complete representation, but also as potential heuristic sources for further development[12].

Inconsistencies should be regarded, then, not as permanent features of 'reality' (whatever that may be) but as temporary flaws of the representations used to model this reality. They may nevertheless be extremely fertile, in a heuristic sense, and it is to this aspect that we now turn.

## VIII. THE HEURISTIC FERTILITY OF INCONSISTENCY

According to Smith, since assenting to a set of statements is usually taken to warrant assent to their deductive closure, dropping the latter in cases of inconsistent theories, means coming up with something weaker than unconditional assent itself to refer to the epistemic attitude which is appropriate in these cases. Thus he introduces the notion of 'entertaining' hypotheses for their 'pragmatic' virtues and states that,

> When we have evidence for the truth of each of two incompatible claims, it is quite rational to *entertain* both. However, the fact that they are inconsistent means that we must mentally flag them to guard against indiscriminate future use. One or the other is false. At best, both can be 'approximately true', 'partially true' or 'true under some disambiguation' etc. (Smith 1988a, p. 244)

Obviously we would agree with this last line but would question why an epistemic attitude different from that of acceptance needs to be marked out for these cases.

The answer is that for Smith, inconsistent theories do not lie within the domain of acceptance but rather occupy the realm of pursuit, or, as he calls it, 'discovery' (1988b). Thus, he explicitly acknowledges that his chief concern is with the question of how scientists move from a particular inconsistent theory to a consistent successor (*ibid.*, p. 435, fn. 20). The heuristic role played by inconsistent theories clearly stands in need of further elucidation. As Post has noted, an inconsistency is usually regarded as a 'formal flaw' which points the way to the construction of a consistent successor (Post *op. cit.*, pp. 221–225). However, not all inconsistencies are fertile in this sense. Bohr's 'programme' began to degenerate, in Lakatosian terms, as further inconsistencies were piled on to the original model in an increasingly desperate attempt to deal with the "undigested" anomalies that began to proliferate (Lakatos 1970, p. 154). Such inconsistencies failed to extend the 'empirical character' of the theory and can thus be regarded as sterile (*ibid.*, p. 154).

That, nevertheless, inconsistent theories may be enormously fertile, heuristically speaking, is something which a 'conservative' view of scientific progress fails to acknowledge. According to this view, the inconsistency must be removed before any further progress can be made, since it is irrational to build upon inconsistent foundations (*ibid.*, p. 144). The opposing 'anarchist' position, on the other hand, elevates inconsistency, at least in the 'weak' sense of involving standard interpretations of certain characteristic terms (*ibid.*, p. 143, fn. 3) to the status of a fundamental property of nature. It is from this view, of course, that the approach of Priest *et al.* draws its strength. If nature were truly inconsistent in some sense, then the appropriate logic of science would be paraconsistent. The argument that such an approach is incompatible with the goal of science being truth (Lakatos *op. cit.*, p. 143) begs the question, since it asserts the existence of true contradictions. Nevertheless, as we mentioned above, such a view fails to do justice to the doxastic attitudes of scientists themselves. For this reason the 'rational' position may be preferred, according to which the heuristic fruitfulness of an inconsistent theory may be exploited without a concomitant claim that it is the last word on the matter (*ibid.*, p. 145).

A crucial role in this process of heuristic exploitation is played by the General Correspondence Principle, used heuristically in the same sense as discussed above. The general strategy identified here is that of 'projection', in the sense that the scientist

> . . . uses the original [inconsistent] proposal along with the confirming evidence available for various parts of that proposal to give a schematic "projection" of what the consistent replacement theory (or some fragment of that theory) will look like. (Smith, 1988b, p. 438)

The same strategy is exemplified in Bohr's use of the correspondence principle and provides a rational procedure for the heuristic use of inconsistency[13].

In common with de Oliveira, Smith emphasises that those parts of an inconsistent theory which are retained in the successor theory contain an 'element of truth', in the sense of enjoying some degree of empirical confirmation (1988b., p. 432; see also p. 435). However, he too rejects talk of acceptance in this context since,

> Acceptance of a set of statements warrants their unrestricted use in all reasoning processes. At least one member of an inconsistent set of statements must be false. So, without additional information, we cannot rationally assent to the *unconditional* use of any one of the statements from a set known to be inconsistent. (*ibid.*, p. 432)

Leaving aside the point that his characterisation of inconsistency begs the question against paraconsistent logics, Smith seems to have in mind precisely the notion of acceptance which we are rejecting in this essay. No scientific theory, past or present, receives such warrant; there are always hedging conditions, acknowledgments of limitations, approximations etc., which are manifestations of the scientists doxastic attitudes towards the theory. And these attitudes are inadequately and incorrectly characterised in terms of belief in the (correspondence) *truth* of the theory. The present work constitutes an attempt to elaborate an alternative view which attempts the difficult trick of accommodating these weaker attitudes while retaining the connection between acceptance and belief of some form. We prefer to stick to 'acceptance' rather than 'entertainment', since even in the case of inconsistent theories there is still the element of commitment which we take to be the hallmark of the former (witness Bohr's elaboration and defence of his theory).

## IX. CONCLUSION: DISCOVERY, JUSTIFICATION AND PARTIALITY

What all this amounts to is a blurring of the line that is (still) often drawn between justification and discovery[14]. The distinction is typically made by tying acceptance to belief in the (correspondence) truth of the theory, and consigning discovery to the realms of psychology and sociology with 'pursuit' inhabiting the 'nether' region somewhere in between (see Whitt 1990). The approximation between the two proceeds from both directions: first of all, heuristic guidelines are admitted, of the sort delineated by Post, lifting discovery, at least in part, back into the domain of rationality (the distinction between discovery and pursuit evaporates as well) and then acceptance is unbound from truth, allowing for that descriptively important element of fallibility[15]. There is then no need for an alternate term such as 'entertainment' since acceptance understood in terms of partial truth is sufficiently broad to do the job in all cases.

To sum up, then, the view we are advocating here contains the following elements: acceptance differs from factual belief in that the former involves a voluntary act of commitment, whereas the latter does not. It is, however, tied to a representational belief in the partial truth of what is accepted and the commitment is to the use of the representation or model concerned. Both inconsistent and strictly false theories may be regarded as partially true and accepted in this sense. And in both cases, the element of truth such theories possess, which ultimately justifies their acceptance, is not lost but is incorporated within the structure of their successors in accordance with the General Correspondence Principle.

We conclude with another quote from Post's 1971 paper, which expresses the core understanding of the position defended here:

> In the heuristic situation of empirical science, truth does not descend deductively from the (highly fallible) most general laws, but ascends inductively from the more modest lower reaches of the deductive structure of the theory held at any one time. Even that infra-structure has to be pruned, and the remaining core is never true for certain, but the modifications forced on us still allow us to refer to that part of the theory as having a likeness to truth. (Post 1971, p. 254)

*Dept. of Philosophy,*
*University of Sao Paulo,*
*Sao Paulo, Brazil*
and
*Dept. of Philosophy,*
*Southeast Missouri State University,*
*Cape Girardeau, MO, U.S.A.*

## NOTES

1. de Oliveira's paper triggered our original discussions of this problem.
2. This should not be taken as a criticism of science but merely as an argument to the effect that it would be unwise to believe in the literal truth of its theories.
3. According to the former interpretation, '$x$ believes that $p$' is taken to be true if $x$, who is supposed to be candid and after due explanation where necessary, overtly professes to accept $p$ as true: we ask $x$ if she does believe $p$, and the issue is decided by $x$'s answer. The latter construction considers that '$x$ believes that $p$' obtains if either $x$ gives her assent to $p$, in accordance with the preceding interpretation, or $x$ is committed to giving assent to $p$ in case $p$ constitutes a logical consequence of propositions believed by $x$. For further details see Rescher (1968).
4. "Standard model theory has concentrated on complete models, corresponding to complete constellations of facts, the limiting cases where any additions would introduce inconsistencies. A certain tradition in the philosophy of language has been cited as one influence leading to the exclusive emphasis on totality: The meaning of a sentence is equated with its truth conditions, and the truth conditions are taken to correspond with states of the whole world in which the sentence is true." (Langholm 1988, p. 1)
5. The logic(s) of truth and acceptance, as understood within such an approach, may, however, be non-classical; in particular, they may be paraconsistent, as we suggest here.
6. Cf. Kyburg's notion of 'strands' in Kyburg (1987 and 1974).
7. Kyburg, again: "It is not the strict inconsistency of the rational corpus that leads to trouble – it is the imposition of deductive closure." (1987, p. 147).
8. Norton's own solution to the problem of inconsistency in this case is to argue that a consistent sub-theory can be constructed from which Planck's law can be recovered. That this is possible with hindsight is irrelevant to the question of what attitude should be adopted towards inconsistent theories before such consistent reconstructions have been identified (Smith 1988b, fn. 20; cf. Brown *op. cit.*, p. 292). Norton's claim does, however, undercut Priest's position and provides further support for the view, to be discussed shortly, that inconsistent theories may be regarded as pointing the way to a consistent successor.
9. Cf. Harman's comment that, in the event of finding ourselves holding inconsistent beliefs, " . . . it is rational simply to retain the contradictory beliefs, trying not to exploit the inconsistency." (Harman 1986, p. 17) Unfortunately, however, no general prescription is given as to *how* we are supposed to avoid exploiting the inconsistency.
10. Likewise, in Brown's account, " . . . acceptance in the sense I am describing (like belief) is closed under combination of contexts." (p. 287); it is this which distinguishes a 'realist' form of acceptance from the 'instrumental' variety, where there is no such closure.
11. Cf. Sperber 1982, p. 170.
12. Sperber makes a brief remark to this effect (*ibid.*, p. 171)
13. "The projected theory serves as a guide for the search for the material content of the proper replacement for the original inconsistent proposal." (Smith 1988b., p. 443).
14. The collapse of the discovery-justification distinction has been much discussed recently, although it too was anticipated in Post's work (1971, p. 217).
15. As Stalnaker points out, " . . . a person may accept a proposition for the moment without the expectation that he will continue to accept it for very long. If a person expects a particular one of his *beliefs* to be overturned, he has already begun to lose it, but an assumption he makes may be quite explicitly temporary, and he may presume that something is true even when expecting to be contradicted." (Stalnaker 1987, p. 80) – by "beliefs" in this context we, of course, understand *factual* beliefs.

## BIBLIOGRAPHY


Arruda, A.I. (1980), 'A Survey of Paraconsistent Logic', in A.I. Arruda, R. Chuaqui and N.C.A. da Costa (eds.), *Mathematical Logic in Latin America*, North-Holland, pp. 1–41.

Arruda, A.I. (1987), 'Aspects of the Historical Development of Paraconsistent Logic', in G. Priest, R. Routley, and J. Norman (eds.), *Paraconsistent Logic: Essays on the Inconsistent*, Philosophia, Munich.

Brown, B. (1990), 'How to be Realistic about Inconsistency in Science', *Studies in History and Philosophy of Science*, 21, pp. 281–294.

Cherniak, C. (1984), 'Computational Complexity and the Universal Acceptance of Logic', *The Journal of Philosophy*, 81, pp. 739–758.

Cohen, L.J. (1989), 'Belief and Acceptance', *Mind*, 98, pp. 367–389.

da Costa, N.C.A. (1974), 'On the Theory of Inconsistent Formal Systems', *Notre Dame Journal of Formal Logic*, 11, pp. 497–510.

da Costa, N.C.A. (1982), 'The Philosophical Import of Paraconsistent Logic', *Journal of Non-Classical Logic*, 1, pp. 1–12.

da Costa, N.C.A. (1989), 'Logic and Pragmatic Truth' in J.E. Fenstad *et al.* (eds.), *Logic, Methodology and Philosophy of Science VIII*, Elsevier, pp. 247–261.

da Costa, N.C.A. and Chuaqui, R. (forthcoming), 'The Logic of Pragmatic Truth'.

da Costa, N.C.A. and Dubikajtis, L. (1977), 'On Jaskowski's Discussive Logic' in A.I. Arruda, N.C.A. da Costa and R. Chuaqui (eds.), *Non-Classical Logics, Model Theory and Computability*, North-Holland, pp. 37–56.

da Costa, N.C.A. and French, S. (1988), 'Pragmatic Probability, Logical Omniscience and the Popper-Miller Argument', *Fundamenta Scientiae*, 9, pp. 43–53.

da Costa, N.C.A. and French, S. (1989a), 'Pragmatic Truth and the Logic of Induction', *The British Journal for the Philosophy of Science*, 40, pp. 333–356.

da Costa, N.C.A. and French, S. (1989b), 'On the Logic of Belief', *Philosophy and Phenomenological Research*, 49, pp. 431–446.

da Costa, N.C.A. and French, S. (1989c), 'Critical Study of *In Contradiction*', *Philosophical Quarterly*, 39, pp. 498–501.

da Costa, N.C.A. and French, S. (1990a), 'The Model-Theoretic Approach in the Philosophy of Science', *Philosophy of Science*, 57, pp. 248–265.

da Costa, N.C.A. and French, S. (1990b), 'Belief, Contradiction and the Logic of Self- Deception', *The American Philosophical Quarterly*, 27, pp. 179–197.

da Costa, N.C.A. and French, S. (forthcoming), 'A Model Theoretic Approach to 'Natural Reasoning'', *International Studies in the Philosophy of Science*.

da Costa, N.C.A. and Marconi, D. (1988), 'An Overview of Paraconsistent Logic in the '80s' in *Logica Nova*, Akademie-Verlag, Berlin.

French, S. (1989), 'A Peircean Approach to the Realism-Empiricism Debate', *Transactions of the Charles S. Peirce Society*, 25, pp. 293–307.

French, S. (1991), 'Rationality, Consistency and Truth', *The Journal of Non-Classical Logic*, 7, pp. 51–71.

Harman, G. (1986), *Change in View*, MIT Press, Cambridge.

Horwich, P. (1991), 'On the Nature and Norms of Theoretical Commitment', *Philosophy of Science*, 58, pp. 1–14.

Jaskowski, S. (1969), 'Propositional Calculus for Contradictory Deductive Systems', *Studia Logica*, 24, pp. 143–157.

Kaplan, M. (1981), 'Rational Acceptance', *Philosophical Studies*, 40, pp. 129–145.

Koertge, N. (1968), *A Study of Relations Between Scientific Theories: A Test of the General Correspondence Principle*, Ph.D. Thesis, University of London, London.

Kyburg, H. (1974), *The Logical Foundations of Statistical Inference*, Reidel, Dordrecht.

Kyburg, H. (1987), 'The Hobgoblin', *The Monist*, 70, pp. 141–151.

Lakatos, I. (1970), 'Falsification and the Methodology of Scientific Research Programs' in I.

Lakatos and A. Musgrave (eds.), *Criticism and the Growth of Knowledge*, Cambridge University Press, Cambridge, pp. 91–195.

Langholm, T. (1988), *Partiality, Truth and Persistence*, Center for the Study of Language and Information, Stanford.

Laymon, R. (1985), 'Idealizations and the Testing of Theories by Experimentation' in P. Achinstein and O. Hannaway (eds.), *Experiment and Observation in Modern Science*, MIT Press, Boston, pp. 147–173.

Laymon, R. (1988), 'The Michelson-Morley Experiment and the Appraisal of Theories' in A. Donovan *et al.* (eds.), *Scrutinizing Science*, Kluwer, Dordrecht, pp. 245–266.

Mikenberg, I. , da Costa, N.C.A. and Chuaqui, R. (1986), 'Pragmatic Truth and Approximation to Truth', *Journal of Symbolic Logic*, 51, pp. 201–221.

Norton, J. (1987), 'The Logical Inconsistency of the Old Quantum Theory of Black Body Radiation', *Philosophy of Science*, 54, pp. 327–350.

Oliveira, M.B. de. (1985), 'The Problem of Induction: A New Approach', *British Journal for the Philosophy of Science*, 36, pp. 129–145.

Post, H.R. (1971), 'Correspondence, Invariance and Heuristics', *Studies in History and Philosophy of Science*, 2, pp. 213–255.

Priest, G. (1987), *In Contradiction*, Martinus Nijhoff, The Hague.

Priest, G. and Routley, R. (1984), 'Introduction: Paraconsistent Logics', *Studia Logica*, 43, pp. 3–16.

Priest, G., Routley, R. and Norman, J. (1989), *Paraconsistent Logic: Essays on the Inconsistent*, Philosophia, Munich.

Rescher, N. (1968), *Topics in Philosophical Logic*, Reidel, Dordrecht.

Rueger, A. (1990), 'Independence from Future Theories: A Research Strategy in Quantum Theory' in A. Fine, M. Forbes and L. Wessels (eds.), *PSA 1990*, Vol. I, P.S.A., East Lansing, pp. 203–211.

Schotch, P.K. and Jennings, R.E. (1980), 'Inference and Necessity', *Journal of Philosophical Logic*, 9, pp. 329–340.

Smith, J. (1988a), 'Scientific Reasoning or Damage Control: Alternative Proposals for Reasoning with Inconsistent Representations of the World', *PSA 1988* , Vol. I, P.S.A., East Lansing, pp. 241–248.

Smith, J. (1988b), 'Inconsistency and Scientific Reasoning', *Studies in History and Philosophy of Science*, 19, pp. 429–445.

Sperber, D. (1982), 'Apparently Irrational Beliefs' in M. Hollis and S. Lukes (eds.), *Rationality and Relativism*, M.I.T. Press, Cambridge, pp. 149–180.

Stalnaker, R.C. (1987), *Inquiry*, MIT Press, Cambridge.

Suppe, F. (1989), *The Semantic Conception of Theories and Scientific Realism*, University of Illinois Press, Urbana and Chicago.

Ullian, J.S. (1990), 'Learning and Meaning' in R.B Barrett and R.F. Gibson (eds.), *Perspectives on Quine*, Blackwell, Oxford, pp. 336–346.

van Fraassen, B.C. (1980), *The Scientific Image*, Oxford University Press, Oxford.

Whitt, L.A. (1990), 'Theory Pursuit: Between Discovery and Acceptance' in A. Fine, M. Forbes and L. Wessels (eds.), *PSA 1990*, Vol. I, P.S.A., East Lansing, pp. 467–483.

AVINASH K. PURI

# Tales from the Classroom: The See-Saw

The following is a rational reconstruction of what must be admitted to have been a minor afternoon in the history of science. It tells of how a group of students came to make – were cajoled, goaded, induced to make – an utterly unoriginal scientific discovery. The discovery was however new to the students, and therein, somewhere, lies the point of the tale.

Such afternoons, a familiar enough aspect of school or college life, hold an obvious and well-established interest for an educationalist. Historians and philosophers of science on the other hand seem loath to turn away from the strides of the great to the stumbles in the classroom. Their reluctance is understandable but also unfortunate – for they miss something vital and that something vital misses them – and the following account aims to counter that reluctance. The account hovers around the notion that science is a process – not a mere product – and that in this sense the science of the apprentice and of the master shares the same intellectual space. This is of course not to deny the profound difference between the two activities, but then the difference too is worthy of attention.

## I. THE SETTING

As part of a first course in *philosophy* the students were about to be introduced to 'The Scientific Method' and the teacher had decided that a session of practical scientific investigation would be a useful preliminary. The students were known to have had a very limited schooling in science and the teacher had anticipated correctly that they had not already come across the discovery planned for the afternoon. (He had a back-up Plan B, so to speak, but it was unnecessary to resort to it). This same teacher is now recounting the story. The characters, apart from the teacher, may be described as young adults. (The setting is an inner-city Institute of pre-university education during the Spring term).

The investigation involved a small see-saw: a flat strip of plywood as long as an arm, pivoted in the middle and marked with lines equally spaced. (In



*S. French and H. Kamminga (eds.), Correspondence, Invariance and Heuristics*, 159–169.

terms of this spacing between the lines, the see-saw was 16 units long). The only other item of apparatus was a box containing a number of identical flat square brass tokens, each with that nice weight of an old British penny. There was no need, during the investigation, to refer to any standard unit of length or weight: one had merely to count the brass tokens or lines as necessary. The teacher had it in mind to nudge the students, by degrees, towards the rule by which a see-saw comes to a balance. How much help these 'pseudo-scientists' would need was, at the outset, an uncertain factor, but the teacher did not intend to be a neutral observer. Given that the process of discovery is liable to go wrong, his secret hope was that what would go wrong would be just enough and not too much.

## II. MAKING THE DISCOVERY

*Teacher*: Last week it was suggested that we might do a bit of science before going on to analyse 'the scientific method'. It is often said that philosophers pontificate about science without ever having done it. Indeed to misquote my old professor: strictly speaking, one should be barred from philosophy of science unless one has, at the very least, won a Nobel Prize or two (for physics of course, and, preferably, for having made a genuine contribution), and by then one should have more sense than to do philosophy of science . . . [1]

I think however – and I suspect you might agree – that this view is a touch too Spartan. One can after all talk intelligently about how birds fly without being able to fly oneself. One may be able to appreciate a good dish of pasta and yet be unable to cook it. Still, I suppose it is no bad thing to be able to cook pasta properly or for that matter to be able to fly. Nor do I think it would be harmful to try to do a bit of science. And when we come to talk about science we may well benefit from having done it. Not that scientists are any good at philosophy, any more than a fish is a hydrodynamicist.

. . . Here, then, is some apparatus I have brought along with me. I have here a see-saw which you can see is balanced as it is.

And when I put one weight on the left-hand side, it becomes unbalanced.

[*A note to the reader*: Throughout the session the teacher places one *single load* only on each side of the see-saw, the load (brass weights set one on top of the other) being located in such a way that its centre of gravity coincides with one of the marked lines on the see-saw. The numerical values he deploys for the loads and distances are intended to minimise arithmetical difficulties on the part of the students. The teacher knows well that a shockingly large computation gap exists between a theory and its consequences for many a modern student (as it does apparently for the modern scientist). But of course, as he does not know for certain what rule the students might come up with, his choice of numerical values could prove unfortunate: what is a simple computation using the teacher's formula, dating back to Archimedes, may be intractable using the student's formula.

It may be added here also that the see-saw used was sensitive to even small imbalances and thus it was necessary to establish a notion of approximate balance. The see-saw was said to be in balance if by moving the weight only a little the direction of imbalance was reversed. When by chance the see-saw decided to display its balanced state by coming to rest in a horizontal position, a thrill of satisfaction was clearly discernible in the room. Mostly, however, one had to make do with what may be called a theory-laden notion of balance.

One important warning. The readers, especially those versed in statics, should disassociate themselves from the following sentiment expressed by one friend with whom this article was discussed: "Surely, everyone knows how to balance a see-saw. How can people be so ignorant?"]

*Teacher*: As I was saying before I was interrupted by the narrator, the see-saw is unbalanced. The task is to bring it back to balance by putting another brass token on the see-saw.

*Primus*: The weight must go on the other side, the right hand side.

*Secundus*: I don't know why it is that Primus always has the first say, but he is right. The weight must go on the other side and of course in the same place. Easy.

An excellent start. The teacher has done the experiment before with other groups and knows that such a good start is not to be assumed lightly. There is the hope also that the rivalry between Primus and Secundus may bear early fruit.

The teacher goes to place the weight on the other side. He can hear another student – Quarta – who apparently had opted for philosophy to get away from science – moaning about how she "hates science, especially physics – and this is physics isn't it? – and science is evil and anyway is this going to be in the philosophy exam? . . . blah . . . blah blah . . . ". Topic for another day, thinks the teacher and puts the weight on the spot indicated by Secundus.

*Teacher*: Yes. The see-saw balances. Very good. Let me next put a weight of 2 tokens on the left hand side of the see-saw on this spot here . . . and I would like you, please, to place a counterweight of 2 tokens yourself. Who will have a go?

Tertia takes up the offer. The teacher has quietly placed his weight 4 units away from the pivot line and expects Tertia to place her weight on the other side at the same distance. *Tertia, however, measures the distances involved from the ends of the see-saw, NOT from the pivot.* Her method leads in this case to the correct placement and a consequent smile of satisfaction. No student comments on Tertia's way of counting in from the ends of the see-saw. The teacher too says nothing and wonders whether her success possibly might mislead others towards her method of counting. He makes a mental note of referring to this later as an example of a *possibly* wrong path taken in the progress of science.

[*Note:* The fact that someone physically counts in from the ends of the see-saw leaves open the possibility that the distance she is trying to measure is from the pivot. In any case the orthodox theory of static equilibrium does not demand that one choose the pivot as the reference point. It is simply that in making that choice one avoids having to know the reaction force at the pivot.]

*Teacher*: Right then. It seems that we know how to balance the see-saw when the weights on the two sides are the same, even if we may be using *different* rules to guide us. What I want to do next is to use *different* weights on the two sides of the see-saw. Look: I place a weight of 4 tokens on this line here and I want you to put a weight of 2 tokens on the other side to balance it. Here, come on Quarta. Don't be shy . . .

Teacher has quietly placed the weight on the spot which, from his perspective, can be described as being 2 units away from the pivot. Quarta participates unwillingly and puts the weight down on the wrong place. She adjusts it by moving it in a direction which makes the imbalance worse! (True, I assure you.)

*Teacher*: Bad luck!

A gaggle of giggles indicates that some students don't think that luck was entirely to blame. The teacher decides not to bring this up later. Perhaps Quarta has never been on a see-saw. Or, maybe, the see-saw teaches you less about statics than might seem to be the case. Could just be panic, thinks the teacher. But it could also be that she was distracted by her hunt for a pattern. Ah yes! Perhaps it could after all be brought up later[2].

*Secundus*: Let me try. The weight of 4 tokens is 2 lines away from the pivot. So the weight of 2 tokens on the other side must be put 4 units away.

The teacher notes with interest Secundus' way of measuring the distance and watches him place the counterweight. The wooden strip oscillates, uncertain whether to nod to the right or to the left. Yes it looks like a good one! Secundus has hit the spot right on and the dithering gives way to a delicate equipoise. Noticeable, in contrast, is the liveliness of the class conversation. A number of ideas are being floated around, it becoming generally accepted that the balance depends on the size of the load and its distance from the pivot to the centre of the brass token. Quarta is still standing apart from the others and even wonders whether there is a rule at all. "Because I know you,

teach", says she, "you are a cheat. I wouldn't be surprised if there ain't no rule". The others say there must be a rule.
*Secundus*: The rule is this: the smaller the weight the bigger the distance; the bigger the weight the smaller the distance!
*Teacher*: Excellent. Excellent. The rule sounds okay, but let us see . . . Can you use your rule to work out this problem: I put a weight of 5 units at a distance of 3. Where should a weight of 3 tokens go on the other side? Make your prediction before you try it out.
*Primus*: The distance must be bigger because the weight is smaller. I say it should go at a distance of 5.

Primus sets to test his prediction but makes a simple MISCOUNT and places the weight on the wrong place. The resulting imbalance makes him doubt his theory. The teacher will use this as an example of how a theory can be mistakenly taken to be refuted. The trivial mistake is corrected. Primus is happy. Quarta still does not understand what the game is all about and goes to her desk in defiance of science, philosophy, and the teacher's authority.
*Teacher*: Okay then. Now what I want is for someone to give me the precise rule. How can we calculate the answer? What is the mathematical rule? What about you, Tertia. You have been a bit quiet. Be bold and have a guess. We are amongst friends, after all. Come on someone. What is the rule?

The response is not immediate. Here there is a look of anxiety, there a look of concentration. The teacher reminds himself that to be bold is not to be rash. He reminds himself also that mathematics is a difficult subject and waits. Quarta, who has been gazing out of the window, returns to the fray and asks the teacher, yet again, whether there is a rule at all. "Maybe it is just chance like throwing a coin. Or like falling in love? . . . blah . . . blah blah . . . ". The teacher is unsure how to counter this epistemological pessimism – seductive and subversive. Remembering the dictum "weak point, shout!" he declares dogmatically that there is indeed a rule, a mathematical pattern, a law of nature, full stop.
*Secundus*: Look, everyone. I have been making a record of the results so far. And this is what they are.

| LEFT | RIGHT | RIGHT | LEFT |
|---|---|---|---|
| weight | weight | distance | distance |
| 1 | 1 | 6 | 6 |
| 2 | 1 | 2 | 1 |
| 4 | 2 | 4 | 2 |
| 3 | 5 | 3 | 5 |

The layout of the Secundus table (slightly tidied up for the reader's sake) surprises the teacher. He would have placed the left side weights and distances next to each other and similarly for the right hand side.

*Secundus*: And if I may carry on, the rule is really very simple! It is all a matter of seeing the pattern. Can you see it? What about you Primus? You are supposed to be smart . . .

Primus, it turns out, has also discovered a pattern.

*Primus*: The rule is: the weight plus the distance on the right hand side should be equal to the weight plus the distance on the other side.

*Secundus*: No! The difference between the two weights is equal to the difference between the two distances.

A squabble ensues.

*Teacher*: How interesting. I must admit I had not quite expected this little outcome and as a matter of fact both your rules fit the data so far. But you should be able to see that the two rules are in fact equivalent. It sometimes happens in science that two people find what they think are different rules but which in fact turn out to be the same. I must one day tell you the story of Schrödinger and Heisenberg who came up with two independent theories of quantum mechanics which were later shown to be equivalent. And we have a similar situation here. You should be able to see that if $W_L + D_L = W_R + D_R$ then logically $W_L - W_R = D_R - D_L$.

Useful names those, Schrödinger's and Heisenberg's, if one ever wants to bluff one's way through quantum mechanics, the teacher finds himself thinking . . .

*Primus*: I see what you mean, teach. I mean about my rule being the same as the rule of Secundus. Still, I said the rule first.

*Secundus*: I thought of it first.

*Teacher*: Perhaps you two can share the Nobel prize, if it comes to that. But first it remains to be seen whether your rule works for other values . . .

*Primus*: It won't work for *every* weight, because the see-saw might collapse. But, otherwise, yes. For instance if one puts a weight of 6 tokens on the right hand side at a distance of 3 units, then a weight of 3 tokens should be put on the other side at a distance of 6 units. 6 + 3 = 3 + 6. Look: I place the weights as indicated and there you have it: balance. Why are you smiling, teach?

*Teacher*: I am sorry. I went for a walk in my head for a moment. I was thinking about how it has been said – by a very famous philosopher of science – that one can *only* learn from one's mistakes; but what if one is unlucky enough not to make any mistakes? . . . Your choice of numbers, Primus, may have been unfortunately lucky, so to speak. Try instead the following: a weight of 6 and a distance of 6 on one side; on the other side a weight of 12. Where should it go?

*Primus*: 6 + 6 = 12 . So on the other side 12 + 0 = 12. The answer is zero. Ummm . . .

*Teacher*: Ummm indeed. Your rule says to put the counterweight on the zero line. Does that make sense?

The teacher expects the answer to be a unanimous "no". The issue he thinks will be decided by a thought experiment. But dear reader, that was not the case!

The class couldn't give up a good rule – the best lousy rule they had – as easily as that. The thing had to be practically tried and refuted.
*Secundus*: Now that I come to think of it, I can see that the rule must be wrong. Look: if in the previous example the counterweight is specified as 13 tokens (rather than 12), the associated distance would have to be negative! And, obviously, negative distances don't exist. Or do they?
*Teacher*: Bravo! That does seem to be a formal flaw of the theory[3]. We surely can't permit negative distances, but you never know with these scientists.
*Primus*: I am still not convinced. I admit that the rule does not work in these *absurd* extreme situations. But the rule might work as long as one does not go beyond the limit (of 12 tokens in this particular case). For any lower value, the rule might work.
*Secundus*: Sorry Primus. No joy. I too was thinking along the same lines, But consider this: the rule tells us in the present example that a counterweight of 11 should be placed at a distance of 1 (11 + 1 being equal to 6 + 6). But this has to be wrong because I have been quietly trying out the 12 token counterweight and have noticed that it needs to be at a distance of 3 units. So something is wrong with the rule anyway because 12 + 3 is not equal to 6 + 6. But furthermore, if a 12-weight needs to be 3 units away for balance, an 11-weight can't possibly have the same effect by being nearer the pivot.
*Teacher*: Exactly, Secundus. For, as we know from our earlier discussion, the lighter the weight the bigger the distance. And that rule still applies. Does it not, Primus? I mean our old qualitative rule may not be the whole truth, but we don't want to throw it away. Do we?

Primus accepts the argument but is not convinced. He has practically to try out the weight of 11 at a distance of 1 to see that it does not work.
*Teacher*: Right then. Where have we got to now? Back to square one? Or can we go on to build a new rule which works? I tell you what, you lot . . . I have been thinking that maybe my presence is a jinx. So why don't I go off to have a cup of tea and you mess around with the see-saw and look at the results that you have obtained so far and see if you can come up with something. Maybe your method of discovery will work better while I am out having a fag – what they call a cigarette in America.

As he is leaving, he hears Secundus thinking aloud: "It is to do with ratios. We should divide. It is halves . . . If you double the distance, one should halve . . . I am sure . . . "

*There it was.* He had it – all in a one big rush of discovery.

## III. AFTER TEA

On the teacher's return Secundus was ready and announced the rule that the weight on the one side divided by the distance on the other side should be the same as the weight on the other side divided by the distance on the *other* other side. A little more work led to the following symbolic formulation:

$$W_L/D_R = W_R/D_L$$

And as this was not readily accessible to everyone, a little further tidying up led to the final more elegant version:

$$W_L \times D_L = W_R \times D_R$$

*Teacher*: That then *is* the rule. Well done, Secundus. Well done everyone for indeed our scientific progress has been the result of a combined effort. But let me now ask the company a question: what about the old formula that says $W_L + D_L = W_R + D_R$? Can we see why it worked when it did work? And would you still use the formula, or do you think we can throw it away now that we have the new formula?

The class with some mathematical assistance from the teacher could see that the older formula worked whenever the equalities $W_L = D_R$ and $W_R = D_L$ happened to be true. The consensus was that there was no great point in retaining the old formula because it had been successful only by chance[4].

*Teacher*: Gentlemen and Ladies, sorry to interrupt your line of thought. But goodness (gracious) me. I have just realised something. Today is Thursday . . .

*Chorus*: And so?

*Teacher*: Well, Thursday is my lucky day and I am afraid the formula we have just found may not work on other days . . .

*Primus*: Don't worry, teach. Thursday is my unlucky day. So that just about cancels out your good luck!

There was a chorus of laughter. The teacher then asked whether there were other factors besides time that might or might not be irrelevant. The class agreed for instance that the colour of the see-saw or of the tokens would not make any difference.

*Teacher*: But here is another thing or two to worry about: our investigation has been conducted on this bench here. I wonder whether the see-saw will behave in a different way if I move to that bench over there. Or, do we have place invariance, as the phrase goes? I wonder also if it matters to the see-saw that we are all watching it. And what would happen if we were not? Maybe it would do its own thing – play, dance and that sort of thing – or maybe simply not exist, if you remember your Berkeley. Or, indeed, quantum mechanically, it might go into some transcendental superposition of states.

*Chorus*: Ha! Ha!

*Quarta*: Oh! I just love that kind of weird nonsense . . . 'transcendental superposition' . . . sounds great. But about this place variance business, yes, the place does make a difference. The experiment won't work on the moon because there is no gravity on the moon. You see the moon is in space and there is no gravity in space because there is no atmosphere in space. And when there is no gravity the weights are weightless, so the thing will be balanced no matter what. Get it?

An appropriate moment, dear reader, to close the curtain.

* * * * * * *

The teacher went on to conduct a philosophical analysis of the session. Very briefly, everyone agreed that truth was hard to discover although they were worried that they might have been a little dim. The teacher told them the truth: they were not. He asked them whether they had used a scientific method or whether they were going about the business like clueless rats. On the whole the students preferred the former alternative but could not pinpoint the method. The teacher asked: which came first – the observation or the conjecture? Like any good scientist they could not remember. However, like bad philosophers, one said observation always comes first, another gave priority to conjecture.

The teacher went home and talked about the day to his wife. He told her about the rule, weight plus distance on one side equalling the weight plus distance on the other side. She said that such a rule could not be the true rule. For, said she, if the formula worked in inches it would not work in centimetres. Here then was a request for a law to be invariant under scale transformation. The teacher thought that was good and told his wife that she was good. She then asked him about the beads. She said, these beads, were they really amber? Because somebody – Uncle Tom – had told her that amber was electrostatic. And when she had tried it out it had not been electrostatic. Great Aunt Elsie had brought the beads from America some sixty years ago and had said they were real amber. "You think the electrostatic might have worn off . . . , teach?" It was only then that the teacher thought that maybe he had a tale to tell.

[The writer's experience as a teacher is that the scientific views expressed in the above should not be lightly dismissed as indicative of a remarkably ignorant, peculiar bunch.]

## POSTSCRIPT

The following week the students were given the teacher's written account of the afternoon (a rough version of what has been presented here). Their reactions briefly were as follows:

1. Their first interest was to identify the characters in the story.
2. They thought the story was very true.
3. They were still unhappy with the mathematics involved.

Why is mathematics so unreasonably disaffective?

ACKNOWLEDGEMENTS

This tale would have been impossible without the students – Dawn, Adam, Melanie, Chinwe and Fatilda. It would not have been written had (a) Harmke Kamminga not kept on bullying me and (b) Elizabeth McGrath not kept on correcting mistakes. My thanks are due also to Ann Massing for her graphic help.

*Hackney College*
*London, U.K.*

NOTES

1. One advantage of using the device of 'rational reconstruction', falling as it does into the legal category of fiction, is that one can take liberties. I hope that my old Professor will forgive the misquote and the fact that in the rest of the article, not only have I stolen many of his ideas, but probably also misunderstood them.
2. The point is important: how one tries to tackle a problem is very much determined by the route by which one arrived at it. The following example, which the teacher is fond of using, illustrates nicely how an utterly trivial puzzle can be made difficult by first posing a distractor. Although the reader unfortunately has been given too much of a clue to be able to participate personally, he or she may invite a friend to carry out the following two tasks:
   i) Divide the following three-quarters of a square into four congruent parts.

   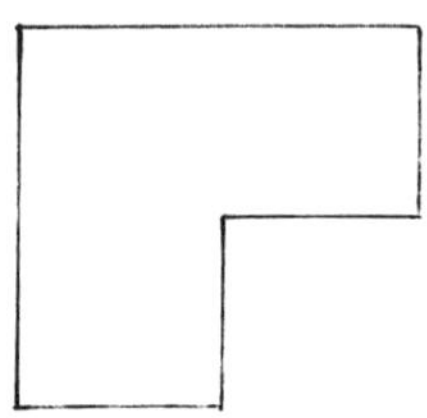

   It is important that the second part of the problem is intimated only after this first task has been completed.
   ii) Divide a complete square into FIVE congruent parts.
   The order in which the two tasks are posed makes an interesting difference to the success rate.
3. Here the teacher is mistaken. The flaw is more informal than formal. It is true that the students' theory is inconsistent with the Archimedean notion that, regardless of the counterweight, a see-saw, if long enough, can always be brought back to balance. The latter notion, however, was more in *his* head than in the students'.
4. Heinz Post used to say – and still does – that good scientists are conservative, taking – I suspect – some delight in the fact that the claim in its aural representation would be misheard with a capital 'C'. Conservatism, however, seems not as yet to have invaded the London Borough of Hackney – scene of the see-saw investigation.

   For the real import of Post's dictum see his article 'Correspondence, Invariance and Heuristics: In Praise of Conservative Induction' (Post 1971). Post claims that in the progress of science one never throws away $S^*$, the well-confirmed part of an older theory even though there might exist an $L$, a new larger theory, which explains $S^*$. However, consider the following $S^*$ from our episode: If (a) weight plus the distance on one side equals the weight plus distance on the other side and if (b) $W_L = D_R$ and $W_R = D_L$, then the see-saw is balanced. $S^*$ here is well-confirmed and indeed follows from *L* and cannot be "thrown away" i.e. denied while maintaining *L*, the later theory discovered by the students. $S^*$ was, however, of no use once *L* had been discovered. $S^*$ was merely an accidental truth en route. The particular counterexample to Post's dictum is trivial in itself but it does serve to raise the question whether there may not be weightier counterexamples from the history of science. To end on a note of agreement with Post: $S^{**}$ ('the bigger the weight, the smaller the distance') was conserved by the students.

## BIBLIOGRAPHY

Post, H.R. (1971), 'Correspondence, Invariance and Heuristics: In Praise of Conservative Induction', *Studies in the History and Philosophy of Science*, 2, pp. 213–255.

GIORA HON

# The Unnatural Nature of the Laws of Nature: Symmetry and Asymmetry

*Symmetry signifies rest and binding, asymmetry motion and loosening, the one order and law, the other arbitrariness and accident, the one formal rigidity and constraint, the other life and freedom.*

Dagobert Frey (quoted by Weyl 1973, p. 16)

*We must relate knowing to characteristic features of life.*

Ilya Prigogine (1980, p. xv)

It has been a recurring scene: Heinz Post, sitting attentively in the front row, would raise his hand at the end of the talk, turn his head half way towards the audience and respond to the speaker with the following words: "I have this naive question which may have bothered other members of the audience." He would then proceed to ask what seems on the face of it a naive question, but with some reflection it would transpire that the question is perspicacious, throwing light both on the kernel of the talk and on its shortcomings. This so-called naive question would invariably reveal where the argument had broken down. That was a true lesson in philosophy; it was a lesson which I have striven to learn and emulate.

I wish to pose a problem here which I hope qualifies as naive in Post's sense. The problem concerns the role of symmetry and asymmetry in forming the ground for studying and understanding nature. Essentially, I claim that the overriding concern with symmetry (and its breaking) has divorced science from its object of study: Nature. Nature is not symmetrical: all phenomena are asymmetrical[1]; they can be discerned precisely because they are asymmetrical. Symmetry renders nature transparent, allowing for the construction of idealized structures, whereas asymmetry reveals concrete phenomena.

Asymmetry is meant here in both a spatial and a temporal sense. I begin with a discussion of the richness of spatial asymmetry whose apprehension by observing nature closely and intently revolutionized art. I shall then stress the crucial role that temporal asymmetry must play if we are to conceive of



*S. French and H. Kamminga (eds.), Correspondence, Invariance and Heuristics*, 171–187.

a physics which reflects our experience of nature. I shall conclude with a clarion call for a new physics – a physics which is valid for what happens in nature, a physics which is directly fastened to the experience of natural phenomena. The call is based on a wedge which is driven between two types of explanation. On the one hand there is the explanation which deploys fundamental theoretical entities to render phenomena intelligible; on the other hand there is the explanation that regards events, and in particular irreversible change, as fundamental, irrespective of the entities that are doing the changing. It appears that the deductive use of theoretical terms has led physics to a new form of scholasticism, to a conception of laws of nature which is unnatural. The question then arises as to whether a different, phenomenological, approach, might not prove fruitful in bringing back to the center of the scene the very experience of nature (cf. Cartwright 1983, Introduction).

In his classic book, *Symmetry*, Hermann Weyl states at the outset that he thinks with Plato that mathematical laws governing nature are the origin of symmetry in nature, and the intuitive realization of this idea in the creative artist's mind is the origin of symmetry in art. However, Weyl admits that the bilateral symmetry of the human body in its outward appearance has acted in the arts as an additional stimulus (Weyl 1973, p. 8). Wherever God or Christ are represented as symbols for everlasting truth and justice they are presented in the bilateral symmetric frontal view, not in profile. There is thus no surprise that buildings and houses of worship, whether they are Greek temples or Christian basilicas and cathedrals, are bilaterally symmetric (*ibid.*, p. 16).

Bilateral symmetry has become the corner stone of paintings and architecture: an extension of the external symmetry of the human body. In the Byzantine tradition and the early Middle Ages divinity was thought to reveal itself in the symmetry of geometric forms and in the richness of the material. The glitter of gold, the gleam of pearls and precious stones were regarded as a reflection of the divine light. It was however understood that these reflections of true beauty could not be realized in material form; aesthetics was thus reduced to formalism (Oertel 1968, p. 91).

The new art of the Renaissance emerged from a fundamental rethinking of this formalism. Whereas in the earlier traditions artists represented sacred images in an unworldly supranatural formal state, they now portray them as real beings, though raised to solemn eminence. A contemporary of Dante, Giotto retained the formal ideal inherited from the past while paving the way to a new reality in which the divine was brought down to earth, in the midst of men, instead of being left in solitary splendour (Venturi 1907).

The paintings of Giotto do not complement the liturgical hymn; they convey, rather, the poetry of Dante. Boccaccio said of Giotto that

> . . . there is nothing in nature . . . that he with his stylus and pen or paintbrush did not depict so true to life that it seemed to be, not a likeness, but a product of nature; so much so that very often men's eyes

> have been deceived in the things he painted and the painted image taken for reality[2].

Giotto became very adept at imitating nature; indeed, he revived the practice of painting people from life (Vasari 1568). In his new art of spatial realism, Giotto illuminated his figures from a single source of light, but his chiaroscuro takes on added depth and subtlety, giving his figures a more sculptural quality. His colour schemes are novel, intense and daring, intending to throw the form into relief and give it a tactile impression[3].

Of Giotto's surviving panel-paintings, the most celebrated is the *Madonna* for Ognissanti (Uffizi) which is usually dated about 1310 but may be earlier. Giotto placed the representation of visual phenomena at the service of a searching interpretation of the human condition. In the *Ognissanti Madonna* the very world of heaven is thus made flesh (Smart 1978, pp. 14, 57–58, 88).

What is the secret of the *Ognissanti Madonna*? How is it made flesh? How can it be taken for reality, as Boccaccio put it. I suggest that the key to this innovative painting is the inherent asymmetry of the Madonna's face: the eyes of the Madonna are not bilaterally symmetric, they irradiate different looks. The fish-shaped right eye is twisted to the right so that it appears to be looking in a different direction from the more realistic left eye. Moreover, the pupil of the right eye is half covered, dimming its focus and suggesting inner awareness, in contrast to the clear focusing of the left eye which fixes our attention and heightens our feeling of being caught in the Madonna's level gaze. Since the space below the Madonna provides the observer with a figurative path of access, we are directly engaged with the Madonna. Yet the right eye is unaware of the spectator's existence (Martin 1965, pp. 9–10). Giotto had perceived what scientific physiognomists observed hundreds of years later: the asymmetry of the face can convey spiritual asymmetry[4].

> A painter's work will not be of any great worth if he takes that of another artist for his model. However, if he learns from nature, his work will indeed bear fruit[5].

This claim of Leonardo da Vinci is borne out by those painters of the Middle Ages who all imitated one another with the result that art gradually deteriorated. Giotto, by contrast, was not content merely to imitate the work of his master, Cimabue. Having been born in a wild mountainous area and being exposed to nature from early childhood, so Leonardo argued (*ibid.*), Giotto had begun to draw directly from nature.

There is thus no surprise that Leonardo admired Giotto. Indeed, Leonardo's famous *Self Portrait* of 1510 (Royal Palace, Turin) in red chalk exhibits precisely the kind of spiritual asymmetry that Giotto presented in his paintings. While Leonardo's right eye, turned aside and strangely blank, belongs to the abstract thinker, dreamer and introvert, the left eye, watchful and keen, belongs to the empiricist, the extrovert and scientist (Martin *op. cit.*, p. 10).

Leonardo was a follower of Giotto in observing nature closely and intently. The movements of water, for example, held great fascination for Leonardo. He regarded water as the driving force of the universe and thought that he might fathom the mysteries of creation by studying the motion of water. Drawings and memoranda scattered over manuscripts of different dates show that the subject absorbed his attention more or less continually. We can see him walking, notebook in hand, along the sea-shore contemplating the ebb and flow of the tides, the winds as they trouble the surface of the water, the surge of the waves, the drift of the sands; or he might be standing by the riverside watching the currents and eddies and inspecting the deposits on the banks; or lingering by a stagnant pond looking at the reflections, the luster on plants and the play of the fish beneath. Whereupon he would throw a stone into the still water and compare the ever-widening circles on the surface with waves of sound ringing through the air. Another time he may be walking up a mountain valley in order to trace a brook to its source while observing the waterfalls and the ceaseless grinding of rocks and pebbles (Richter 1952, p. 17)[6].

These meticulous observations of the movements of water reappear in Leonardo's paintings: eddies, whirlpools, turbulences and currents, are recast in a lock of hair, a curl, the falling of a drape, the shape of a dress – the principles of the motion of water being the rules that govern representations of matter subject to a force. Drawing water in motion which takes the form of hair, Leonardo instructed:

> Observe the motion of the surface of the water, how it resembles that of hair, which has two movements – one depends on the weight of the hair, the other on the direction of the curls; thus the water forms whirling eddies, one part following the impetus of the chief current, and the other following the incidental motion and the return flow. (*ibid.*, p. 25)

Leonardo intimated that his "intention is to consult experience first and then with reasoning show why such experience is bound to operate in such a way." Although "nature begins with the cause and ends with the experience, we must follow", Leonardo declared, "the opposite course, namely, begin with the experience, and by means of it investigate the cause." This was in his view the true rule by which those who analyse the effects of nature must proceed (*ibid.*, p. 6)[7].

Recalling that for Heracleitus motion of water is the paradigm of change, of flux – the principal case of Becoming – we might be able to appreciate the challenge which Leonardo had set himself. Following the new methods of science, we are accustomed to look not where the problems lie, but rather where there is light, where, in Post's apt phrase, 'inevitable research' abounds (Post 1974, pp. 7, 12). Leonardo, by contrast, had groped in the dark: knowledge should be directly fastened to the experience of natural phenomena; an experience which is alas in a constant flux. How are we to understand this experience? How are we to grasp change? The explanation of change, as Post remarks, is after all the fundamental problem of science (Post 1968, p. 225).

It appears to be a deep-rooted conviction of human reason that a rational explanation of phenomena has been reached in so far as apparent differences are discovered to be reducible to an underlying identity. The demand for a rational explanation seems to be satisfied when it is shown that an apparently new fact was really there all the time; when one reduces the apparent difference between what was there at one moment and what was there at another to a real identity (Field 1925–1926, pp. 125–126, 128)[8]. Atomism is a good example; its programme is, in Post's words, "to explain everything, but everything, in terms of a denumerable number of identical invariant units, or at least units of limited variety, i.e. of a small number of species". As Post remarks further,

> . . . we are looking for invariant units more or less hidden behind the world of change . . . We are looking, then, for invariant factors of explanation . . . The deeper atomistic programme is to 'diagonalise' the whole universe into completely independent factors. (Post 1975, pp. 20–21)

However, 'diagonalising' the whole universe may be very costly; it may be achieved at the expense of the richness of the contingent which is associated directly with experience. Should we then follow the doctrine of atomism and annul the off-diagonal elements – the very elements which contain information of the rich contingent? My reply is *no*. I suggest here a transition from the abstract beauty of Platonic symmetries to an Aristotelean belief in the richness and variety of the concrete and particular – the asymmetrical. "Things are made to look the same only when we fail to examine them too closely." (Cartwright *op. cit.*, p. 19)

Permanence, order and unity are the mark of rationality in the classic tradition; they are eulogistic predicates. Classical philosophy and modern science speak much about unity and little about unreconciled diversity; much about the eternal and permanent and little about change; much about necessity and little about contingency; much about the comprehending universal and little about the recalcitrant particular. It appears that the conception of the eternal exercises an hypnotic influence. The permanent gives peace; the variable, the changing, presents by contrast a constant challenge. Indeed, the permanent satisfies genuine emotional, intellectual and practical needs. Hence the method of dealing with the variable and precarious by means of the stable and constant. Change is to be resolved into combinations of the permanent: one introduces units of limited variety, preferably all alike, and searches, as Schiller put it, for "the pole at rest in the world of fleeing appearances" (Post 1975, p. 25)[9]. Every change is interpreted as a displacement not affecting the atoms themselves; they remain eternally the same. Thus atomism seeks to explain all change in terms of unchanging, invariant, ultimate units. In practice, invariants are a necessity for bringing the phenomena in flux to pass as mathematical functions. However, when such terms are employed there is ground for suspecting that a simplification – an idealization, a 'diagonalisation' – of phenomena has been performed (Dewey 1968, pp. 26–27, 41–42).

Notwithstanding the yearning for the permanent, one may feel, like Sagredo – a participant in Galileo's *Dialogue Concerning the Two Chief World Systems* – great astonishment to hear it attributed as a prime perfection of the heavenly bodies that they are invariant, immutable, inalterable, etc., while on the other hand it is considered a great imperfection to be alterable, generable and mutable. "For my part", Sagredo remarks,

> I consider the earth very noble and admirable precisely because of the diverse alterations, changes, generations, etc. that occur in it incessantly. If, not being subject to any changes, it were a vast desert of sand or a mountain of jasper, or if at the time of the flood the waters which covered it had frozen, and it had remained an enormous glove of ice where nothing was ever born or ever altered or changed, I should deem it a useless lump in the universe, devoid of activity and, in a word, superfluous and essentially nonexistent. (Galileo 1974, pp. 58–59)

The static Parmenidean world may be knowable, but in Sagredo's view it would be vacuous. It may be 'diagonalised', but then it will lack the off-diagonal elements – the asymmetrical contingent. I discern here two sets of contrasting concepts. To the 'diagonal' belong symmetry-elements which yield invariances that have the status of necessity. The 'off-diagonal' elements represent, by contrast, the asymmetrical: rich contingent elements which change all the time.

Experience and knowledge of the contingent – the 'off-diagonal' elements – are, I claim, an essential requisite for developing a physics that is valid for what happens in nature. This is the message of Mach's most influential book, *The Science of Mechanics*. "The aim of my whole book", writes Mach,

> . . . is to convince the reader that we cannot make up properties of nature with the help of self-evident suppositions, but that these suppositions must be taken from experience. (Mach 1974, p. 27)

Mach therefore opens his book with a didactic illustration: how symmetry can lead astray; how the delight in mathematical demonstration based on symmetry arguments can lure the natural philosopher into begging the question. The case is the law of the lever; the culprit is Archimedes and the message is clear: the general law of the lever could not be deduced from the equilibrium of equal weights on equal arms. Mach seeks to show where the experience that already contains the general law of the lever is covertly introduced. The deduction from simple and almost self-evident theorems may charm the natural philosopher who has an affection for Euclid's method, but it may be nonetheless logically invalid.

In his treatise, Archimedes starts from the following assumptions which he regards as self-evident:

1. Magnitudes of equal weight acting at equal distances (from their point of support) are in equilibrium.
2. Magnitudes of equal weight acting at unequal distances (from their point of support) are not in equilibrium, but the one acting at the greater distance sinks.

According to Mach, Archimedes deduces from these assumptions the following proposition:

> Commensurable magnitudes are in equilibrium when they are inversely proportional to their distances (from the point of support) (*ibid.*, p. 14).

If the spectator were to place himself or herself in the plane of symmetry of the arrangement in question, the first proposition manifests itself as a highly imperative intuitive perception – a result determined by the bilateral symmetry of our own body. However, as a matter of fact, the conclusion that the equilibrium-disturbing effect of a weight $P$ at the distance $L$ from the axis of rotation is measured by the product $P \times L$ (the so-called statical moment), is covertly or tacitly introduced by Archimedes and all his successors. It is obvious that if the arrangement is absolutely symmetrical in every respect, then equilibrium is obtained on the assumption of *any* form of dependence, whatever the disturbing factor on $L$, or, generally, on the assumption $P \times f(L)$. Consequently, the *particular* form of dependence, $P \times L$, cannot be inferred from the equilibrium. Furthermore, we might suppose that this was self-evident, entirely apart from any experience, according to the so-called principle of sufficient reason; that in view of the symmetry of the entire arrangement there is no reason why rotation should occur in the one direction rather than in the other. But Mach reminds us that we may forget that

> . . . a great multitude of negative and positive experiences is implicitly contained in our assumption; the negative, for instance, that dissimilar colors of the lever-arms, the position of the spectator, an occurrence in the vicinity, and the like, exercise no influence; the positive, on the other hand, (as it appears in the second assumption), that not only the weights but also their distances from the supporting point are decisive factors in the disturbance of equilibrium. (*ibid.*, p. 15)

Assuming that *only* the weights and their distances from the point of support are decisive, the first proposition of Archimedes really incorporates a large amount of experimental evidence and is eminently qualified to be made the foundation of further investigations. But notice that from the mere assumption of the equilibrium of equal weights at equal distances, the *inverse* proportionality of weight and lever-arm is derived! How is that possible? It is possible since Archimedes both assumes and knows that distance from the fulcrum is determinative. This procedure is not permissible: the deduction contains the proposition to be demonstrated. It is an instance of *petitio principii*. The central methodological assumption, the principle of symmetry, has betrayed Archimedes[10].

Symmetries are transformations that leave all relevant structure intact, hence the expectation, or rather the demand, that laws of nature should be universally symmetric. It transpires, however, that this demand cannot be fulfilled. According to Weyl,

> . . . if nature were all lawfulness then every phenomenon would share the full symmetry of the universal laws of nature as formulated by the theory of relativity. The mere fact that this is not so proves that *contingency* is an essential feature of the world. (Weyl *op. cit.*, p. 26)[11]

In other words, the laws of nature do not determine uniquely the one world that actually exists! The laws should have told us not merely what is and does happen, but what must be and must happen (van Fraassen 1989, p. 288). This is however not the case: the laws of nature are not universally symmetric; they have as it were failed to accomplish their principal metaphysical objective. This failure of the laws of nature originates in the separation between the initial conditions of the system and its dynamical development. Initial conditions are understood as conditions which fix the integration constants, that is, they select from the set of all possible solutions of the laws those ones which correspond to the processes found *contingently* in nature (Houtappel, van Dam and Wigner 1965, pp. 596–600; Zeh 1989, p. 2).

Yet the strong tradition of classicism encourages the geometrical mind to discern symmetries and to determine their associated invariants. In this tradition, where Reason and Nature are used interchangeably, there is place neither for the contingent nor for its experience. It is no accident that Pascal's *Pensées,* written at the height of classicism but seeking to undermine it, should begin with a distinction between the geometrical and the intuitive mind. Pascal's phrase for the latter is *esprit de finesse*; it means the ability to distinguish and deal with concrete things, with living beings, as against the geometrician's ability to manipulate abstractly axioms and definitions. The geometrician's universe is articulate, distinct and colourless; the *esprit de finesse* by contrast sees colours and hues, continuity and indefiniteness: the flux of water, the formation of a curl, the asymmetry of the face (Barzun 1975, pp. 39–40).

It has been claimed that modern scientific knowledge tends to increase its formal simplicity (Einstein 1960); but this simplicity is won at the cost of a growing gap between the fundamental hypothesis of the theory, on the one hand, and the directly observed phenomena, on the other hand. Indeed, as Einstein remarked,

> . . . theory is compelled to pass more and more from the inductive to deductive method, even though the most important demand to be made of every scientific theory will always remain: that it must fit the facts. (*ibid.*)

Modern physics thus leads us away from the very experience of nature. The achievements of the geometrician provides formal explanations but this understanding is detached from experience, it becomes less and less natural.

It appears that the contrasting views of Heracleitus and Parmenides are still with us today as they were in antiquity: the former acknowledges the experience of "everything flows" but despairs of explanation, whereas the latter considers change an illusion and despairs of reference to empirical data. A possible way out from this dilemma is to combine the two horns in the spirit of Democritus – the programme of atomism which Post supports (Post 1975, pp. 24–26). But as I have indicated, this may prove unsatisfactory. I therefore would like to propose a different approach by reconsidering critically the role of symmetry and in particular by focusing attention on asymmetry.

The attentive natural philosopher should combine intuitive knowledge with a great power of abstract formulation: the geometrical mind should coalesce with the *esprit de finesse*. According to Mach, everything which we experience in nature imprints itself *uncomprehended* and *unanalysed* in our precepts and ideas, which, then, in their turn, mimic the processes of nature in their most general and most striking features. In these accumulated experiences we possess a treasure-store of which only the smallest portion is embodied in clear articulate thought. It takes a great power of abstraction to tap this reservoir of intuitive knowledge and bring it to fruition (Mach *op. cit.*, pp. 35–36).

This process of imprinting, the way nature imprints itself in our precepts and ideas, has the characteristics of a measuring process – the sense-organs being instruments which yield 'readings', albeit only qualitative ones. It is thus by its very nature an irreversible process. Indeed, the possibility of confronting a scientific theory physically with experience in the attempt to test it, presupposes a domain where irreversible processes occur. The question of 'consciousness' or 'observer' does not arise here; rather, following Bohr, the existence and functioning of a device should be stressed: the blackening of a grain of silver bromide emulsion, the blip of the Geiger counter, the triggering of a photodetector, indeed, the function of the retina of the eye – a device capable of "an irreversible act of amplification". This act brings the measuring process to a 'close', thereby defining a 'phenomenon' (Wheeler and Zurek 1983, pp. 184–185, 207). These are the kinds of processes that have their imprints recorded and stored. But to understand this process of imprint, or, what comes to the same thing, to formulate a theory of irreversible processes, one must have a clear understanding of the concept of time[12]. Physical time has been conceived in classical physics as well as in the quantum theory as a parameter, an independent real variable (Bunge 1967, p. 94). But so far no clear understanding of this concept has emerged (see for example Horwich 1987 and Savitt 1991). I suggest that the concept of time and in particular the common experience of its asymmetry hold the key for further developments in physics.

Shortly before his death, Einstein wrote in a letter of condolence to the family of his life-long friend Michael Besso that

> . . . for us who are convinced physicists, the distinction between past, present

> and future is only an illusion, however persistent. (quoted by Prigogine 1980, p. 203)

This remark of Einstein should be taken literally: he did believe in a timeless universe. Indeed, the concept of the Now worried him seriously. According to Einstein,

> . . . the experience of the Now means something special for man, something essentially different from the past and the future, but . . . this important difference does not and cannot occur within physics.

That this experience cannot be grasped by science seemed to Einstein a matter of painful but inevitable resignation. He concluded that

> . . . there is something essential about the Now which is just outside the realm of science. (Quoted by Zeh *op. cit.*, pp. 149, 150, 151; cf. Prigogine *op. cit.*, pp. 202–203)

This position is not surprising since Einstein held that

> . . . *according to our present knowledge*, all elementary processes are reversible. (Einstein 1969, p. 688, emphasis in the original.)

Time, in this conception of physics, is indeed just a real variable. Physics simply does not offer any conceptual means of characterizing a present which might objectively separate the past from the future. Notice however that Einstein qualified his remark. He did not exclude the possibility of a physics which would render elementary processes irreversible.

Since the fundamental laws of dynamics are invariant under time-reversal, they are compatible with reversed phenomena in which for example concentrically focussing waves would eject a stone out of a pond. However, such a solution of the dynamical laws has never been observed in nature. Similarly, of the two types of solutions which the Maxwell equations yield for the wave equation, the retarded and advanced potentials, only the retarded field seems to have a physical meaning, in the sense that only this solution has been observed in nature. One therefore argues that the advanced solutions would require improbable initial conditions, or one claims intuitively and somewhat mystically that "for reasons of causality", or "for physical reason", one observes in nature only the solution based on the retarded potential and its linear combinations. It is worth referring here to Dingle's warning caricature which Post considered fair: there is this school arithmetic problem which allows the solution "number of workmen = minus 3". A modern physicist would conclude that there *are* three negative workmen (Post 1971, p. 222, fn. 23).

Clearly, the time directed notion of causality cannot be derived from the fundamental dynamical laws themselves. However, the possibility arises here of qualitatively formulating a structure which could be added to determinism and thereby define a direction for time (Zeh *op. cit.*, p. 12). This is precisely what Ritz did at the beginning of this century. He attempted a radical solution to the problem by requiring the exclusive existence of retarded waves by law of nature. Under such a stipulation the electromagnetic field would not possess any degree of freedom, and its boundary conditions would be globally fixed. This suggestion led to a controversy with Einstein who favoured the opposite point of view. In 1909 they published the following notice:

> While Einstein believes that one may restrict oneself to this case without essentially restricting the generality of the consideration, Ritz considers this restriction as not allowed in principle. If one accepts this point of view, experience requires one to consider the representation by means of the retarded potentials as the only possible one, provided one tends to assume that the fact of the irreversibility of radiation processes has to be present in the laws of nature. Ritz considers the restriction to the form of the retarded potentials as one of the roots of the Second Law, whereas Einstein believes that the irreversibility is exclusively based on reasons of probability. (Quoted by Zeh *ibid.*, p. 13)

Given the underlying time reversal invariance of the fundamental laws of dynamics, why don't we find just as many systems with their entropic orientation in time in one direction as in the other (Sklar 1986, p. 216)? The asymmetry of our experience of time deeply affects our own form of existence. If physics is to justify the claim that its laws are valid for everything that happens in nature, it should be able to explain or describe this fundamental experience. Surprisingly the very laws of nature are in pronounced contrast to this fundamental asymmetry (Zeh *op. cit.*, p. 1). Ritz's suggestion is an attempt to address this stark contrast[13].

Whence then does the temporal character of reality, or for that matter its *illusion*, come from? Could we dare to follow Parmenides consistently in espousing his unambiguously clear and simple denial of time and change? Could we accept Hermann Weyl's affirmation that "the objective world simply is, it does not become", and that it appears to become only to our 'blind-folded consciousness' (*abgeblendete Bewusstsein*) (quoted by Capek 1951, pp. 25, 41)? Capek is right in claiming that

> . . . even when change and succession are declared to be illusions, they, being illusions, still remain at least *psychologically* real. Chased from the realm of objective reality, change and succession take refuge in the world of our inner states. No matter how illegitimate it is to project the psychological succession outside of our private 'stream of consciousness', the stream itself of our mental states remains irreducibly changing and successive. (*ibid.*, p. 27)

Succession and change possess undeniable reality, at least in the realm of consciousness, otherwise the very illusion of temporality would be impossible. Yet even an illusion must have a reason. To have then the *illusion* of time, the impression of succession must be somehow connected to a certain ontological locus (*ibid.*, p. 30). But whatever that locus may be, it is clear that the physics we know today does not contain the concept of the flow of time that we have the impression of experiencing. Does this subjective, real, experience induce an arrow of time which is absent from the objective, fictitious, physical reality?

This is a question of conformity: is there a conformity between nature and our thought? Hertz, in his optimistic approach to knowledge, argued that "experience teaches us that . . . such a conformity does in fact exist". In his view,

> . . . we form for ourselves images or symbols of external objects; and the form which we give them is such that the necessary consequents of the images in thought are always the images of the necessary consequents in nature of the things pictured.

We thereby solve, according to Hertz, the most direct, and in a sense the most important, problem which our conscious knowledge of nature should enable us to solve: the anticipation of future events, so that we may arrange our present affairs in accordance with such anticipation (Hertz 1956, p. 1). The case of the asymmetry of time does not seem to fall under Hertz's happy claim, unless we start doing physics differently. In the physics we see today there is incompatibility of temporality and necessity: time emerges precisely in the act by which it is denied (Capek *op. cit.*, p. 30).

For quite a while physics has been concerned with understanding the behaviour of elementary particles that are believed to be the fundamental building blocks of nature. Atomism, as we have seen, states that the behaviour of macroscopic bodies, indeed the universe at large, can be understood purely in terms of these elementary particles and the interactions between them which are governed by reversible laws. On this view, macroscopic phenomena must also be reversible, and the apparently irreversible changes must be approximations, or illusions, resulting partly from our observations being conducted over too short a time scale. However, the laws of classical physics were set up on the unquestioned assumption that, although events may be reversible, it is always possible to talk about what has happened and to learn from past events. Why can we not consider this assumption explicitly and take its implications seriously? If we were to follow Leonardo da Vinci, Mach and Hertz, and pursue the view that we should learn our ways of thinking about natural phenomena from the way that nature behaves, then we should take as reality those processes in the physical world that are actually observed: the blackening of a grain of silver bromide emulsion, the blip of the Geiger counter, the triggering of a photodetector, the formation of a bubble in the liquid-hydrogen bubble chamber, indeed the cat's death. We should then treat as 'illusion', or at least as an approximation to reality, sub-atomic processes such as the passing of a photon through both slits. Instead of the fundamental laws referring to microscopic reversible processes, with macroscopic irreversible behaviour as an approximation or illusion, it is the irreversible laws that should be taken as fundamental and reversibility as an approximation. This approximation in which microscopic elementary particles move subject to reversible laws is valid only in the very special circumstances where a particle, or particles, are effectively decoupled from their interaction with the rest of the universe. The idea that reversibility is an approximation to irreversibility rather than the other way round is certainly more compatible with our view of the universe as we experience it (Rae 1986, pp. 106–116).

This is Prigogine's suggestion: irreversible changes are the fundamental entities in nature. The fundamental concepts in this approach are the changes rather than the objects that are doing the changing. A model of the physical world that attributes all reality to the changes, while stating that it is impossible to make a consistent description of what it is that is changing, is difficult to accept. But we should recall here that Einstein has already introduced in his theory of special relativity the concept of event as the fundamental entity of physics. As Post put it, "... we are dealing not with things, but with events." (Post 1971, p. 248, fn. 77). Following Prigogine, we go further and construe event as an irreversible change. In other words, one introduces first the second law before even defining the entities involved (Prigogine *op. cit.*; Prigogine and Stengers 1984; Rae *op. cit.*, pp. 106–116)[14].

Post is a fierce critic of all those who use the quantum theory to introduce some form of mysticism to science in general and to physics in particular. The outrageous leap from the measurement problem to the necessary existence of the human soul is just one extreme example. If the irreversible measuring processes were to be taken as the primary reality, then the so-called quantum theoretical explanation of extra-sensory perception and other 'paranormal' phenomena, would come to a nought. This new approach to physics in which the central theme is the idea that time does flow in one direction, seeks to reconcile one's subjective experience of nature and the belief in an objective physical reality. It is truly a scientific approach.

Modern physics was born in the seventeenth century with a principal concern for celestial mechanical phenomena. From its birth up to Carnot in the early part of the nineteenth century, physics neglected irreversible processes. Since the time of Carnot irreversible processes have been increasingly studied, but the fundamental laws have been assumed to be reversible – the irreversible aspects of *actual* processes being ascribed to a host of secondary causes such as statistical effects, interactions, measurement, boundary conditions as well as observation over too short a time scale. The time would appear to be ripe for exploring the consequences of a new physics which takes asymmetry, temporal and spatial, as fundamental (Whyte 1956).

Correspondence, invariance and heuristics are the central concepts of Post's philosophy of science. His faith in these concepts has originated in his observation that "no theory that ever 'worked' adequately turned out to be a blind alley" (Post *op. cit.*, p. 237). But the beliefs that there are possible procedures for going from a weaker to a stronger theory, that these procedures are rational and thus that scientific progress is linear and that heuristics is possible (*ibid.*, p. 251), stem from a much broader philosophy which we may call a philosophy of life. It seems to me that this philosophy consists of a rare mixture of an incisive critical irony which has a revolutionary fervour[15], and a strong belief in a true conservatism which praises tradition. What makes this philosophy of contradictory features work is the notion of restriction: *in der Beschränkung zeigt sich erst der Meister.* "In science, as in art," remarks Post, "work within imposed restrictions is more fertile than the mere assertion of the admitted right to be different." (Post *ibid.*, pp. 218, 250–251). My call for a new physics is precisely in this spirit, except that I call for greater restrictions. Since *actual* processes are undoubtedly irreversible: no grown-up man has ever turned into a child or an oak into an acorn, the fundamental laws are insufficient to represent the most general of all conditions of the real world (Post 1968, p. 229, citing Ostwald approvingly). We have to look then for fundamental laws which do conform to 'characteristic features of life'[16]. In other words, to follow the correspondence principle and its concomitant heuristics which Post advocates, a new *L*-theory must be formulated so that

> . . . if *S* is classical mechanics, which is symmetric under time reflection $t \to -t$, $S^*$[17] may be a degenerate form of a *strictly irreversible L-theory*, whose footprints show up (as flaws in *S*) in the many irreversible processes in nature. Very commonly, certain degeneracies in the *S*-theory are removed by the more precise *L*-theory. (Post 1971, p. 243; my emphasis)

Thus, the new physics I call for is this very *L*-theory.

I conclude, as Post habitually does, with a moral: "The fish does not profit from studying hydrodynamics", remarked Lakatos, to which Heinz Post replied, " . . . the hydrodynamicist, . . . *does* profit from learning how to swim." (Post 1974, p.3; emphasis in the original). It would appear to be high time for the hydrodynamicist to take swimming lessons.

## ACKNOWLEDGEMENT

I wish to express my debt to Martin Carrier, Steven French, Harmke Kamminga, Saul Smilansky and Gereon Wolters. I wish further to acknowledge the generous assistance of the *Humboldt Stiftung* and the *Zentrum Philosophie und Wissenschaftstheorie*, Konstanz University.

*Department of Philosophy*
*Haifa University, Mt. Carmel*
*Haifa, Israel*
and
*Humboldt Fellow*
*Zentrum Philosophie und Wissenschaftstheorie*
*Konstanz University, Germany*

## NOTES

1. "It is now evident that the world is chirally asymmetric at all scales, from the scale of elementary particles upward." (Hegstrom and Kondepudi 1990, p. 101)
2. Boccaccio, G., *Decameron*, circa 1350; quoted by Martindale and Baccheschi 1969, p. 9.
3. On the tactile quality of Giotto's paintings see Berenson 1971, pp. 62–65, 69–74.
4. Abraham (1934). Abraham produced photographs of faces consisting of two right halves and two left halves. The difference between the resulting two facial expressions and the original face is astounding.
5. Leonardo da Vinci, *Codex Atlanticus*, circa 1500, 141; quoted by Martindale and Baccheschi 1969, p. 10; cf. Richter 1952, pp. 226–227.
6. Hence Leonardo's conclusion that streams are the chief agents in shaping the earth's surface (Richter, 1952, p. 17).
7. Similarly, Newton remarked that the " . . . whole burden of philosophy seems to consist in this – from the phenomena of motions to investigate the forces of nature". This, in Post's view, is the business of scientists. (See Post 1971, p. 215, fn.1)
8. This is one of the principal theses of Meyerson's philosophy. According to Meyerson, causality is the law that rules the human mind and that law explicates itself by seeking identities or invariant aspects in the processes of change. "Scientific laws are the result and not the cause of the mind's natural tendency to seek causality in terms of identity." (Biagioli, 1988, p. 15). See Meyerson 1976.
9. "*Sucht den ruhenden Pol in der Erscheinungen Flucht.*" (Schiller 1943, p. 264 (*Elegie*, line 138)). Cf. Post 1975, p. 20.
10. For a critique of Mach's analysis, see Goe 1972. According to Goe, " . . . the negative verdict passed by Mach on the validity of the Archimedean proof of the lever principle has often been accepted in the literature uncritically." (*ibid.*, p. 330). Goe claims that

    > Mach misrepresents Archimedes . . . Under the guise of reproducing Archimedes' reasoning as it applies to the cases indicated only "in general outlines", Mach departs from it in an essential way, by eliminating all reference to centres of gravity and treating only of weights suspended from a lever, each at one point. (*ibid.*, pp. 340–341)

    However, Goe agrees that Mach is right in pointing out that the particular expression $P \times L$ for the disturbing factor cannot be deduced from the principles stated. (*ibid.*, p. 342) On his part, Mach informs the reader in the later editions of the *Mechanics* that a paper by Vailati came to his attention in which the author took issue with Mach's exposition and analysis. Vailati, the Italian mathematical philosopher, argues against Mach's treatment much along the lines that Goe has followed, that is, criticizing Mach's failure to see that Archimedes derives the lever principle indirectly via considerations about centres of gravity. Mach, however, does not deny that criticism; he stresses rather the philosophical import of his discussion:

    > . . . if the reader has derived some usefulness out of this discussion, I am not very particular about maintaining every word I have used. (Mach, 1974, p. 28)

    Mach's Archimedes may have been a strawman, but the lesson is clear.
11. Emphasis in the original. Quoted by van Fraassen 1989, p. 287.
12. "We cannot even state a theory of irreversible processes unless the time concept is at hand". (Bunge 1967, p. 97)
13. For other possibilities and their respective criticisms see Sklar 1986, p. 215. In this review, Sklar argues that

    > . . . although enormous advances have been made in understanding the explanatory ground on which the familiar statistical posits of statistical mechanics rests, the final understanding of the basic principle of temporal asymmetry still eludes us. (*ibid.*, p. 210)

14. For criticisms of Prigogine's suggestion see, e.g., Denbigh 1982, 1985; Earman 1986, p. 229ff.; Sklar 1986, pp. 222ff.; Verstraeten 1991.
15. Against Ideolog*ies*: 'Ideologies are either vicious or redundant.' (Post 1974, p. 3)
16. See the second motto.
17. $S^*$ is the core of $S$, and must be invariant under at least all the transformations under which the new theory $L$ is invariant.

## BIBLIOGRAPHY


Abraham, P. (1934), 'Une Figure, Deux Visages,' *Nouvelle Revue Française*, 42, pp. 409–429, 585–614.

Barzun, J. (1975), *Classic, Romantic, and Modern*, The University of Chicago Press, Chicago and London.

Berenson, B. (1971), *Italian Painters of the Renaissance*, World Publishing, New York.

Biagioli, M. (1988), 'Meyerson: Science and the "Irrational"', *Studies in History and Philosophy of Science*, 19, pp. 5–42.

Bunge, M. (1967), *Foundations of Physics*, Springer, Berlin.

Capek, M. (1951), 'The Doctrine of Necessity Re-examined', *The Review of Metaphysics*, 5, no.1, pp. 1–54.

Capek, M. (ed.) (1976), *The Concepts of Space and Time*, Boston Studies in the Philosophy of Science, Vol. XXII, Reidel, Dordrecht.

Cartwright, N. (1983), *How the Laws of Physics Lie*, Oxford University Press, Oxford.

Denbigh, K. (1982), 'A Review of Prigogine 1980', *The British Journal for the Philosophy of Science*, 33, pp. 325–329.

Denbigh, K. (1985), 'A Review of Prigogine and Stengers 1984', *The British Journal for the Philosophy of Science*, 36, pp. 352–354.

Dewey, J. (1968), *Experience and Nature*, Open Court, Lasalle, Illinois.

Earman, J. (1986), 'The Problem of Irreversibility', *Proceedings of the Philosophy of Science Association*, Vol. II, PSA, East Lansing, pp. 226–233.

Einstein, A. (1960), 'Field Theories, Old and New', *New York Times*, 3 February 1929, Readex Microprint Cor., New York.

Einstein, A. (1969), *Philosopher-Scientist*, The Library of Living Philosophers, Vol.VII, Open Court, La Salle, Illinois.

Field, G.C. (1925–1926), 'Ancient Philosophy and Modern Science', *Proc. Arist. Soc.*, 26, pp. 117–134.

Fraassen, Bas C. van. (1989), *Laws and Symmetry*, Clarendon Press, Oxford.

Galileo, G. (1974), *Dialogue Concerning the Two Chief World Systems*, S. Drake (trs.), University of California Press.

Goe, G. (1972), 'Archimedes' Theory of the Lever and Mach's Critique', *Studies in History and Philosophy of Science*, 2, pp. 329–345.

Grünbaum, A. (1974), *Philosophical Problems of Space and Time*, Boston Studies in the Philosophy of Science, Vol. XII, 2nd ed., Reidel, Dordrecht.

Hertz, H. (1956), *The Principles of Mechanics Presented in a New Form*, Dover, New York.

Houtappel, R.M.F., van Dam, H. and Wigner, E.P. (1965), 'The Conceptual Basis and Use of the Geometric Invariance Principles', *Review of Modern Physics*, 37, pp. 595–631.

Hegstrom, R.A. and Kondepudi, D.K. (1990), 'The Handedness of the Universe', *Scientific American*, January, pp. 98–105.

Horwich, P. (1987), *Asymmetries in Time*, MIT Press, Cambridge, Massachusetts.

Mach, E. (1974), *The Science of Mechanics*, Open Court, Lasalle, Illinois.

Martin, F.D. (1965), 'Spiritual Asymmetry in Portraiture', *The British Journal of Aesthetics*, 5, pp. 6–13.

Martindale, A. and Baccheschi, E. (eds.) (1969), *The Complete Paintings of Giotto*, Weidenfeld and Nicolson.
Meyerson, E. (1976), 'The Elimination of Time in Classical Science' in Capek (ed.), *op. cit.*, pp. 255–264.
Oertel, B. (1968), *Early Italian Painting to 1400*, Thames and Hudson, London.
Post, H.R. (1968), 'Atomism 1900', *Physics Education*, 3, pp. 225–232, 307–312.
Post, H.R. (1971), 'Correspondence, Invariance and Heuristics: In Praise of Conservative Induction', *Studies in History and Philosophy of Science*, 2, pp. 213–255.
Post, H.R. (1974), *Against Ideologies, Inaugural Lecture*, 28 November 1974, Chelsea College, University of London.
Post, H.R. (1975), 'The Problem of Atomism', *British Journal for the Philosophy of Science*, 26, pp. 19–26.
Prigogine, I. (1980), *From Being to Becoming*, Freeman, San Francisco.
Prigogine, I. and Stengers, I. (1984), *Order out of Chaos*, Bantam Books, New York.
Rae, A.I.M. (1986), *Quantum Physics: Illusion or Reality?*, Cambridge University Press, Cambridge.
Reichenbach, H. (1956), *The Direction of Time*, University of California Press, Berkeley.
Richter, I.A. (ed.) (1952), *Selections From The Notebooks of Leonardo da Vinci*, Oxford University Press, Oxford.
Savitt, S.F. (1991), 'Critical Notice: a review of Horwich's book, *Asymmetries in Time*', *The Canadian Journal of Philosophy*, 21, pp. 399–417.
Schiller, F. (1943), *Schillers Werke*, Nationalausgabe, 1. Band, Gedichte 1776–1799, Weimar.
Sklar, L. (1986), 'The Elusive Object of Desire: In Pursuit of the Kinetic Equations and the Second Law', *Proceedings of the Philosophy of Science Association*, Vol. II, P.S.A., East Lansing, pp. 209–225.
Smart, A. (1978), *The Dawn of Italian Painting 1250–1400*, Phaidon, Oxford.
Vasari, G. (1568), *Le Vite*, in Martindale and Baccheschi (1969) p. 10.
Venturi, A. (1907), *Storia dell'arte italiana*, in Martindale and Baccheschi *op. cit.*, pp. 12–13.
Verstraeten, G. (1991), 'Some Critical remarks Concerning Prigogine's Conception of Temporal Irreversibility', *Philosophy of Science*, 58, pp. 639–654.
Weyl, H. (1973), *Symmetry*, Princeton University Press, Princeton, New Jersey.
Wheeler, J.A. and Zurek, W.H. (eds.) (1983), *Quantum Theory and Measurement*, Princeton University Press, Princeton, New Jersey.
Whyte, L.L. (1956), 'One-way Processes in Physics and Biophysics', *The British Journal for the Philosophy of Science*, 6, pp. 107–121.
Zeh, H.D. (1989), *The Physical Basis of the Direction of Time*, Springer Verlag, Berlin.

ALAN CHALMERS

# Galilean Relativity and Galileo's Relativity

*Look for a flaw or footprint in the existing theory. Treat any flaw as a footprint, and build your . . . theory up from it.*

H.R. Post

Many of the students who studied History and Philosophy of Science under Heinz Post, even those who did not continue their studies professionally afterwards, would claim that the main benefit they derived from his teaching was an increased liking for and appreciation of physics. His success in that regard indicates that contemporary physics can best be illustrated by the complexities of accurate history of science rather than the Orwellian rewrites and mythical simplifications that continue to adorn science texts and popular histories. In order to capture the distinctive epistemological status of physics it is not necessary to characterise its significant advances as coming about at the drop of a hat or two from the leaning tower of Pisa. More accurate accounts of the experimentation and argumentation constituting physics serve the task better. History of Science need not be rated X[1]. Encouraged by these reflections, I would like to think that the following remarks on the kind of relativity involved in Galileo's physics will serve to illuminate contemporary physics as well as its seventeenth century antecedents.

In his article 'Correspondence, Invariance and Heuristics' (Post 1971) Heinz Post stresses the extent to which new theories grow out of problems in the old, and attempts to characterise typical relations that hold between a theory and its successor. He attempts to exploit those relations to construct heuristic guidelines for the construction of successful theories. In the following I will argue that key elements of Galileo's physics did indeed grow out of problems in Aristotle's physics. However, a consequence of this debt that Galileo owed to Aristotle is that Galileo's physics must be situated closer to Aristotle on the path from Aristotle to Newton than is typically appreciated. In particular, Galilean invariance is not to be found in Galileo's writings, nor could it possibly have occurred there.



*S. French and H. Kamminga (eds.), Correspondence, Invariance and Heuristics*, 189–205.

## I. GALILEAN RELATIVITY

There is not much room for confusion about what the term 'Galilean relativity' has come to denote in contemporary physics. It denotes the kind of relativity implied by Newtonian mechanics. If an inertial reference frame is defined as a reference frame in which Newton's law of inertia holds, then Galilean relativity involves the assertion that the behaviour of a system is unaffected by its position, orientation or uniform velocity with respect to such an inertial frame. This is reflected in the fact that the form of the laws of Newtonian mechanics is invariant with respect to a linear or angular displacement of the co-ordinate system and to a translation of the co-ordinate system at uniform velocity in a straight line. This set of transformations has become known as the Galilean transformations. In this article we shall not be concerned with displacements of frames of reference but only with their linear translation at uniform velocity.

The essentials of Galilean invariance are frequently brought out by the following kind of thought experiment. We imagine a laboratory which is considered to be at rest in an inertial frame and a second laboratory moving at uniform velocity in a straight line relative to it. We imagine further that the laboratories are isolated from each other and from any other external system. That is, the laboratories have no windows and provide no other means of access to one another. Then, according to Galilean relativity, no mechanical experiment can be performed that can serve to reveal in which of the two laboratories it was conducted. Motion in a straight line at constant velocity has no observable consequences in itself. Uniform rectilinear motion of a system only has observable consequences relative to some other system. Newton himself presented a statement of Galilean relativity as Corollary V to his laws of motion.

> The motions of bodies included in a given space are the same among themselves, whether the space is at rest, or moves uniformly forwards in a straight line without any circular motion. (Cajori 1962, p. 20)[2]

The use of the terms 'Galilean relativity' and 'Galilean transformation' as I have described them is widespread in contemporary physics. The terminology strongly suggests of course, that it was Galileo who first formulated and defended Galilean relativity, and claims to that effect are often made explicit, not only by physicists but also by reputable historians of science. Thus Landau and Lifshitz, whom Professor Post respected enough in this teaching days to criticise whenever he could, write:

> Thus there is not one but an infinity of inertial reference frames moving, relative to one another, uniformly in a straight line. In all these frames the properties of space and time are the same, and the laws of mechanics are the same. This constitutes *Galileo's relativity principle*, one of the most important principles of mechanics. (Landau and Lifshitz 1969, p. 6; italics in original)

Ludovico Geymonat, a respected biographer of Galileo, writes in similar vein:

> Galileo put forth what is now usually called the 'Galilean principle of relativity'. This states that on the basis of mechanical experiments performed within a system, it is impossible to decide whether the system is at rest or in uniform rectilinear motion. (Geymonat 1965, p. 119)

I shall argue that Galileo neither formulated nor defended Galilean relativity and that he could not possibly have done so given the means at his disposal. Before I embark on that task I must admit that the attribution of Galilean relativity to Galileo is not without some basis in his writings. One of the most striking passages is the following, from the *Dialogue Concerning the Two Chief World Systems*, which I quote at length, and which Geymonat also quotes in support of his attribution of Galilean relativity to Galileo:

> Shut yourself up with some friend in the main cabin below decks in some large ship, and have with you there some flies, butterflies, and other flying animals. Have a large bowl of water with some fish in it; hang up a bottle that empties drop by drop into a vessel beneath it. With the ship standing still, observe carefully how the little animals fly with equal speed to all sides of the cabin. The fish swim indifferently in all directions; the drops fall into the vessel beneath; and, in throwing something to your friend, you need throw it no more strongly in one direction than another, the distances being equal; jumping with your feet together, you pass equal spaces in every direction. When you have observed all these things carefully (though there is no doubt that when the ship is standing still everything will happen in this way), have the ship proceed with any speed you like, so long as the motion is uniform and not fluctuating this way and that. You will discover not the least change in all the effects named, nor could you tell from any of them whether the ship was moving or standing still. In jumping, you will pass on the floor the same space as before, nor will you make larger jumps towards the stern than towards the prow even though the ship is moving quite rapidly, despite the fact that during the time that you are in the air the floor under you will be going in a direction opposite to your jump. In throwing something to your companion, you will need no more force to get it to him whether he is in the direction of the bow or the stern, with yourself situated opposite. The droplets will fall as before into the vessel beneath without dropping towards the stern, although while the drops are in the air the ship runs many spans. The fish in their water will swim toward the front of their bowl with no more effort than toward the back, and will go with equal ease to bait placed anywhere around the edges of the bowl. Finally the butterflies and flies continue their flights indifferently toward every side, nor will it ever happen that they are concentrated toward the stern, as if tired out from keeping up with the course of the ship, from which they have been separated for long intervals by

keeping themselves in the air. And if smoke is made by burning some incense, it will be seen going up in the form of a little cloud remaining still and moving no more to one side than the other. (Galileo 1967, pp. 186–187)

This can certainly be read as a strong statement of Galilean relativity, illustrated by a thought experiment akin to the one described above. Other, shorter but equally forceful, statements can be found in the *Dialogue* to reinforce the case. Just one of several possible examples reads as follows:

> Motion, in so far as it is and acts as motion, to that extent exists relatively to things that lack it; and among other things which all share equally in any motion, it does not act and is as if it did not exist. (*ibid.*, p. 116; for similar passages, see pp. 171 and 248)

In spite of the apparently convincing case that can be mounted by a selection of quotations such as these, I propose to argue that those readings which take them to be expressions of Galilean relativity are misreadings.

## II. GALILEO'S RELATIVITY I: PERCEPTUAL RELATIVITY

One version of relativity that is formulated and defended by Galileo is what I shall refer to as perceptual relativity. According to this principle, it is not possible for humans to detect with the senses motions in which they themselves participate. Galileo uses this principle in the *Dialogue* to argue that the motion of the earth will not be detectable by terrestrial observations since observers will share that motion. Galileo illustrates his case, for example, by comparing observations on a moving ship with those on a moving earth. A passenger on a ship who gazes at the sail yard will not thereby detect its motion because it does not move relative to the observer's eyes.

> And this comes about because the motion which the ship confers upon the sail yard, it confers also upon you and upon your eyes, so that you need not move them a bit in order to gaze at the top of the sail yard, which consequently feels motionless to you. (*ibid.*, p. 249)

If a stone is dropped from the mast of a uniformly moving ship the vertical fall of the stone is perceived because it is not shared by the observer, whereas the horizontal motion, possessed by ship, stone and observer, remains unperceived. Galileo then likens this situation to that in which a stone is dropped from a tower on a moving earth.

> Now transfer this argument to the whirling of the earth and to the rock placed on top of a tower, whose motion you cannot discern because in common with the rock you possess from the earth that motion which is required for following the tower; you do not need to move your eyes. Next, if you add to the rock a downward motion which is peculiar to it

and not shared by you, and which is mixed with this circular motion, the circular portion of the motion which is common to the stone and the eye continues to be imperceptible. The straight motion alone is sensible, for to follow that you must move your eyes downward. (*ibid.*, p. 250)

Galileo certainly illustrates and defends perceptual relativity in a forceful and vivid way[3]. However, it must be noted that perceptual relativity was by no means novel in Galileo's time. It appears in Euclid and in medieval optics and was used by Buridan and Oresme in the fourteenth century and by Copernicus in the sixteenth to explain why a steady motion of the earth would go undetected by the senses[4]. It is also important to note that perceptual relativity is much weaker than Galilean relativity. This is evident as soon as we distinguish two aspects of the tower experiment that figures so prominently in Galileo's defence of the diurnal rotation of the earth. The first aspect concerns the question of whether the horizontal motion of the falling stone would be observed by a terrestrial observer on the assumption that the stone stays with the tower. Galileo employed perceptual relativity to argue that it would not. The second aspect concerns the question of whether the stone will indeed stay with the tower or get left behind. If it were to get left behind then, of course, the tower experiment would be capable of providing positive evidence that the earth moves. Galileo argues that the stone will indeed stay with the tower just as, he argues further, a stone dropped from the mast of a uniformly moving ship will remain alongside the mast. It is Galileo's discussion of this matter, together with his espousal of perceptual relativity, that has provided scope for his interpreters to attribute Galilean relativity to him.

## III. GALILEO'S RELATIVITY II: INERTIAL MOTION

To defend the claim that uniform motion cannot be detected by dropping stones or by similar experiments Galileo introduces what can be regarded as an example of inertial motion, if we understand such a motion to be one that persists indefinitely without a cause, whether that cause be internal to a moving object, as Galileo, following Aristotle, imagined weight to be, or an external push or pull[5]. It is vital for my discussion of the distinction between Galileo's relativity and Galilean relativity to appreciate the difference between the inertial motions that Galileo invoked and those characterised by Newton's first law of motion.

It is quite inappropriate to attribute a *principle* of inertia of any generality to Galileo, and this becomes quite clear once we examine the contexts in which examples of inertial motion occur in Galileo's writing and the ways in which he argues for them. Galileo's chief example of a motion that persists without a cause is that of a heavy object moving in a circle with the earth as centre. Galileo's first published discussion of this motion, in his *Letters on Sunspots* (1613) is also, I think, the best one. It reads as follows:

> For I seem to have observed that physical bodies have physical inclination to some motion (as heavy bodies downward), which motion is exercised by them through an intrinsic property and without need of a particular external mover, whenever they are not impeded by some obstacle. And to some other motion they have a repugnance (as the same heavy bodies to motion upward), and therefore they neither move in that manner unless thrown violently by an external mover. Finally, to some movements they are indifferent, as are these same heavy bodies to horizontal motion, to which they have neither inclination (since it is not towards the centre of the earth) nor repugnance (since it does not carry them away from that centre). And therefore, all external impediments removed, a heavy body on a spherical surface concentric with the earth will be indifferent to rest and to movements toward any part of the horizon. And it will maintain itself in that state in which it has been placed; that is, if placed in a state of rest, it will conserve that; and if placed in movement toward the west (for example), it will maintain itself in that movement. Thus a ship, for instance, having once received some impetus through the tranquil sea, would move continually around our globe without ever stopping; and placed at rest it would perpetually remain at rest, if in the first case all external impediments could be removed, and in the second case no external cause of motion were added. (Drake 1957, pp. 113–114)

The claims made here, and the way in which they are argued for, appear in much the same form both in the *Dialogue* and in *Two New Sciences* (see Galileo 1967, pp. 145ff and Galileo 1974, p. 243). They reveal, I suggest, that Galileo had traversed less of the path from Aristotle to Newton than is typically appreciated. A review of some of Aristotle's expressed views on nature and natural motion will prepare the way for a substantiation of that assertion.

Aristotle's idea of nature, and of natural motions, as he introduced them in Book II of the *Physics*, involved two inter-related notions. The first is between naturally occurring and artificially produced things. Plants and animals are natural. Material objects, such as stones, are also natural insofar as they are material objects. A bed is a natural object insofar as it is wooden, and hence heavy, but not insofar as it is constructed in the form of a bed. The second notion concerns natural, as opposed to forced or violent changes. Natural objects,

> have within themselves a principle of movement (or change) and rest – in some cases local only, in others quantitative, as in growth and shrinkage, and in others again qualitative, in the way of modification. (*Physics*, II, 1, 192b, 10–15)

The growth and reproduction of plants and animals and the fall of heavy and the rise of light objects are all natural insofar as the principal cause of those changes resides in the nature of the changing items.

These two aspects of Aristotle's conception of nature were for him intimately related. A naturally occurring thing has a natural motion characteristic of it being a natural thing of that kind whereas a manufactured object does not have a natural motion characteristic of it being a manufactured object of that kind. Thus a bed will have a natural tendency to fall not insofar as it is a bed, but insofar as it is wooden. Further, if a bed made of freshly hewn wood were to be planted, and were it to take root, a tree would grow, not a bed.

It is not difficult to think of examples that cut across these two aspects of nature and natural change that Aristotle presents as necessarily linked. A watch is certainly a manufactured object, but the elasticity of the spring is natural to it, so that once wound up the watch will run of its own accord. To the possible objection that the fact that the watch needs to be wound up renders its motion forced rather than natural, I reply that even heavy objects need to be raised, and impediments removed, before they will fall, and this fact is not sufficient to disqualify falling as a natural motion for Aristotle (see *Physics*, VII, 4, 255b). In order to appreciate why examples such as the running of a watch are out of place in a discussion of nature and natural motions from an Aristotelian point of view, we must consider a third aspect of Aristotle's position. This concerns the extent to which he regarded nature as a whole, as an ordered, self-perpetuating system. As Ross put it,

> nature is here to be thought of . . . as a collective term, for the natures of all 'natural bodies' working harmoniously together. (Ross 1945, p. 68)

Monstrous births, for example, count as natural according to the two criteria discussed above, but there is another strong sense in which they are not natural. They happen 'by nature' but not 'according to nature' where, in the latter case, nature is considered as an orderly system and natural motions or changes are those contributing to or serving to maintain that order.

The third, teleological, aspect of Aristotle's conception of nature is centrally involved in his understanding of the natural motion of material objects. In order to specify the directions of these motions it is necessary to invoke the notion of natural place. The four elements have a natural place in the universe conceived of as an ordered whole, with earth at the centre, fire at the periphery of the terrestrial region just beneath the moon's orbit and water and air in between. The natural place of a material object will be a location between the centre of the universe, which is also the centre of the earth, and the extremity of the terrestrial region, depending on the relative amounts of the four elements it contains. Ideally all objects will occupy their natural place, and they tend to move naturally straight up or down toward their natural places whenever they are distant from them. The natural motions of material objects contribute to, and help to maintain, the order of the universe as a whole, and cannot be specified in the Aristotelian system without invoking the ordered whole. Although Aristotle often talks of the teleology of nature in an anthropomorphic way (e.g. "nature does nothing in vain") it is best understood as what Ross referred to as "a *de facto* teleology" (*ibid.*, p. 79). Nature is, as a matter of fact, ordered and natural motions serve to maintain, or where necessary restore, that order.

Galileo certainly departed from Aristotle's conception of nature and of natural motion in significant respects. Some of those departures were necessitated by Galileo's adoption of Copernican astronomy. Nevertheless, Galileo retained key features of the Aristotelian conception, to an extent that a case can be made for regarding his views as a modification and extension of it rather than involving its wholescale rejection. Galileo retained and utilised the distinction between natural and unnatural motions and he retained the idea that the role of natural motions is to retain or restore the order of the Universe.

For Galileo, the downward fall of heavy objects was natural, in the Aristotelian sense, insofar as the principal cause of the motion resides in the nature of the falling object. As is explained by Galileo in the passage from the *Letters on Sunspots* quoted above, the downwards fall of heavy bodies "is exercised by them through an intrinsic property and without need of a particular external mover". Not only would such a statement be at home in Aristotle's *Physics* or *De Caelo* but one can see how the way in which Galileo exploited it to arrive at his conception of horizontal inertia arises in a natural way when certain ambiguities or problems in Aristotle's writings on natural motion are confronted. What is more, some scholastic writers, when they did confront these problems, were moved to adopt a position that comes close to Galileo's own, and Galileo was familiar with their efforts in this regard. It is worth documenting these points in a little detail.

It is perfectly clear from Aristotle's discussion in *De Caelo* and elsewhere that the natural motion of heavy objects is vertically downwards, towards the centre of the earth. However, there is an ambiguity concerning which motions of such bodies are to count as unnatural. In Book I, unnatural motion is presented as the opposite to natural motion. This is quite explicit, for example, in the passage that reads,

> if it is the downward movement which is unnatural, the upward motion will be the natural; and if it is the upward which is unnatural the downward will be natural. (*De Caelo*, I, 3, 270A, 1–2)

Thus the unnatural motion of a heavy object, and the only unnatural motion of such an object, is vertically upwards. This narrow construal of unnatural motion is necessary for some of the details of Aristotle's *a priori* argument for the structure of his earth-centred universe. Elsewhere, unnatural motion is construed by Aristotle in a broader sense to mean any motion that is not natural, so that any motion of a heavy object other than its natural vertical fall will be unnatural.

> Now an unnatural movement presupposes a natural movement which it contravenes, and which, however many the unnatural movements, is always one. For naturally a thing moves in one way, while its unnatural movements are manifold. (*De Caelo*, III, 2, 24–26)

A reasonable attempt to remove the contradiction would be to understand a motion of a heavy object to be unnatural insofar as its distance from the centre of the earth increases and natural insofar as it decreases.[6] But this raises the question of how to characterise the motion of a heavy object that neither recedes from nor approaches the centre of the earth. It would appear to be neither natural nor unnatural. Our consideration of Aristotle's own texts has brought us to the brink of Galileo's horizontal motion, that is neither natural nor forced, although Galileo was not the first to travel that path, as we shall see.

A second path leading to the singular character of horizontal motion can also be traced to Aristotle's own writing. In *Meteorologica* (341a, 1ff)[7] Aristotle claimed that the element fire, and also the upper reaches of the air, when they have risen high to their natural places, participate to some extent in the circular rotation of the heavens. If this is the case, is that circular motion of fire natural or unnatural? The question was raised by Albertus Magnus in the thirteenth century and pursued further by Martius Vitelleschi and Ludovicus Rugerius at the Collegio Romano late in the sixteenth century. William Wallace's study of Galileo's early notebooks has shown that Galileo was familiar with, and drew heavily on, those sources (Wallace 1981, pp. 270–271, 283–284, 334–335). In the context of the problems posed by the circular motion of fire, Vitelleschi suggested that it is only the exact opposite of natural motion, motion vertically downwards for fire and vertically upwards for earth, that is strictly speaking unnatural or violent, in keeping with Aristotle's expressed view in *De Caelo*, Book I. He portrayed horizontal motion as intermediate between natural and unnatural and referred to it as "neutral". He remarked that the Aristotelian view that no unnatural motion can be perpetual should be applied only to unnatural motion in the strict sense. This enabled Vitelleschi to accommodate the perpetual character of the neutral motion of fire in the upper regions of the terrestrial region. Once again, we are brought to the brink of Galileo's horizontal inertia by a consideration of, and an attempt to extend and improve, Aristotle's text. It is significant here that when the special character of horizontal motion first made its appearance in Galileo's writings, that motion was referred to by Galileo, in a marginal note, as "neutral" (Galileo 1960, p. 67n).

In Galileo's first public exposition of horizontal inertia, quoted above, Galileo referred to horizontal motion as "indifferent" rather than "neutral". But the fact that the indifference is with respect to natural downwards and violent upwards motion demonstrates the extent to which Galileo has retained those categories. We do not have to look beyond Aristotle to find versions of Galileo's expressed views that falling heavy bodies accelerate downwards, that they do so because of their nature so that no external driving force is necessary and that heavy bodies must be forced to move upwards by an external mover and that they decelerate once that force is removed. As for the indifferent character of horizontal motion, this too emerges in a natural way from Aristotle's writings, and did emerge in the work of commentators on Aristotle with quite different concerns than those of Galileo.

Turning now to that important aspect of Aristotle's conception of natural motions that saw them acting 'in accordance with nature' and thereby contributing to the order of nature as a whole, we find that that conception plays a role in Galileo's deliberations too. Early in the *Dialogue* we find Galileo, through Salviati, agreeing with Aristotle that the world "is of necessity most orderly, having its parts disposed in the highest and most perfect order amongst themselves", which assumption, Salviati tells Simplicio, "I do not believe to be denied either by you or by anyone else" (Galileo 1967, p. 19). Galileo proceeds, after a digression concerning the motion of bodies down inclined planes, to explain how the circulation of the stars and planets and the fall of heavy bodies to the earth serve to maintain or restore the ideal order.

> I therefore conclude that only circular motion can naturally suit bodies which are integral parts of this universe as constituted in the best arrangement, and that the most which can be said for straight motion is that it is assigned by nature to its bodies (and their parts) whenever these are to be found outside their proper places, arranged badly, and are therefore in need of being restored to their natural state by the shortest path. (*ibid.*, p. 32)

Here Galileo appropriates the idea that appears in Copernicus' *De Revolutionibus*[8] according to which objects fall to the ground in order to restore the earth to its ideally spherical shape, and he applies the same consideration to other heavenly bodies such as the sun and moon.

> Now just as all the parts of the earth mutually cooperate to form its whole, from which it follows that they have equal tendencies to come together in order to unite in the best possible way and adapt themselves by taking a spherical shape why may we not believe that the sun, moon and other world bodies are also round in shape merely by a concordant instinct and natural tendency of all their component parts? If at any time one of these parts were forcibly separated from the whole, is it not reasonable to believe that it would return spontaneously and by natural tendency? (Galileo *ibid.*, pp. 33–34)

Galileo's account of why objects fall (and why the heavenly bodies as a whole circulate) involves a *de facto* teleology no less than Aristotle's does, albeit a different one.

Having noted crucial respects in which Galileo can be said to have incorporated Aristotelian notions into his physics, I do not wish to overstretch the point. I do not wish to emulate the learned doctor, rightly scorned by Galileo, who claimed that the invention of the telescope should be accredited to Aristotle, citing in his defence Aristotle's explanation of why stars can be seen in the daytime from the bottom of a well (see Galileo 1967, p. 109 and *De Generatione Animalium*, V, 1, 780b, 21). Galileo's conception of nature and of natural motions differed from Aristotle's in significant respects, and some of those differences marked Galilean innovations and important steps towards the Newtonian view. Aristotle's conception of the natural motion of terrestrial bodies involved his notion of natural places, which in turn involved the four elements. Galileo dispensed with this. For Galileo all bodies, whatever their composition, have a natural tendency to fall and do so unless they are forced upwards by a denser medium in which they are immersed. What is more, Galileo specified that all bodies, whatever their weight or density, fall with the same uniform acceleration. Galileo also abandoned the Aristotelian idea of the centre of the earth coinciding with the centre of the Universe and constituting the point towards or away from which the natural motions of terrestrial objects are directed and on which the natural motions of celestial objects are centered. As we have seen, Galileo adopted an alternative teleology to accommodate that certain kind of acceleration that nature does employ for descending heavy things[9]. The reason I stress the Aristotelian features that remained in Galileo's physics is to help me stress the extent to which Galileo's conception of inertia, and hence his relativity, fell short of, and differed from the Newtonian versions.

Galileo certainly held the view that unimpeded horizontal motions are inertial, in the sense that they persist indefinitely without an external cause. The reason why they are inertial, for Galileo, is the fact that, in this special case, the centre of the earth is neither approached nor receded from. It is a special case, and for that reason it is quite inappropriate to attribute anything so grandiose as a *principle* of circular inertia to Galileo. This point is strengthened by the fact that there is another example of inertial motion to be found in Galileo's physics, which is a natural consequence of his views on natural fall and which also involves a neutralisation of that tendency, but in this case the inertial motion is linear. The example, which occurs in a text on percussion which Galileo probably intended as a fifth 'Day' to be added to the four that constitute *Two New Sciences*, involves two equal weights hanging vertically from each end of a rope that passes over a frictionless pulley. In this case the tendency of one of the weights to fall is exactly countered by that of the other. Consequently, if one of the weights is set in downwards motion, with a corresponding upwards motion of the other, then the motion persists without the action of a persisting cause. Galileo himself likens the situation to that of a motion on a horizontal plane (Galileo 1974, pp. 296–297). For Galileo, inertial motions can result in special circumstances as a consequence of the natural tendency of objects to fall. There is no general principle of inertia in Galileo's physics, circular or otherwise.

The limited character of Galileo's conception of inertial motion notwithstanding, it was sufficient for him to be able to accommodate the tower argument, stones dropped form the mast of a moving ship and the like. A stone will drop at the foot of the tower of a uniformly rotating earth and at the foot of the mast of a uniformly moving ship because any horizontal velocity that it has will be conserved. Galileo's discussion of horizontal inertia in the *Dialogue* occurs in the context of the tower experiment and culminates in his description of the null effects of experiments to detect the uniform motion of a ship, reproduced in Section I. This leads me to the conclusion that Galileo's relativity consists of perceptual relativity plus horizontal circular inertia and nothing more. The latter ensures that horizontal motions that are shared will remain shared in the absence of countervailing causes, whilst the former ensures that those motions that are shared will remain unperceived.

Near the beginning of Day 2 of the *Dialogue* Galileo has Salviati set up the discussion of the possible effects of a rotating earth as follows:

> Then let the beginning of our reflections be the consideration that whatever motion comes to be attributed to the earth must necessarily remain imperceptible to us and as if nonexistent, so long as we look at terrestrial objects; for as inhabitants of the earth, we consequently participate in the same motion. (Galileo 1967, p. 114)

That "reflection" is not completely borne out by the ensuing discussion, as Galileo was aware and which serves to emphasise the fact that Galileo's relativity should not be interpreted in too general a sense. Quite apart from his theory of the tides (which had its faults, to be sure, but inconsistency with his conception of relativity was not one of them), there are terrestrial experiments by means of which the rotation of the earth could in principle be detected on the assumption that Galileo's physics is correct. Two points on the earth's surface at differing distances from the equator will not share the same velocity by virtue of the earth's rotation since they will describe circles of differing radii in equal times. This opens the possibility of an observer at one of the points detecting the motion of the other. As Galileo points out, in principle cannon shots directed at targets due North should fall East of the target because the eastward motion of the target will be somewhat less than that of the cannon. Of course, in practice "this difference is insensible because of the small distance from the cannon to the mark", bearing in mind that repeated shots at a target "vary by yard" due to unavoidable inaccuracies (*ibid.*, pp. 179 and 182).

## IV. THE LACK OF GENERALITY OF GALILEO'S PHYSICS

One way to appreciate the extent to which Galileo's relativity lacked the generality and fell short of Galilean relativity is to pose the question of what would happen to Galileo's experiments below decks on a moving ship if the ship were to move uniformly in a straight line tangential to the earth's surface. According to Galileo's physics, and in fact, the outcome of Galileo's experiment will be such as to reveal the motion of the ship. The further the ship moves the greater will the downwards direction towards the centre of the earth diverge from a line parallel to the walls of the cabin. Consequently, drops from the bottle will eventually miss the vessel beneath and equal jumps will no longer take one equal distances in any direction because jumps towards the bow will become increasingly uphill as the ship departs ever further from its starting point on the earth's surface.

A modern reader will be inclined to dismiss my thought experiment as irrelevant to a discussion of relativity since, because of the earth's attraction, the ship is not an isolated system in the sense required. The motion of the ship relative to the earth's gravitational field can be detected within the cabin below decks. But Galileo could not possibly have made use of a thought experiment involving a system isolated from gravitational effects. Terrestrial objects tend to fall, for Galileo, because of their nature. Remove that tendency and one is not discussing the motion of terrestrial objects at all. Galileo is far from considering gravity as an extraneous effect due to the earth's attraction as is evident from his scorn for the similar idea of the tides being caused by the moon's attraction. Koyré was right to insist that Galileo was "unable to abstract from gravity" (Koyré 1978, p. 187) provided we interpret that remark, not as a psychological limitation on Galileo's part, but as referring to a basic feature of his physics. As we have seen, the key premise of his physics was that terrestrial objects have a natural tendency to fall, so that their natural downwards motion is accelerated and any upwards motion is retarded. As a natural extension of this we have the indifference to, and hence conservation of, horizontal motion. Take away gravity and none of these claims are stateable. Take away gravity and Galileo has no physics left, and certainly no materials from which to fashion any kind of relativity principle that goes beyond perceptual relativity.

The extent to which Galileo's physics depended on his notion of the natural fall of terrestrial objects is evident from the paucity of Galileo's attempt to offer an explanation of planetary motions. It is clear that as soon as we turn to the motion of the earth or a planet as a whole, the idea that the parts of the earth or a planet tend to return to the whole to restore it to a spherical form is irrelevant. When Galileo considers such motions early in the *Dialogue* he makes no attempt to exploit his own science of motion. It was clearly of no help. Rather, he invokes the idea of a perfectly ordered universe and argues that the celestial bodies, as wholes, must move in circles because, "only circular motion can naturally suit bodies which are integral parts of the universe as constituted in the best arrangement" (Galileo 1967, p. 32).

A comparison of Newton's mechanics with Galileo's serves vividly to highlight the high degree of generality of the former. Newton's laws refer to bodies in general, but to no bodies in particular. *A body* moving uniformly in a straight line free of forces will continue to do so indefinitely, whilst *any two bodies* will attract each other with a force proportional to the product of their masses and the inverse of the square of their separation. By contrast, Galileo's mechanics is applicable to terrestrial bodies in the vicinity of the earth and its principles cannot be stated without reference to a specific body, namely, the earth. Further, whilst Galileo's mechanics functions in a region centered on the earth, Newton's laws treat bodies as contingencies in an infinite featureless space. The straight lines referred to in the first law are situated in such a space.

The modern reader will find no difficulty in appreciating the thought experiment referred to in Section I involving laboratories isolated in space. But it is important to stress that locating systems in space in this kind of way is unproblematic, and even natural, only after something like the Newtonian view of the world has become commonplace. Once we accept the infinite, featureless space figuring in that view, Newton's first law, for instance, becomes not only unproblematic but virtually tautologous. For, after all, why should the uniform motion of an object moving freely in space change? The fact that acceptance of and familiarity with something like Newton's world view is a precondition for the plausibility and intelligibility of the law of inertia is brought out by the fact that predecessors of Newton did contemplate analogues of that law only to find them absurd. For instance, Aristotle, in his *Physics*, argued against the notion of a void space by pointing to some of the unacceptable consequences of such a postulate. One of those consequences he spelt out as follows:

> Further, no one could say why a thing once set in motion should stop anywhere; for why should it stop *here* rather than *here*? So that a thing will either be at rest or must be moved *ad infinitum*, unless something more powerful get in its way. (*Physics*, IV, 215a, 19–23)

That is, if there is a void space then Newton's first law holds. But Newton's first law is absurd. Therefore, space is not infinite. Similar statements occur in later authors. Consider, for example, the following discussion of the matter:

> Besides, straight motion being by nature infinite (because a straight line is infinite and indeterminate), it is impossible that anything should have by nature the principle of moving in a straight line; or, in other words, towards a place where it is impossible to arrive, there being no finite end. For nature, as Aristotle well says himself, never undertakes to do that which cannot be done, nor endeavours to move whither it is impossible to arrive.

I think that this amounts to the claim that Newton's first law is absurd. And it suits the argument of my paper very well to point out that this passage is due to Galileo himself; not a young Galileo but the mature Galileo of the *Dialogue*, whilst it is uttered, not by Simplicio the Aristotelian, but by Salviati, who can be taken to be expressing Galileo's own view here (1967, p. 19).

The fact that a horizontal line over a limited distance on the earth's surface differs very little from a straight line makes it possible to minimise the difference between Galileo's and Newtonian inertia from a practical point of view. After all, Galileo was able to demonstrate the parabolic trajectory of a projectile by assuming the horizontal to be straight. The following passage, due to E.J. Dijksterhuis, is an example of how the difference between Galileo's horizontal, and hence circular, inertia and Newton's linear inertia can be played down.

> The situation is thus as follows: according to the Galilean law of inertia, proper, a particle that is free from external influences (note that gravity is

> not included among them) perseveres in a circular motion having the centre of the earth for its centre. Over short distances this motion is considered rectilinear; subsequently the limitation to short distances is forgotten, and it is said that the particle would continue its rectilinear motion on a horizontal plane surface if no external factors interfered. Thus what might be called the circular view of inertia of Galileo gradually developed into the conception that was formulated in the first law by Newton. (Dijksterhuis 1961, p. 352)

I hope that my discussion has demonstrated that, in spite of the fact that in some circumstances the difference between Galilean and Newtonian inertia is of no practical consequence, there is a very large conceptual gap between the two. Moreover, it is precisely that gap that accounts for the corresponding gap between Galileo's and Galilean relativity.

Heinz Post intended his study of the relationship between theories and their successors to apply to science "of the last three hundred years" (Post *op. cit.*, p. 250) and so Galileo's physics lies outside its range. Because the aims of Aristotelian theory differed from those of modern science, and because the scientific revolution in which Galileo participated saw a shift from pre-science to science rather than a shift from one scientific theory to another, we should not try too hard to fit Galileo's innovations into Post's scheme. Nevertheless, several of the themes stressed by Post are evident in the story I have told above. Galileo was led to his new physics of the motion of heavy bodies at least in part by starting with the Aristotelian view and attempting to repair its deficiencies in the area of the problems with which he was concerned. Galileo discovered a footprint of his new theory in the old, but problematic, Aristotelian orthodoxy. Insofar as he was prepared to work within Aristotelian physics and adapt it to his purposes, rather than build an entirely new framework, Galileo was conservative, more so than Descartes, for example, who did construct a new theoretical framework. It is in keeping with Post's thesis that it was conservative Galileo, not revolutionary Descartes, that met with significant scientific success. Galileo's physics had its own limitations and flaws. Galileo's successors were able to see in some of them the footprints of Galilean relativity, whilst many contemporary scientists and historians have come to mistake the footprint for the real thing.

ACKNOWLEDGEMENT

I am grateful to Dr. M. Shortland for some helpful criticisms of an early draft of this article.

*Dept. of History and Philosophy of Science,*
*University of Sydney,*
*Sydney, Australia*

## NOTES

1. Stephen Brush discusses the problematic tension between actual history of science and the accounts typically presented to scientists in 'Should the History of Science be Rated X' (Brush 1974).
2. Strictly speaking, Galilean relativity only follows from Newton's laws if we exclude forces that themselves vary from one inertial frame to another.
3. For further evidence of this, see, for example, his striking example of an artist drawing a sketch on board a moving ship (Galileo 1967, pp. 172–3), and his extension of the discussion to the sense of touch (*ibid.*, 254–5).
4. For example, according to Copernicus,

   > every observed change of place is caused by a motion of either the observed object or the observer, or of course, by an unequal displacement of each. For when things move with equal speed in the same direction, the motion is not perceived, as between the observed object and the observer, I mean. It is the earth, however, from which the celestial ballet is beheld in its repeated performance before our eyes. Therefore, if any motion is ascribed to the earth, in all things outside it the same motion will appear, but in the opposite direction as though they were moving past it. (Copernicus 1978, pp. 11–12)

   Some of the precedents for what I call perceptual relativity are noted by Paul Feyerabend (1975). I have criticised his account of Galileo's introduction of relativity, and especially his claim that Galileo needed to resort to "propaganda" and "psychological tricks" in Chalmers (1986).
5. Whilst the term 'inertial motion' is a convenient one for me to use here, it must be noted that it was not a term that Galileo employed.
6. This, in effect, is the way of defusing the contradiction suggested by J.L Stocks in a note on the quoted passage in this translation of *De Caelo*; D. Ross (ed.), *The Works of Aristotle*, Vol. 2. My quotations from *De Caelo* are taken from this source.
7. This is presumably the passage Galileo had in mind when he referred to Aristotle's views on this matter in the *Dialogue* (1967, p. 141).
8. "Hence the statement that the motion of a simple body is simple holds true in particular for circular motion, as long as the simple body abides in its natural place and with its whole. For when it is in place, it has none but circular motion, which remains wholly within itself like a body at rest. Rectilinear motion, however, affects things which leave their natural place or are thrust out of it or quit it in any manner whatsoever. Yet nothing is so incompatible with the orderly arrangement of the universe and the design of the totality as something out of place; therefore rectilinear motion occurs only to things that are not in proper condition and are not in complete accord with their nature, when they are separated from their whole and forsake its unity." *On the Revolutions*, p. 17.
9. Here I have adapted the terminology used by the Galileo of *Two New Sciences* to indicate that the views I have attributed to him are present even in his most mature work; see Galileo 1974, p. 153.

## BIBLIOGRAPHY

Brush, S. (1974), 'Should the History of Science by Rated X', *Science*, 183, pp. 1164–72.

Cajori, F. (1962), *Sir Isaac Newton's Mathematical Principles of Natural Philosophy and his System of the World*, University of California Press, Berkeley.

Chalmers, A. (1986), 'The Galileo that Feyerabend Missed' in J.A. Schuster and R.R. Yeo (eds.), *The Politics and Rhetoric of Scientific Method*, Reidel, Dordrecht, pp. 1–31.

Copernicus, N. (1978), *On the Revolutions* (trans. E. Rosen), Macmilllan, London.

Dijksterhuis, E.J. (1961), *The Mechanization of the World Picture*, Oxford University Press, Oxford.

Drake, S. (1957), *Discourses and Opinions of Galileo*, Doubleday Anchor, New York.

Feyerabend, P.K. (1975), *Against Method*, New Left Books, London.

Galileo Galilei (1960), *On Motion and Mechanics* (trans. I.E. Drabkin and S. Drake), University of Wisconsin Press, Madison.

Galileo Galilei (1967), *Dialogue Concerning the Two Chief World Systems* (trans. S. Drake) University of California Press, Berkeley.

Galileo Galilei (1974), *Two New Sciences* (trans. S. Drake), University of Wisconsin Press, Madison.

Geymonat, L. (1965), *Galileo Galilei*, McGraw Hill, New York.

Koyré, A. (1978), *Galilean Studies* (trans. J. Mepham), Harvester, Brighton.

Landau, L.D. and Lifshitz, E.M. (1969), *Mechanics*, Pergamon, Oxford.

Post, H.R. (1971), 'Correspondence, Invariance and Heuristics', *Studies in History and Philosophy of Science*, 2, pp. 213–55.

Ross, W.D. (1945), *Aristotle*, 4th revised edition, Methuen, London.

Wallace, W.A. (1981), *Prelude to Galileo*, Reidel, Dordrecht.

CLIVE KILMISTER AND BARRIE TONKINSON

# Pragmatic Circles in Relativistic Time Keeping

## I. INTRODUCTION

Remarks like "moving clocks go slow" would never pass Heinz Post's lips, so it is with diffidence that we present some conclusions about clocks in relativity. Our confidence was somewhat restored only when we felt we were able to propose a method for clarifying a philosophical puzzlement which we think he would like: we call it the Pragmatic Circle. We want to start by considering space and time in Newton's theory and in Einstein's theory of special relativity (SR) in the context of Post's Correspondence Principle. In doing this we will see that we need to consider what constraints should be applied to the spacetime associated with inertial frames. This question and the fact that there are many inertial frames, all equally good for describing nature, leads to a time keeping problem which did not exist in the Newtonian world. Our aim is to discuss the question of whether and in what circumstances clocks measure proper time and also to comment on the logical status of the answer.

We approach this question by revisiting Einstein's theory with the intention of clarifying foundational issues and then moving on to the consideration of accelerated frames of reference. This work needs to be extended into General Relativity, which we will refer to, but our main concern will be with SR. With this more secure background we will propose an application of the pragmatic circle method to time-keeping.

The General Correspondence Principle:

> Roughly speaking, this is the requirement that any acceptable new theory $L$ should account for the success of its predecessor $S$ by 'degenerating' into that theory under those conditions under which $S$ has been well confirmed by tests. (Post 1971, p. 228)

On p. 232 of his 1971 paper Post gives an example of 'inconsistent correspondence' between SR and classical mechanics as $v/c \to 0$. We would like to take more general guidance from Post's heuristic criteria and try to see how we might reinterpret the Newtonian theory ($S$) in terms of Einstein's theory



*S. French and H. Kamminga (eds.), Correspondence, Invariance and Heuristics*, 207–225.

($L$). This endeavour is important for, rather than completely abandon Newton's concept of time, there is a sense in which all clocks in SR tell the time in the same sort of way. It is only because of this that we are able to make comparisons between clocks and finally, through the pragmatic circle approach, say what we mean by "good clocks keep proper time".

In the Newtonian world we speak of time with the meaning that a particular instant, or time, extends across the whole of Newtonian space – a global time. In SR we do not abandon this notion of a time which applies to the whole of a space, rather reinterpret it as frame time. Global time becomes what we might call 'localized global time' to the extent that each frame has its own extensive time. When we consider what meaning we might attach to time in a particular inertial frame we start by considering the time at a point. Confining ourselves to operational definitions, the time at a point is the time given by a clock situated at the point, and we now ask how we might "spread time across our local inertial space". We do this with the realization that an operational method of relating distant instants and local instants drives us to the need for a definition of simultaneity. The Newtonian Global instant becomes an Einsteinian frame simultaneity set. 'At the same time as' becomes the equivalence relation 'is simultaneous with'. Thus we are able to talk about a frame time somewhat in the spirit of a global time only once we have an infinity of inertial frames each keeping their own localized global time.

Having formed the translator for Newtonian time points, by which they are mapped onto Einsteinian time points, we naturally try to do the same for space. We want to retain the extensive properties of Newtonian space, homogeneity, isotropy etc., and since a 3+1 view is perfectly legitimate in the Einsteinian world we may be tempted to take Newtonian space across into the new theory. We are, however, aware that we must apply constraints to the space associated with our inertial frames; how are we to find a criterion for doing this? Our criterion will be such that the Lorentz transformations hold, i.e. that spacetime is flat enough. Thus the spatial part of the translation, due to the need to apply constraints, involves a reinterpretation in terms of the higher dimensional entity of spacetime.

Having arrived at spacetime we might look back at the Newtonian side of the mapping where we see that our earlier considerations about time only involved the points of Newtonian and Einsteinian time, the topological properties rather than the metrical properties. The metrical properties of time, both for Newton and Einstein, are characterized operationally by the difference between readings on a clock and so can be incorporated in the full translation. To sum up: Newtonian temporal points translate into Einsteinian simultaneity sets. Newtonian space plus the metrical properties of time translate into spacetime.

In the Newtonian universe we imagine clocks distributed through space all reading the same time at a Newtonian instant. Once we have a set of inertial frames in uniform relative motion we feel less secure about what readings we might attribute to the various clocks. Since all frames are equivalent for the description of nature we hypothesize that all clocks keep proper time, but we seek a methodology for comparing clocks such that we can do an experiment to check that a particular clock is keeping proper time. Our method will employ an iterative process in which we adjust our clocks and theory until we are satisfied that we have experimentally established the consistency of clock readings in disparate inertial frames. We call this the method of the pragmatic circle. It was by reflecting on foundational issues in Einstein's special theory of relativity that we found ourselves trapped inside the circle, so we consider these next.

In 1905 Einstein's first paper, on what later was known as special relativity, appeared. He contrasts the ease with which time measurements are made (e.g. by means of a watch) of local events with the difficulty of giving time for distant events. Then he suggests "a much more practical determination" for establishing such a distant time "by definition" as the average of the times of reception and transmission of a signal sent to the distant event.

Before discussing this definition of Einstein's we need to make one qualification. He does not put it forward for an arbitrary frame of reference; instead, an observer who is called stationary – a choice of words that subsequent confusions have shown to be unfortunate – is at rest in a frame of reference in which "the equations of Newtonian mechanics hold good"; what we would call an inertial frame. Here Einstein initiates a practice, followed since in many expositions, where one uses the word 'observer' instead of 'frame'. There is no harm in this provided there are no overtones of human consciousness and that the difference between what might be called an observed value and the corresponding reduced value is appreciated – e.g. an observed frequency and the value after reduction when Doppler shift has been applied.

Returning to Einstein's use of the word 'definition'; this may seem unexpected in this context and so, in our clarification of the foundations of the theory, we need to say something about it. It has already been used by Einstein in the preceding discussion in the same paper in talking about local time. A definition of local time, or a prescription for finding it, is given in terms of 'the position of the small hand of my watch'. If we want to consider what we might mean by distant time we must similarly give a definition or prescription for finding it. We are going to do for the difficult distant case what has already been done for the easier local case; might we borrow from the local example to help us in this endeavour? In the local case we have that elapsed time, or interval, is the difference between readings on a single clock at A, giving 'A- time' (similarly 'B- time').

We now establish, by definition, that the difference between a reading at A and a reading on a distant clock at B will also be called a time interval. In this way we make the crucial move which enables us to 'spread time across space'; this then is where we first need to use the term 'definition'. Secondly we complete our definition of distant time by setting the hands of our clocks according to the synchrony condition:

> Similar clocks at points A, B of space are synchronised if, when a light signal leaves A at time $t_A$ by A-time and is reflected at B at time $t_B$ by B-time and returns to A at $t'_A$ then:
>
> $$t_B - t_A = t'_A - t_B$$

Thus we have established what might be described as a global rule, for a particular frame, in that we are able to assign the time to any event observed in that frame.

Had 'definition' been omitted its absence would not have been too easy to spot and there would perhaps have been less objection on the grounds that use of the term sounds somewhat cavalier. Einstein (1905) – see translation in Kilmister (1970) p. 190 – says "we assume that this definition (the synchrony definition) is free from contradictions, and possible for any number of points". Leaving aside conventionalist considerations we believe that this statement is supportable, very briefly, on the grounds that a consequence of light being the fastest signal is that time does not elapse on the signal, so it carries no intrinsic means of giving an elapsed time during its transit in a single direction. Therefore, on this account, there are no grounds for conflict in the setting of a distant clock.

Lying rather more deeply behind the use of the term 'definition' are some philosophical considerations. We need to be aware that Einstein does not intend, at this stage, to become involved in ontology; he is only going to concern himself with pointer readings – operational definitions. Perhaps this is why (Electrodynamics of Moving Bodies 1905) 'time' invariably seems to carry inverted commas; 'time' is an instant – a single pointer reading, or an interval – the difference between pointer readings – according to context.

There is a little more to be said about why the idea of a definition is used by Einstein; for this we need to consider his choice of a signal. Because the paper is intended to be about electrodynamics he takes the signal, without more ado, to be light. But surely it was in his mind that this was the only choice, because light was not only the fastest signal known, but also involved with the various puzzling experiments like Fresnel's. Einstein discussed the choice of light signals in 1909:

> Thus the theory of relativity changes our views on the nature of light, in that light does not enter in connection with any hypothetical medium but as something that exists independently, like matter.

Then in 1922,

> The consequence of the Maxwell-Lorentz equations, that light is propagated with the velocity $c$, at least with respect to a definite inertial system k, must therefore be regarded as proved. According to the principle of special relativity we must also assume the truth of this principle for every other inertial system. (Einstein 1922/1946)

From the same reference, and apparently in order to correct misconceptions, Einstein continues:

> The theory of relativity is often criticized for giving, without justification, a central theoretical role to the propagation of light, in that it founds the concept of time upon the law of propagation of light. The situation, however, is somewhat as follows. In order to give physical significance to the concept of time, processes of some kind are required which enable relations to be established between different places. It is immaterial what kind of processes one chooses for such a definition of time. It is advantageous, however, for the theory, to choose only those processes concerning which we know something certain. This holds for the propagation of light in vacuo in a higher degree than for any other process which could be considered, thanks to the investigations of Maxwell and H.A. Lorentz. From all of these considerations, space and time data have a physically real, and not mere fictitious, significance; in particular this holds for all of the relations in which coordinates and time enter. (Einstein 1922/1946)

What are we to make of this? In the first instance it sounds as though Einstein does not believe that the use of a light signal is essential. Then he seems to confirm the worst fears of those who believe that his setting up of the theory is somewhat arbitrary when he says, "It is advantageous, however, for the theory, to choose . . . "

We believe the interpretation to be as follows. Einstein means what he says when he says "It is immaterial what kind of processes one chooses . . . ". This is because we can only know about time, in the sense of distant time, by processes which enable relations to be established between different places. Now those processes which give physical significance to the concept of time (our knowledge of time) must all be consistent; so we can use what we will, provided always that the processes used have such content as to properly constrain the theory. In the future the possibility exists of using processes which are not electromagnetic waves. "It is advantageous, for the theory, to choose only those processes concerning which we know something certain", then "from all of these considerations space and time data have a physically real, and not mere fictitious, significance . . . ". Here we find exemplified the deep relationship Einstein sees between theory and the physical world. Thus, while a theory is a free construct of the mind, such a construct is disciplined by fundamental physical considerations which encapsulate our knowledge of the real world. The greater our certainty about physical processes the better it will be (more advantageous) for our theory. When it comes to establishing relations between different places we have the greatest certainty about the propagation of light in vacuo and must therefore use this in the construction of our theory. Rather than be over critical, we should see the wording of the 1905 paper as a rather low key introduction to Einstein's brilliant insight. This little phrase, "a much more practical determination", in the first instance needs to be taken at face value. Secondly it introduces the need for a prescription for time and finally it is the modest introduction to an epoch making idea.

## II. THE LORENTZ TRANSFORMATION AND ONTOLOGY

Einstein's definition of synchrony seems eminently reasonable, but so far only one reference frame has been considered. The extent of the changes needed in our ideas about time become apparent as soon as one considers two inertial observers, the second one being in uniform motion relative to the 'stationary' one first considered by Einstein. The respective global frame times constructed by these two observers can not agree, as one can see at once by the following simple argument: Suppose that the first observer A sends a light signal to the second B at time $t_1$ and that the reflected signal from B is received by A at time $t_2$. A then reflects the signal again to be reflected by B and finally received back at A at time $t_3$. Now if B is synchronised with A according to A's time, the definition gives, for the time of the two reflection events at B, the average of $t_1$ and $t_2$ and the average of $t_2$ and $t_3$ respectively. Now if this synchronisation agrees with that in the inertial frame in which B is at rest, then the reflection at A will, by applying the same definition, be the average of the two times at B. This average is

$$\frac{t_2}{2} + \frac{(t_1 + t_3)}{4}$$

and this should be $t_2$, but this is the case only if $t_2 - t_1 = t_3 - t_2$; that is, exactly when the two observers are at rest in the same inertial frame. An important consequence of Einstein's definition is therefore a change of ontology. In the pre-Einstein world there were one-place functions ('times') from events to real numbers:

$$E \rightarrow t = f(E)$$

These have been replaced by two-place functions:

$$E \rightarrow t = f(E,F)$$

where $F$ denotes the inertial frame of the observer who, armed with a watch, is sending and receiving light signals.

The ontological change is not without effect on the view which it is convenient (to say no more) to hold about time. If one takes Leibniz's relational standpoint, that the world simply contains an ordering relation between events which is reified into numerical form by means of the rational or real number field, then the change is easy to accommodate. It is only necessary that the order of perceived events is not changed by a change from one inertial frame to another. Notwithstanding the well-publicised changes in the order of some events in special relativity, events which are in communication with each other cannot be so changed in order. If one is a realist about time, the ontological change calls forth rather more trouble. Such a person could remain a realist and simply let the operationalists get on with their definitions as an epistemic matter, but the realist should now consider shifting allegiance to spacetime. Paraphrasing Einstein:

> Just as from the Newtonian standpoint, time is absolute and space is absolute, so for special relativity we must say that the spacetime continuum is absolute – meaning physically real and independent in physical properties. (Einstein 1922/1946).

The formal statement of this ontological change is the Lorentz transformation. The path originally followed by Einstein to derive this is a complex one and involves appeal to rigid coordinate systems. It will clarify our argument if we do not follow it here, but appeal instead to the equivalence of two inertial observers and to Einstein's definition which we have been discussing. This way of deriving the transformations, implicit in Einstein's paper, is used in a different context by Milne (1938) and has been popularised by Bondi under the title *k*-calculus. It is convenient to represent the events under discussion on a space-time diagram, (Figure 1) with the time increasing vertically upwards as is usual in this subject. The diagram is drawn from the point of view of Einstein's arbitrarily designated 'stationary' observer A so that he is represented by the vertical line OA. The other observer is supposed to leave him at time $t$ = O (on which they agree) and to move to the right with constant speed. He is represented by the line OB. The dotted lines in the figure represent light (or radar) signals. If a signal is sent at time $t$ (in his global time) by A, the time of reception by B (in his global time) will be proportional to $t$. Calling the constant of proportionality $k$, so that B measures the reception of the signal at time $kt$ in his global time, the constant $k$ simply measures the rate of separation of the observers. The next step in the argument appeals to the equivalence of the two inertial observers: the signal returned by B at

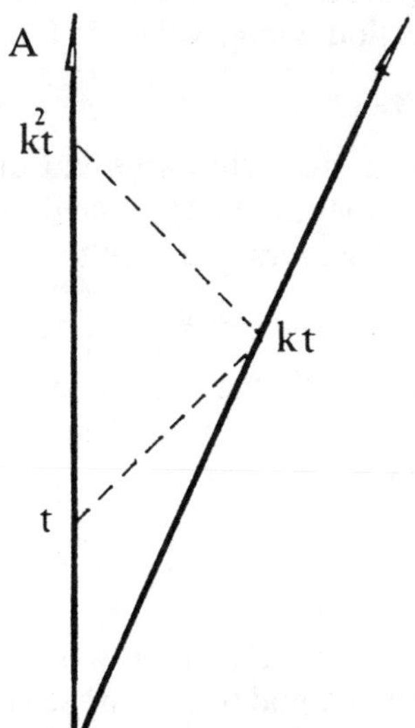


*Fig. 1.* Time-keeping between two equivalent observers.

time $kt$ will be measured by A to arrive at time $k(kt) = k^2t$. In A's global time, then, the time of reflection of the signal at B will be

$$\frac{(t + k_2t)}{2} = \frac{t(1 + k_2)}{2}$$

and since $1 + k^2$ is always greater than $2k$ (because $(1 - k)^2$ is greater than zero) A's time measure of this event is greater than B's.

In the misleading form 'moving clocks run slow' this result has become part of the accepted theory. But this shortened version, and even Einstein's original, convey a wholly incorrect suggestion: that the motion of the clock in some way produces an effect on its rate; whether this is dynamical, kinematical, or possibly an incorrect functioning is unclear (discounting any incorrect functioning which would be produced in, for example, a pendulum clock). The notion that the rate of a moving clock is changed has perhaps been one of the greatest points of misunderstanding in special relativity, but this is not what the theory says. The correct interpretation is, rather, that a moving clock goes at exactly the same rate and only reads less because it has experienced less time. This is not at all puzzling so long as it is realised that, because of the change in ontology, different amounts of time may be measured, in different frames, between a given pair of events.

If A sends a signal at time $t_1$ and receives it back at time $t_2$ then, in addition to Einstein's prescription of $(t_1 + t_2)/2$ for the time of the reflection event, A can argue that since the light has travelled twice the distance to this event (once in each direction) the distance of this event from him is $(t_2 - t_1)/2$, remembering that we take the velocity of light to be unity. One can then refer to A's construction of a (one dimensional) coordinate system, giving time and space coordinates $t$, $x$ to the reflection event, where it follows at once that

$$t_1 = t - x, \; t_2 = t + x$$

If now one imagines A and B surveying an event distant from both of them (but in the same straight line for simplicity), one derives from Figure 2, using the equivalence of the two observers, the results

$$t' - x' = k(t - x), \; t + x = k(t' + x')$$

and these two equations are exactly the usual one-dimensional Lorentz transformations in a slightly unfamiliar notation. The factor $k$ is easily seen to be related to the velocity of separation by

$$v = \frac{(k^2 - 1)}{(k^2 + 1)}$$

The earlier use of the term 'definition' is now seen as extended to embody the idea of applying a uniform method of synchronisation for all inertial frames. This application of a uniform method brings us back to the use of light signals. The quotation above from Einstein (1922) continues:

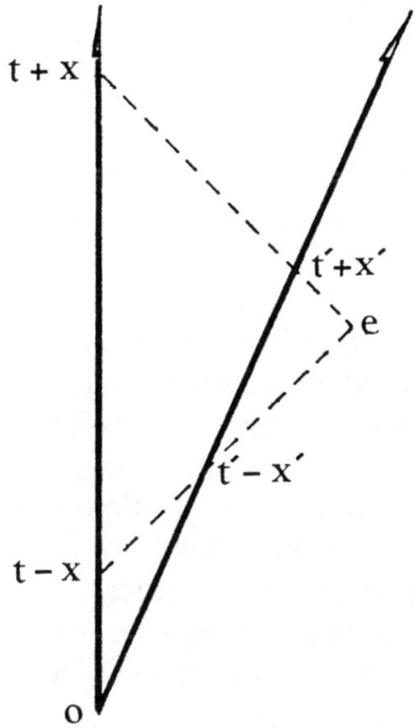


*Fig. 2.* Two equivalent observers set up the Lorentz transformation.

> There is, therefore, sense in asking . . . what the true equations of transformation are by which we pass from one inertial system $K$ to another $K'$, moving relatively to it. It may be shown that this is uniquely settled by means of the principle of the constancy of the velocity of light and the principle of special relativity. (Einstein *ibid.*)

We need such a method because the use of light signals brings in the uniformity of the frames – our inability to distinguish one from the other. We come full circle in the realisation that light signals must be used, to the extent that equality, once recognised, is precisely the sort of characteristic which would be "advantageous" to the theory if it is to have wide applicability – be a high level theory in the sense of Hempel.

## III. THE BEHAVIOUR OF CLOCKS

The Lorentz transformations between inertial frames have two important properties. The first of these is indeed assumed in the derivation. It is that the two frames are equivalent as far as all mechanical phenomena are concerned (and that includes timekeeping), so that matters may be described equally from either point of view. The second property of the transformations is slightly more technical. The expression given above for the Lorentz transformation leads at once, by differentiation to

$$dt' - dx' = k(dt - dx),\ dt' + dx' = (1/k)(dt + dx)$$

and multiplying these together shows that

$$dt'^2 - dx'^2 = dt^2 - dx^2 = ds^2 \text{ (say)}$$

Apply this to the primed frame, so that for all events at the clock/origin, $dx' = 0$, and therefore the invariant $ds = dt'$. That is, the coordinate increment $dt'$ equals the corresponding increment $ds$ of the invariant quantity, which is accordingly referred to as the *proper time*.

Now in Minkowski spacetime the coordinates, which are those we have been constructing, have metrical significance, meaning that increments in coordinate distance can be identified with corresponding calibrations on a measuring rod, and increments in coordinate time correspond with calibrations on a clock. Of course we refer only to very accurate measuring rods and very accurate clocks. This brings us to an important question: how can we be sure we have a very accurate, or so called good, clock? For the moment we will make do with what has been called 'the clock hypothesis' – that good clocks in inertial frames measure proper time (e.g. see Rindler 1969/1977).

The two properties of being equivalent and of clocks measuring proper time are characteristic in the sense that it is only between inertial observers that both hold. It is (almost) universally assumed by workers in relativity that arbitrarily moving clocks also measure proper time. What we can say at this point is that it is a logical consequence of the foundations of the theory that, in an inertial frame, we can write $ds = dt$, but that the question of how we might go about making the clock hypothesis secure, both for inertial and non-inertial frames, will be an important consideration later in this paper.

A well-known example of an investigation in which it is assumed that the clock measures proper time, although the observers are not independent, is that of uniformly accelerated observers. This is used in the usual sense of "uniform relativistic acceleration" (Born 1909), but for our purposes it is unimportant what is meant by this; it is simply another motion, but it happens to be one with which it is easy to deal. It is usually specified in the text-books by

$$x = [a^2 + t^2]^{1/2} - a, \; a = 1/g,$$

where $g$ is the acceleration. Since $2x = t_2 - t_1$, $2t = t_2 + t_1$, this may be rewritten as

$$1/t_1 - 1/t_2 = g$$

Now make the assumption that a moving clock measures proper time. The proper time is given by

$$ds^2 = dt_1 \cdot dt_2$$

which is easily integrated to give

$$t_1 = a(1 - e^{-gs}), \; t_2 = a(e^{gs} - 1)$$

For a general event, as in Figure 3

$$t_1 = a(1 - e^{-gs_1}),\ t_2 = a(e^{gs_2} - 1)$$

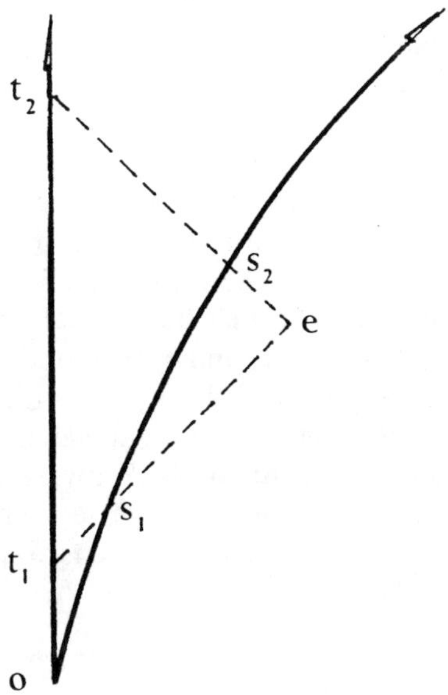


*Fig. 3.* Setting up the transformation between an inertial observer and an accelerated one.

so that

$$ds^2 = dt_1 \cdot dt_2 = e^{2gx'}(dt'^2 - dx'^2)$$

The measured (proper) time is then not $dt'$ except for the events at the origin, ($x' = 0$). But in the case when $s_1 = s_2$,

$$t = \frac{(t_1 + t_2)}{2} = a\sinh gs$$

and so $t/s = (\sinh gs)/gs > 1$ is the clock paradox again. And this form shows that the direct effect of acceleration (as distinct from its indirect effect in producing velocity) is nil, since

$$\frac{dt}{ds} = \cosh gs$$

but

$$\frac{dx}{ds} = \sinh gs$$

so that the velocity is tanh $gs$ and the factor from the velocity, $(1 - v^2)^{-1/2}$ is exactly cosh $gs$ as well. It is very easy to see that this is not an effect that depends on the acceleration being constant; the same argument serves to prove it for any motion.

In the argument above we assumed, without drawing attention to it, that the accelerated observer would use Einstein's rule for setting up a coordinate system, although that rule was enunciated for inertial observers. Now one knows, from Einstein's own remarks about coordinate systems in general relativity, that as soon as one leaves the realm of inertial observers there is a certain arbitrariness in coordinates; so that the accelerated observer beta, say, is at liberty to define his coordinate system in this way (one may ask in passing what other possibilities are open to him?). All that the critic may ask is, how good is such a coordinate system? The only objection to it seems to be that it does not cover the whole domain that the inertial observer alpha's covers; for the expressions show that $t_1 < a$, and $t_2 > a$, and these are barriers. But in what alpha describes as a straight line path with uniform velocity the approach to such a barrier will be one with exponential slowing. So they are rather like those in Kruskal's extension of the Schwarzschild metric; moreover if $g$ is small, so that $a$ is very large, the barriers are at a great distance. Rindler (1977), pp 156–164, gives a good discussion of this. However, he uses the metric form

$$ds^2 = X^2 dT^2 - dX^2$$

which can be derived from ours by a coordinate transformation:

$$(gX)^2 = e^{2gx'} - 2e^{gx'}\cosh gt' + 1$$

$$(e^{gx'}\cosh gt' - 1) \tanh T = e^{gx'}\sinh gt'$$

In this argument it has been assumed that proper time is measured by a moving clock, and no assumption has been made about the equivalence of observers. So we now take up the question of whether beta describes alpha as uniformly accelerated relative to him, which would restore the equivalence of the observers, and would reintroduce the clock paradox in an acute form. Since beta has a non-Euclidean form for his metric, there is a temptation to assume that beta will use the resources of the tensor calculus and differential geometry to reduce his observations. If he did so, it is easy to calculate that he simply finds alpha to be unaccelerated; and this need occasion no surprise, since this *is* exactly alpha's state of affairs, and all that is done in carrying out beta's calculation is a coordinate transformation. But of course, beta has employed a much more naive approach in setting up his coordinate system, and he must be supposed to persevere in this in observing alpha. From the two expressions

$$t = a(e^{gs_1} - 1) = a(1 - e^{-gs_2})$$

it easily follows that

$$e^{-gx'}\cosh gt' = 1$$

The minus sign results from alpha moving off in the negative direction according to beta, and so in order to compare alpha's view of beta with beta's view of alpha, we have to compare $(1 + z^2)^{1/2} - 1$ with log cosh $z$. These two expressions are different, so beta does not see alpha as uniformly accelerated and there is no clock paradox, but it is worth remarking that they are rather alike in numerical value; indeed they agree within 7% over the whole range of positive $z$. We are not aware of any analytic reason for this, but one can see easily that for small $z$ we are comparing

$$\frac{1}{2}z^2 - \frac{z^4}{8}$$

with

$$\frac{1}{2}z^2 - \frac{z^4}{12}$$

and for large $z$ we are comparing

$$z + \frac{1}{2z} - 1$$

with

$$z + e^{-2z} - \log 2$$

Thus we can be sure that alpha and beta are not equivalent, but in experiments to limited accuracy they may well not be able to determine this.

Sometimes an appeal is made to general relativity at this stage. This is useless for two distinct reasons. Firstly, that measured time is proper time is an assumption of general relativity; so the theory cannot be called in to validate the assumption. Secondly, the coordinate systems in general relativity (as was realised by Einstein well before 1915) are purely conventional labels of events. No argument which uses the particular time coordinate to say something about the behaviour of clocks can be valid; but if coordinates are not to be used, there remains only the proper time. In any case, the main burden of general relativity is with the gravitational field, and the discussions up to now in this paper have been with the idealised case in which gravity can be neglected.

The issue of accelerated observers was looked at very carefully 25 years ago by Romain (1963) and he gives many references to the preceding fifty years. His method is to look at the most general form without otiose special assumptions such as that clocks measure proper time. In this way he gets very general forms that can be checked with experiment. However, his conclusion, "The final solution of the problem of correctly interpreting experimental evidence is not yet within easy reach." does not need changing at present.

## IV. DIFFICULTIES SUMMARISED

There are a number of difficulties and puzzles about the conventional position which we have just described. In this section we draw them together in roughly the order in which they have occurred in the account in Part 1.

(i) In Einstein's original paper very little attention is paid to the determination of time for purely local events. Einstein has two short comments about the "small hand of my watch". Although this is understandable, for one does not want to get into the technical details of the construction of chronometers, it produces some philosophical discomfort when the theory goes on to make predictions about the behaviour of moving clocks relative to fixed ones. Some of this discomfort is only the result of incorrect phrases like "moving clocks go slow" and is removed by simply abstaining from such language. But there remains a residue of doubt about the nature of the clocks employed.

(ii) Einstein begins with a stationary observer for whom "the equations of Newtonian mechanics hold good" and then considers another observer in uniform motion relative to the first. In other words, the set of inertial frames is privileged by the investigation in a way that is not true of Newtonian mechanics for there the Galilean transformation holds between *any* two frames in uniform relative motion. Why is the set of inertial frames picked out in this way?

(iii) As we set out at some length above, the role of light as the means of signalling is a little mysterious. It is clear that Einstein sees the need for some argument, over and above the fact that it is the fastest signal known, and that it is precisely the paradoxes connected with the observed speed of light that gave rise to the Lorentz transformation. But it cannot be said that the requirement that the theory should "choose only those processes concerning which we know something certain" is really strong enough to bear the weight of choosing the unique signal.

(iv) There is finally the tangled question of which clocks measure proper time. Here there are three situations to be considered, only two of which came up above.

(a) In inertial frames in special relativity it is a consequence of the theory that $ds = dt$; the question of what we mean by "good clocks measure proper time" (the clock hypothesis) is considered below. The proper time is also the coordinate time of the rest-frame of the clock, provided that a Minkowski coordinate-system is used.

(b) For an accelerated clock in special relativity it seems to be widely believed that such a clock measures proper time. If this is so, the treatment given above shows that the clock discrepancy is exactly that (associated with) the velocity, so that the assumption is equivalent to saying that acceleration has no direct influence on what the clock measures. But it remains an assumption, and the empirical evidence for it is in favour, but not so strongly as all that.

(c) In general relativity the assumption is always made that clocks measure proper time, and this is usually recognised as an assumption. It plays an important role in the 'third crucial test' of the theory, the gravitational red-shift, for the theory of that is based on the hypothesis that the vibrating atom in the strong field has the same period of vibration as that in the laboratory, measured in proper terms.

The two cases (b), (c) are independent. This can be seen from the fact that in general relativity it would be possible (if highly unlikely) that clocks did not measure $ds$, but rather the square root of

$$(g_{ij} + R_{iabc}R_j^{\ abc})dx^i dx^j$$

or some such combination. But all such expressions would reduce to $ds$ in special relativity, where $R_{iabc} = 0$. These difficulties and puzzles are connected in a way that we discuss in the next section.

## V. TIME KEEPING

It is easiest to begin with the last of the difficulties just described. It is odd that the question of whether clocks measure proper time should have such a different status for uniformly moving clocks in special relativity, for accelerated clocks and for clocks in an irreducible gravitational field. Before we tackle this it is necessary to say something about time-keeping in general.

We do this following Leibniz, in terms of seeing the basic perception of time as simply that of the ordering of events. (If a realist view of time is taken, the difficulties here will all translate into it.) This order is then reified by assigning numbers to the events so that the ordering of the numbers mirrors the temporal ordering of the events. Ideally any such assigning of numbers is possible and could be called a clock. In practice, however, a dynamical constraint is applied; a time reckoning is chosen which "makes Newton's first law true". This opens all sorts of problems; how can one be sure that the particle chosen as the representative bearer of Newton's first law is truly under no forces, and should a fundamental perception like time depend on an arguably less prior perception of motion? We shall return to these points later. One can summarise the position by saying that, whether or not Leibniz's conception of time is defensible, the time-concept captured by it is only what may be called 'topological time'. It is a time-concept which correctly labels the relative position of perceived events as far as their temporal relations are concerned. There is, in addition, a 'metrical time'; making Newton's first law true says something about this, but the following difficulties over duration show that something more is needed. The rational numbers are evidently adequate for Leibniz's construction, but they are inadequate to describe duration. For all possible events perceived between $t = 0$ and $t = 1$ can be labelled by rationals, and the rational numbers can be enumerated. This enumeration provides a first, second, . . . event. Suppose the first event is centred in a short interval of duration 1/4, the second in 1/8 . . . the rth in $1/2^{r+1}$.

The total of all these durations cannot exceed 1/2, and overlapping will very likely reduce this figure. The total duration cannot therefore be 1 as it should. This shows that the notion of duration cannot be derived from that of the topological time. It is in order to provide something more than topological time that the notion of a clock is introduced; the small hand of Einstein's watch has much to do. One needs, for instance, to give meaning to the idea of adding two equal durations. What we choose to call a clock and how it relates to its environment form a fundamental part of our understanding about space and time.

To return to the problem of what clocks read, we ought to begin by limiting the question to 'good' clocks; for evidently the theory is only considering idealised clocks free of obvious mechanical defects. But what is a 'good' clock? Surely one that measures proper time when it is at rest in an inertial frame. And when is a frame inertial? One need not go on; our considerations enter a circle from which there is no escape. But there is no reason to be fearful, for such circles are common structures in physics. They only seem worrying because we are unable to say which is the cart and which the horse between theory and experiment. It is only that in the present case the circle seems rather daunting because we are dealing with a theory of space and time and it seems natural to want to define time first.

The method we are proposing to make progress, which we call that of the pragmatic circle, is to consider an iterative process. As well as theory and experiment a part is played by technology or engineering. We might proceed as follows: assume a flat space-time in which the Lorentz transformations are supposed to hold. Observe clocks in various frames and compare readings. If it is the case that the readings do not transform according to the Lorentz formulae, then investigate the clocks. A problem may be found; e.g. poor temperature control. There need be no appeal to high level theory; we may decide to improve temperature control, either because we have a satisfactory (low level) explanation of why temperature affects frequency stability, or because we have empirical data which show that it is so. With the improved clocks a new inconsistency is found. We may conjecture that the space-time volume surveyed is too large to be taken as flat, so we constrain observations to a smaller volume. Then we enquire whether a correction to theory is needed to accommodate yet more precise experimental data and so on. The point is that the meaning of such a statement as "a good clock measures proper time at rest in an inertial frame in special relativity" is got through an iterative process in which theory and experiment are variously defined by, and definers of, the terms in use. In saying this we are really only setting down in a rather abstract form the actual procedure in the laboratory or workshop. It is just that our form is less abstract than the 'pure theory' version.

When we go on to consider the cases of accelerated clocks, or clocks in gravitational fields, the argument becomes more difficult; there is no symmetry between accelerated observers as there was between inertial ones. Self-gravity cannot be regarded as eliminated. But the method of procedure is the same.

(In particular, looking for a good clock in a gravitational field will eliminate the use of a grandfather clocks at an early stage; but that is only one small aspect). The point which we seek to make by formulating the pragmatic circle method is simply that the situation in these two cases is, after all, no different in kind from the special relativity case of clocks at rest in an inertial frame. In each case there is an iterative process; the processes are of the same general kind. If there were a significant difference between any two of the three cases, it would merely be an empirical one. One could put it this way: good clocks measure proper time within certain error bounds. Good clocks exist at rest in inertial frames in special relativity, but the corresponding empirical evidence that there exist good accelerated clocks is less strong, and the same goes for clocks in gravitational fields. In any case the three cases stand in exactly the same logical situation.

It may be useful to enlarge the whole investigation to include the field and the determination of a good clock together. Taking the example of a light clock, extended across space, it is not immediately clear where the boundary between the clock and the space (in which it might be considered to be immersed) lies. In performing a Lorentz transformation one would be concerned with whether a clock tells something that transforms as proper time (is invariant) and whether the coordinates transform correctly. According to which is perceived as the best direction to begin, one might now consider the form of the metric, or the goodness of the clock, or ask whether there was some effect on the (otherwise good) clock which had not been accounted for. One would be concerned with establishing the nature of a good clock; whether for example it is the case that a spatial element is necessarily a constitutive characteristic of a clock – as seems to be the case (at least in a restricted sense) even for atomic clocks.

As to the use of possible alternatives to light signals, Einstein has suggested that other physical processes might be used. We have previously noted the unifying role played by the velocity of light as crucial in the formulation of special relativity. Thus, while, in principle, it would be possible to use other processes the choice can never be arbitrary. Philosophical and physical underpinning, such as exists in the unifying power of the invariant velocity of light, would necessarily be needed in any alternative approach. We trust that we have correctly interpreted Heinz Post's teaching in drawing this conclusion.

We have previously referred to the privileged status of the set of inertial frames and the difficulty over the practical realisation of Newton's First Law. We turn again to Einstein for his important commentary on the principle of inertia and further insight into his philosophical disposition.

> The possibility of explaining the numerical equality of inertia and gravitation by the unity of their nature gives the general theory of relativity, according to my conviction, such a superiority over the conceptions of classical mechanics, that all the difficulties encountered in development must be considered as small in comparison.

> What justifies us in dispensing with the preference for inertial systems over all other coordinate systems, a preference that seems so securely established by experiment based upon the principle of inertia? The weakness of the principle of inertia lies in this, that it involves an argument in a circle. (Einstein 1922/1946, p. 58)

Einstein now considers the limitation of the principle of inertia and notes that it is established to a high degree of approximation provided the spatial extent of its applicability is restricted and that small perturbations are neglected. Thus while Einstein identifies a fundamental weakness in the principle of inertia he virtually tells us how to avoid the difficulties it entails. We should note that so called 'Galilean regions' (where special relativity holds) are known only in so far as important constraints have been applied; 'remarkable accuracy' is the best we can finally hope for.

Clifford had already taken up the theme in a paper given before the British Association in 1872:

> The word exact has a practical and a theoretical meaning. When a grocer weighs you out a certain quantity of sugar very carefully, and says it is exactly a pound, he means that the difference between the mass of the sugar and that of the pound weight he employs is too small to be detected by his scales. If a chemist had made a special investigation, wishing to be as accurate as he could, and told you this was exactly a pound of sugar, he would mean that the mass of the sugar differed from that of a certain standard piece of platinum by a quantity too small to be detected by his means of weighing, which are a thousandfold more accurate than the grocer's. But what would a mathematician mean, if he made the same statement? He would mean this. Suppose the mass of the standard pound to be represented by a length, say a foot, measured on a certain line; so that half a pound would be represented by six inches, and so on. And let the difference between the mass of the sugar and that of the standard pound be drawn upon the same line to the same scale. Then, if that difference were magnified an infinite number of times, it would still be invisible. This is the theoretical meaning of exactness; the practical meaning is only a very close approximation; how close, depends upon the circumstances. The knowledge then of an exact law in the theoretical sense would be equivalent to an infinite observation. I do not say that such knowledge is impossible to man, but I do say that it would be absolutely different in kind from any knowledge that we now possess. (Clifford 1872)

We see that the pragmatic circle has been with us for a long time; we believe that the uncertainties that it brings out are common in physics, but particularly pertinent with regard to the fundamental problem of time keeping.

Importantly we see that a certain level of uncertainty is to be expected at each point in the circle, pragmatic acceptance and iterative processes enable us to keep telling the time and doing physics.

ACKNOWLEDGEMENTS

Clive Kilmister wishes to acknowledge the help of the Leverhulme Trust.

*Dept. of History and Philosophy of Science,*
*King's College, University of London,*
*London, U.K.*

## BIBLIOGRAPHY

Born, M. (1909), *Annalen der Physik*, 30, 1.

Einstein, A. (1905), 'On The Electrodynamics of Moving Bodies', *Annalen der Physik*, 17, 891.

Einstein, A. (1922–1946), *The Meaning of Relativity*, Princeton University Press.

Kilmister, C.W. (1970), *Special Theory of Relativity*, Pergamon Press (qv. for translation of Einstein (1905) and key papers in Special Relativity).

Milne, E.A. (1983), *Kinematic Relativity*, Oxford University Press.

Post, H.R. (1971), 'Correspondence, Invariance and Heuristics', *Studies in History and Philosophy of Science*, 2, pp. 213–255.

Rindler, W. (1969–1977), *Essential Relativity*, Springer-Verlag.

Romain, J.E. (1963), 'Time Measurements in Accelerated Frames of Reference', *Reviews of Modern Physics*, 35, pp. 376–389.

Zeeman, P. (1914), 'Fresnel's coefficient for light of different colours', *Proceedings of the Academy of Sciences, Amsterdam*, 17, 445.

HARVEY R. BROWN

# Correspondence, Invariance and Heuristics in the Emergence of Special Relativity

## I. INTRODUCTION

"It is particularly interesting to point to progress in science towards new, better theories, entirely guided by factors internal to the old theory." In his provocative 1971 essay on the 'rationale' of scientific discovery (Post 1971), Heinz Post gave this as one of the reasons for embarking on such a study. In the case of the emergence of special relativity, one of the four scientific revolutions Post discussed in defending his rationale of 'conservative induction', the observation is particularly apt. In the present paper, I want first to elaborate on this theme. Einstein's 1905 postulates have larger "footprints" – to borrow Post's term – in the "old theory" (if by this is meant the assortment of Maxwell-Lorentz ether electrodynamics, classical mechanics and thermodynamics) than is perhaps still widely appreciated.

As far as the precise 1905 light postulate (as opposed to its standard modern variant) is concerned, in calling it the "true essence of the old aether point of view" Pauli was to provide a striking description that is very close to the truth. The light postulate was Einstein's homage to the successes of the Lorentz theory of the electron, in spite of his earlier truly revolutionary hypothesis in 1905 concerning the light quantum (Section II). In adopting the principle of relativity, Einstein was not, as is so often claimed, extending the 17th century principle beyond its supposed 'mechanical' remit into the domain of electromagnetism and beyond. He simply restored to the Galilean-Newtonian principle its original, intended scope. But whereas Newton (in a rare lapse) presented the principle as a simple corollary to his laws of motion, Einstein, with the methodological status of thermodynamics in mind and like Huygens before him, raised it to a postulate (Section III). Now the young Albert Keinstein did the same in 1705 and *derived* Galilean kinematics, thus doing for Galilean/Newtonian space-time what Einstein did for Minkowski space-time two hundred years later. The fable of Keinstein will be told in Section IV, which contains a comparison of his derivation with Einstein's route to the Lorentz transformations, and comments on the non-trivial issue of the correspondence between classical and relativistic kinematics.



*S. French and H. Kamminga (eds.), Correspondence, Invariance and Heuristics*, 227–260.

Special relativity (SR) has of course received a more 'mature' (geometric) formulation since Einstein's 1905 work. But *interpretation* of the geometric formulation has, on occasions, obscured the important footprints of the theory in the 'old' physics, and the consequences have not, in my opinion, always been happy. In the remainder of the paper I will examine the alternative, "surest and clearest" derivation of relativistic kinematics given by Friedman in his acclaimed 1983 study of the foundations of space-time physics. This derivation rests on the postulated metric structure of Minkowski space-time and its 'symmetry group', the role of clocks as hodometers in space-time, and the independent (geodesic) equations of motion for free particles and light. The derivation is ostensibly *dynamics-free*; in particular it does not appeal to the principle of relativity. But though it is clear mathematically, it is physically mysterious.

For the role of the *dynamical* principle of relativity in Einstein's derivation of so-called relativistic *kinematics* is subtle. It emphasises a deep point (partly) understood by Lorentz, but which in the post-Minkowski outlook is sometimes overlooked, although it was stressed by Pauli in 1921, Einstein himself in 1949 and J. S. Bell more recently. The point is that in SR the issue of the origins of length contraction and time dilation – and hence of space-time geometry – cannot be divorced from *dynamical considerations*.

The nature of Friedman's pre-dynamic derivation is summarised in Section V below; Section VI contains a discussion of the difficulties in it. To some extent the discussion here expands on earlier remarks concerning the origins of space-time geometry made by Ehlers, Pirani and Schild in 1972 and Sklar in 1977 (although I am not interested here in the 'conventionalism' issue central to the concerns of the latter). In Section VII it is argued that the logic of Friedman's proof is moreover hard to reconcile with his earlier discussions of 'Newtonian space-time' and 'classical electrodynamics'. (Friedman's treatment of the latter theory is problematic in its own right; I argue that the 'Galilean' kinematics built into 'classical electrodynamics' cannot straightforwardly be identified with the standard kinematics of pre-electrodynamic Newtonian mechanics.) This leads to a brief critical discussion in Section VIII of the modern 'space-time theories' approach to relativity physics currently popular amongst prominent philosophers of space-time physics (although an important apparent difference between the views of Friedman and those of Earman is stressed). In the concluding Section IX, Einstein's 1949 reflections on the role of rods, clocks and geometry in SR are cited. Finally, mention is made of the 1980 geometrical programme of Anandan as a possible alternative to the 'space-time theories' approach – and one more in keeping with the deep heuristics of Einstein's 1905 work, where, as they must be, kinematics and dynamics are intrinsically intertwined.

## II. EINSTEIN'S 1905 LIGHT POSTULATE

Careful commentators of Einstein's 1905 relativity paper have long recognised that the numerical invariance of the light-speed is not a fundamental postulate, but is presented in the paper as a *consequence* of Einstein's postulates. His fundamental 'light postulate' – in the most explicit form given in his paper – is a claim about the behaviour of light in a specific 'resting' inertial frame:

> Any ray of light moves in the 'resting' coordinate system with a definite velocity *c*, which is independent of whether the ray was emitted by a resting or by a moving body. (Einstein 1905, Section I.2)[1]

In 1905 there was no *direct* experimental evidence that an inertial frame exists in which the two-way light-speed is 'constant' in this sense: source-speed independent and isotropic[2]. (Such evidence was certainly not provided by the Michelson-Morley experiment of 1887, which demonstrated only that the two-way light-speed is isotropic when the source is at rest in the laboratory frame; hence the brief post-1905 resurgence of the 'emission' theory of light in the work of Ritz.) Why then did Einstein postulate the existence of such a frame?

It is well-known that SR arose essentially out of the null-results of 19th century experiments designed to detect first- and second-order effects of the 'wind' caused by motion through the electromagnetic ether. What is arguably less widely appreciated is the fact that because in 1905 Einstein already suspected that the Maxwell-Lorentz field equations, which were then commonly accepted as strictly valid at least in the ether rest frame, could have at best statistical (macroscopic) validity in that frame[3], he deliberately refrained from basing his derivation of the Lorentz transformations on them (see the discussion in Brown and Maia forthcoming 1993 and Torretti 1983, pp. 48–49). A derivation of such fundamental importance had to be based on something solid (besides the relativity principle), some *specific* feature of the emerging electromagnetic world picture that was if not incontestable, then securely grounded.

Ever since the advent of the mechanical ether of Faraday and Maxwell, all ether theories of electromagnetism shared the characteristic prediction that the speed of propagation of disturbances in the medium, as measured in its rest frame, is isotropic and independent of the motion of the source of the disturbance[4]. Indeed, it is in the nature of a mechanical luminiferous ether, that the light-speed is determined ultimately only by the elastic properties of this rarefied, isotropic 'ponderable' material. But in the Lorentz theory of the electron, developed from 1890 onwards, the ether lost its mechanical status; it was a new kind of imponderable matter, no*thing* more than the seat of the electromagnetic field (see Hirosige 1969 and Earman 1990, p. 51). Thus Lorentz could provide no obvious explanation for light-speed constancy in the ether rest frame, other than the fact that it was a consequence of his fundamental field equations (this point was made in Saunders and Brown 1991, p. 47, fn. 18).

We tend to forget nowadays just how successful the Lorentz electron theory was. It was not of course only that at the macroscopic level the field equations coincided with Maxwell's. Einstein (1907) himself was particularly impressed with the fact that the Fresnel 'drag' coefficient – fundamental in accounting for null-results in the first-order ether wind experiments and the non-null 1851 Fizeau interference effect – ceased being *ad hoc* and counter-intuitive[5], gaining in Lorentz's treatment a *dynamical* explanation in the context of a wholly static (undraggable) ether. (We shall return to this point in Section VII.) Finally, it should not be forgotten that part of Einstein's sympathy, such as it was, with Lorentz's programme stems from the fact that prior to 1905, he (Einstein) had convinced himself that the alternative 'emission' model of light, in which the light-speed varies with the speed of the source, could not generate reasonable electromagnetic field equations at the macroscopic level[6].

Einstein was to write in 1907 of the constancy of the light-speed:

> It is by no means natural to expect that [it] . . . should be actually satisfied in nature, yet – *at least for a coordinate system of a certain state of motion* – it is made likely by the confirmations which Lorentz's theory, that is based on the assumption of an absolutely resting ether, has obtained by experiment. (Einstein 1907; my emphasis; see Miller 1981, p. 202)

Thus Einstein's 1905 'resting frame' was precisely the frame in which Lorentz's hypothetical ether is said to be at rest – at least before the relativity principle comes into play[7]. Yet one cannot help wondering whether the 'experimental confirmations' that the Lorentz theory enjoyed (and the theoretical difficulties with the emission theory of light) were the whole story in Einstein's mind. He was surely aware in 1905 of the extremely flimsy *explanatory* role that the ether played in the Lorentz theory. Hence he must have seen that adopting the "true essence of the old aether point of view" (Pauli's 1921 description of the light postulate, Pauli 1981, p. 5) in constructing SR – which was designed to render the ether "superfluous"! – was far from incoherent.

Einstein's light postulate constituted then recognition of that element of Lorentz's theory which he thought would survive any developments in the theory of radiation. The degree of robustness he attributed to it is testified by the fact that the most radical existing threat to the wave theory of light was his own. Einstein regarded his earlier 1905 light quantum hypothesis as more revolutionary than anything in his relativity paper; moreover, it was surely more *prima facie* compatible with the emission theory of light than with the 'ether' light postulate (this point was stressed by Hoffman (1982); see also Torretti *op. cit.*, p. 49). That Einstein was nonetheless right to adopt this postulate in his relativity paper is one of the best, early indications of his remarkable physical judgement[8].

## III. THE PRINCIPLE OF RELATIVITY

In his autobiographical notes (1969), Einstein describes his "despair" af reading Planck's 1900 work on blackbody radiation, of discovering the tr laws of mechanics and electrodynamics "by means of constructive efforts bas on known facts". He continued:

> . . . I came to the conviction that only the discovery of a universal forn principle could lead us to assured results. The example I saw before me w thermodynamics.

Thus, Einstein adopted the principle of relativity,

> as a restricting principle for natural laws, comparable to the restricti principle of the non-existence of the *perpetuum mobile* which underlies th modynamics.

The principle implies:

> the independence of the laws (thus specially also of the law of the constan of the light velocity) of the choice of the inertial system . . . (pp. 53–5

And so it was in his 1905 paper; an essential application of the relativ principle in Section I.3 of the paper was to infer from the light postulate tl the light-speed is invariant under boosts. (Another important application of t principle in the derivation of the Lorentz transformations will be highlight in Sections IV and VI.) But what precisely is Einstein's principle of re tivity (PR), and what is its relation to its Galilean-Newtonian predecessor

Let us start with Newton, and Corollary V of the laws of motion ( 'axioms') in the *Principia*. Here it was claimed that the behaviour of a system of bodies, given Newton's laws as valid in the absolute rest frame, independent of the collective state of uniform, rectilinear motion of that syste relative to absolute space. "Clear proof" of the claim, states Newton, is giv by "the experiment of a ship", undoubtedly a reference to Galileo's famo thought-experiment in the *Dialogue Concerning the Two Chief World Systen*

It is noteworthy that Newton's theoretical proof of Corollary V (which reli on what we today call the Galilean transformations) refers only to collisi forces. Yet Newton clearly believed that the relativity principle extend beyond collision dynamics; after all, Galileo's thought-experiment explicit involved objects in free-fall and hence gravitational effects. Indeed, Corolla V is *assumed* by Newton to hold for various kinds of central forces in secti XI of Book I of the *Principia*. And the reason that such a generalisati holds is that Newtonian forces – including gravity, and forces acting on lig corpuscles, as well as the dynamical restitution constant for inelastic collisio – are assumed to be independent of the collective state of absolute moti of the system of bodies in question. But this is also the reason, as Barbo recently pointed out, for the *non-sequitur* in Newton's proof of Corollary such velocity-independence (of the collision force, or any other) is not strict a consequence of his laws (this point appears in Barbour's magnificent study, 1989, pp. 31–32, 577–578). Not that compatibility of velocity-*dependent* forces with the laws of motion *and* the Galilean transformations is assured, as I hope the discussion in Section VII below will show. But Barbour is essentially right: *no simple, definitive* proof of the Galilean invariance of Newton's laws is possible without an extra non-trivial assumption concerning forces (and a related one for masses, too)[9].

Galileo's ship experiment did not explicitly mention optical or electromagnetic effects. But it would be foolhardy to conclude that they constituted for him a probable counterexample to the inertial relativity principle, given the role it played in his defence of Copernicanism. In the case of Newton, the point can be put more forcibly. Nothing in his semi-corpuscular theory of light indicated a possible violation of relativity in optics[10], and as regards magnetism, Newton explicitly referred to it in the *Principia* as just one of the several possible central ('centripetal') forces in nature which come under the sway of his laws of motion (Newton 1934)[11]. Magnetism was in principle as much part of Newtonian *mechanics* as gravity was, and although it had not the same dependence on distance, there is every reason to think that it was likewise to be read as velocity-independent. (Newton's less frequent references in the *Principia* to the then even more obscure phenomenon of electrical activity do nothing to alter this overall picture.)

The classical, Galilean-Newtonian principle of relativity was then almost certainly universal in scope, holding for all the known fundamental forces[12]. Furthermore, neither Galileo's nor Newton's formulation of the principle (as opposed to its '*proof*' in Corollary V) referred, nor needed to refer, to the Galilean transformations. It was essentially the statement that the outcome of any two replica experiments performed in relatively moving 'spaces' would be identical, when the same initial conditions were set up in each space, and when either space is at rest or moving at uniform speed in a straight line relative to the surface of the earth (Galileo) or absolute space (Newton)[13]. Such a claim 's well-defined whether or not standard rods and clocks contract and dilate spectively once in motion relative to such spaces, in other words whatever (linear) coordinate transformations are between chosen coordinate systems hem. (The relevance of this point will be made clear below.)

'he emergence and success of the wave theory of light and the eventual mption of optics within a unified field theory of electricity and mag- in the 19th century of course changed the picture considerably. Since nagnetic waves were originally seen as propagating disturbances in a cal ether, the understanding was roughly that electromagnetism same status *vis-à-vis* the Newtonian relativity principle as the on of sound in a static medium. The equations of propagation of igh some ponderable medium such as air do not of course *refute* (though they do not take the same form in all inertial frames when emains static); they are known *a priori* to be outside its scope y due to the obvious symmetry-breaking nature of the static medium. (Evidence for the static nature of the *bulk* of the luminiferous ether essentially rested on the phenomenon of stellar aberration[14], but variations of that same experiment and others required that transparent bodies in motion dragged part of the ether contained in them according to Fresnel's formula.) By the time we arrive at the Lorentz theory of the electron, however, for reasons we have seen above the *prima facie* role of the ether as a natural symmetry-breaker in this sense is not quite so obvious, despite the fact that it is now postulated to be *entirely* static. Indeed, it transpired that its role in the light of Lorentz's theorem of 'corresponding states' – his attempt to account for the continuing lack of success in experimentally detecting motion through the ether – is that of securing a (so far) *undetectable* standard of rest[15].

In his search for a 'restricting principle' in order to develop a non-'constructive' theory of the nature of space and time, or rather the behaviour of real rods and clocks when viewed to be in motion, and thus eventually establishing the electrodynamics of moving charged bodies, Einstein's first step was to declare that this entity – the standard of absolute rest in electrodynamics – is *intrinsically* unobservable[16]. At the beginning of his 1905 paper, he discusses physical grounds for the "conjecture that to the concept of absolute rest there corresponds no properties of the phenomena, neither in mechanics, nor in electrodynamics . . . "[17] and proceeds to 'raise' the conjecture to a 'postulate'. In Section I.3, he introduces a yet *stronger* claim:

> The laws by which the states of physical systems undergo changes are independent of whether these changes of states are referred to one or the other of two coordinate systems moving relatively to each other in uniform translational motion.

Einstein's 'restricting' principle of relativity (PR) thus incorporates his 'postulate' above that "for every reference system in which the laws of mechanics are valid, the laws of electrodynamics and optics are also valid". But notice that PR is more general than the postulate: it implicitly applies to *all* the fundamental laws of physics. We shall see in Section VI that the derivation of the Lorentz transformations in fact requires this stronger claim.

PR is the statement that in all inertial reference frames, the form of the fundamental laws – which includes the values of the constants appearing in them[18] – is the same if the coordinate systems in each frame are similar. (It is implicit that 'form' here means *simplest* form, since otherwise PR so stated vacuously applies to *any* theory written in generally covariant form; we return to this point in Section VII.) Once again, the statement is not intrinsically tied to a specific form of the (linear) transformations between such coordinate systems. Indeed the whole point of the 'kinematical' part of Einstein's 1905 paper is to reveal the implications of PR *in conjunction with the independent light postulate* for the apparent behaviour of arbitrary, ideal rods and clocks in motion, given also his light-based convention for synchronising distant clocks (a natural choice given the light postulate) and the usual spatio-temporal symmetries in the rest frame. Furthermore, it is hard to see how PR differs substantially in meaning and scope from the classical relativity principle discussed above, given that Newton's first law of motion (the law of inertia) holds in all the relevant reference frames in both cases. In short, by denying Lorentz's ether any hope of eventually fulfilling its stated function, Einstein had rendered it as otiose as Newton's absolute space, and reinforced Newton's (and arguably Galileo's) belief that the fundamental laws relating to *all* physical interactions lie within the domain of applicability of the relativity principle.

## IV. THE FABLE OF KEINSTEIN

In 1705, an obscure young European, largely self-taught natural philosopher named Albert Keinstein anticipated the method Einstein was to use to derive relativistic kinematics two hundred years later. Keinstein already realised that the principle of relativity was not strictly a consequence of Newton's laws and sought to examine its consequences once elevated to the status of a postulate. Like Einstein, he looked to current physics for additional principles to combine with PR.

Keinstein was aware that one possibility (not that it was widely discussed at the time) was the principle that any signal or body propagating with infinite speed in absolute space would do so independently of the motion of the source of that signal. Combined with PR, this entails the existence of an invariant speed, namely the infinite speed. But Keinstein regarded such a notion with caution; the instantaneous action-at-a-distance in Newton's gravitational theory was regarded on the Continent as regrettable, and at any rate it provided no grounds for thinking that a *manipulable* signal with infinite speed was attainable: the principle seemed unphysical. Instead Keinstein looked to Newton's laws, and particularly the second. I now give the argument in full.

Keinstein considered the pair of inertial frames $S$, $S'$ corresponding to Newton's absolute rest frame (in which the laws of motion are postulated) and a moving frame, respectively. (It was known that the 'relative space' in $S$ is Euclidean; that the same holds for $S'$ follows from the principle of relativity.) The coordinate systems adapted to the frames were assumed to be in what is now called the 'standard configuration', with $S'$ being measured to move relatively to $S$ along the positive $x$ -direction with velocity $v$. Keinstein realised that the temporal coordinates in $S$ and $S'$ have no physical significance until the convention is established in each frame which 'spreads time through space', i.e. which determines how distant clocks are synchronised. He naturally considered *clock transport* as the appropriate choice of synchrony procedure. Finally, Keinstein assumed that standard, ideal rods and clocks remain so under boosts, and that the coordinate transformations from $S$ to $S'$ are linear. (We discuss the grounds for this condition in the next section.)

It can readily be shown that if spatial isotropy holds in $S$, the linear boost transformations between coordinates arranged in the standard configuration are given by:

$$x' = a(x - \nu t) \tag{1.1}$$
$$y' = by \tag{1.2}$$
$$z' = bz \tag{1.3}$$
$$t' = dx + et \tag{1.4}$$

where $a$, $b$, $d$, and $e$ are functions of $\nu$. Now for some body moving inertially along the positive $x$-direction in $S$ with velocity $\dot{x}$, the velocity transformation formula resulting from (1) is

$$\dot{x}' = \frac{\{a(\dot{x} - \nu)\}}{\{d\dot{x} + e\}} \tag{2}$$

Furthermore, for some body moving along the positive $x$-direction in $S$ *with acceleration* $\ddot{x}$ the acceleration transformation formula resulting from (1) is

$$\ddot{x}' = \frac{\{a(e + d\nu)\ddot{x}\}}{\{d\dot{x} + e\}^3} \tag{3}$$

where $\dot{x} = \dot{x}(t)$ is the instantaneous velocity of the body as measured in $S$, so that $\ddot{x}'$ is generally a function of $t'$, even when the acceleration as measured in $S$ is uniform[19]. Now suppose, as Newton believed, that both inertial masses and forces are velocity-independent in $S$. From PR and Newton's second law it now follows that the *acceleration of a body is numerically invariant.*

Invariance of acceleration for all values of $\nu$ and $\dot{x}$ implies in (3) that $d = 0$ and $e^2 = a$.

If we now reconsider a uniformly moving body and put $\dot{x} = 0$ in (2), then $\dot{x}'$ will be the 'reciprocal' velocity of frame $S$ as measured to move along the positive $x'$-direction in $S'$, which we denote by $\bar{\nu}$. From (2), we have

$$\bar{\nu} = -(a\nu)/e$$

Now the *principle of reciprocity* dictates that $\bar{\nu} = -\nu$, i.e. that $a = e$. (Einstein followed Keinstein in assuming reciprocity in his 1905 derivation, but was not aware that it is actually a consequence of light-speed invariance, as we see below[20]. But since $e^2 = a$, we have that $a = e = 1$ (assuming that as $\nu \to 0$, $a \to 1$, hence $a \neq 0$) . This leaves only the coefficient $b$ in the transverse transformations (1.2, 1.3). Now PR implies that the inverse transformations from $S'$ to $S$ also take the form (1) with $-\nu$ substituting $\nu$, given reciprocity. (PR essentially implies that the coordinate transformations form a group, in the sense that elements of the group all have the same form: the associated coefficients are the same functions of velocity. Here Keinstein, like Einstein after him, is effectively appealing to the existence of the inverse element of the group. We return to this crucial argument in Section VI.) So we have

$$b(\nu) = 1/\{b(-\nu)\} \tag{4}$$

The isotropy of space in $S$ implies that

$$b(\mathrm{v}) = b(-\mathrm{v}) \tag{5}$$

since otherwise any contraction caused by motion along the $x$-axis in a rod pointing in a perpendicular direction would depend on the direction of that motion. Thus from (4) and (5) we obtain $b = 1$ (assuming again that as $\mathrm{v} \to 0$, $b \to 1$). The standard Galilean transformations have now been derived.

Readers will no doubt be rightly skeptical that anyone in 1705 could erect such a modern-looking argument. The fictitious Keinstein was not only employing technical concepts and insights that were only to gain recognition much later, he was resorting to a form of reasoning that would certainly have surprised his contemporaries. Note however that it was not the use of PR as a postulate or 'restricting principle' that was new in 1705. Huygens and indeed Newton (prior to writing the *Principia*) had used it as such[21]. What is significant is the notion of applying PR to generalised coordinate transformations, to obtain moreover what would probably have seemed an *a priori* given[22]. The point at any rate is that PR can be utilised to *derive* Galilean kinematics in Newtonian mechanics, without postulating the invariance of the infinite speed[23].

We now compare Keinstein's derivation with Einstein's, or rather with a simplified version of it. $S$ is now Einstein's 'resting' frame. Invariance of the isotropic, one-way light-speed $c$ (a consequence as we have seen of PR and Einstein's light postulate; recall that the invariance principle as stated requires clocks to be synchronised according to the Einstein convention) itself can straightforwardly be shown to imply for inertial coordinate systems in the standard configuration (1) that

$$x' = \pm k\gamma(x - \mathrm{v}t) \tag{6.1}$$

$$y' = \pm ky \tag{6.2}$$

$$z' = \pm kz \tag{6.3}$$

$$t' = \pm k\gamma(t - \mathrm{v}x/c^2) \tag{6.4}$$

for $|\mathrm{v}| < c$, where $\gamma$ is the Lorentz factor $(1 - \mathrm{v}^2/c^2)^{-1/2}$, and $k$ is a positive dimensionless function of v. The minus part in the $\pm$ signs in (6) is again removed by requiring that the transformations reduce to the identity transformation in the limit $\mathrm{v} \to 0$. Notice that reciprocity is a consequence of (6); Einstein had no need to introduce it as a separate assumption.

We can now use PR and isotropy as was done in deriving (4) and (5) above, this time for the transverse coordinate transformations (6.2, 6.3), thus obtaining $k = 1$[24]. (Isotropy of time dilation resulting from motion of clocks along the $x$-axis, as well as of length contraction for moving rods aligned along that axis, is also now assured given the fact that the factor $\gamma$ in (6.1) and (6.4) is an even function of v[25]. Again in inferring contraction and dilation effects from the coordinate transformations, we are employing the important assumption above that standard, ideal rods and clocks remain so under boosts.

We return to this point in section VII.) This completes the derivation of the Lorentz transformations[26].

A final word about *correspondence*. It is perhaps surprising how intricate the task is of clarifying precisely how traditional Galilean kinematics re-emerge in the appropriate 'classical limit' of the theory.

First, the condition $v/c \ll 1$ (with v defined as above) is clearly not sufficient; the effect is to render the value of the Lorentz factor $\gamma$ near unity, but the non-classical space-dependent term in (6.4) remains. Thus, it must be supposed that distances are not 'too big': we must further impose $x \ll c^2/v$ (see, for example, Taylor and Wheeler 1966, problem 38, p. 84). But now consider a body moving with velocity $u$ with respect to $S$ (and $u'$ with respect to $S'$ ). For $u \sim c$, it is the difference between the relativistic velocity transformation formula linking $u$ and $u'$ (which follows from the Lorentz transformations) and its Galilean counterpart, even when the above conditions are satisfied, that accounts, for example, for the non-null results in the celebrated 1851 Fizeau interference experiment confirming the Fresnel drag coefficient. (This point is clarified in Section VII below.) Thus we must of course put $u/c \ll 1$ as well. So now the kinematics formally look just like their classical counterpart – but notice that the transformations are still defined with respect to the Einstein synchrony convention! Pre-relativity physicists, on the other hand, would presumably have adopted, had they thought about it, *clock transport* as the method for synchronising distant clocks in defining the boost transformations, as Keinstein did. Thus a *full* account of the matching of traditional Galilean kinematics with those given in SR in the 'classical domain' requires I think a final ingredient. This is the Eddington-Winnie theorem stating that *slow* clock transport synchrony in SR is tantamount to introducing the Einstein convention[27].

## V. FRIEDMAN'S GEOMETRIC DERIVATION OF THE LORENTZ TRANSFORMATIONS

In his 1983 study of space-time physics, Friedman compares the standard derivation above of the Lorentz transformations based on Einstein's 1905 postulates with a more geometrical and supposedly clearer derivation constructed without the principle of relativity (*op. cit.*, Sections IV. 1–2). In his discussion of the former, Friedman gives particular emphasis to the assumption therein of the linearity of the coordinate transformations between inertial frames. He rejects Einstein's obscure justification of it in terms of the homogeneity of space and time, and argues that it is only secured under the premiss of the flatness of four-dimensional space-time, and the assumption that the transformations preserve the flat affine structure (p. 141). This would make Einstein's (and Keinstein's) derivation rely on a surreptitious four-dimensional geometrical assumption, something quite alien to their approach. But in fact no such reliance exists. It has long been known that in order for the frames in question to be *inertial*, so that uniform velocities are transformed into uniform velocities, linearity is required if the finitude of coordinates is preserved under the transformations and the coefficients therein are differentiable up to the third order (see Fock 1959, pp. 12–16 and 377–384 and Torretti *op. cit.*, pp. 75–76)[28]. Finally it is noteworthy that Friedman nowhere connects the requirement in the derivation that the coordinate transformations form a group in the requisite sense with the principle of relativity, which misleadingly appears to play no role in his account of the derivation.

Friedman's own geometrical derivation presupposes a four-dimensional differentiable manifold endowed with a flat affine connection and a compatible indefinite (Minkowski) metric 'field' with signature $\langle 1, 3\rangle$, which allows for a division of the tangent (vector) space at each point into three sectors – let us call them the '$t$-like', '$s$-like' and 'null' sectors. Lengths can then be defined for segments of *inter alia* a '$t$-like' curve (one whose tangent vector field is everywhere in the $t$-like sector). It is now assumed that the motion of a particle – or a clock – always defines a $t$-like curve in the manifold, so let us call such curves 'time-like'. The connection with kinematics is further elaborated *via* the chronometric assumption, which 'interprets' a segment length on such a curve as the actual reading of an ideal clock (the physical constitution of which is not specified), whose trajectory is the curve in question and whose operation covers the segment in question. (Friedman omits to operationally define the lengths of $s$-like, or 'space-like' curves, but the reader can guess what the definition is. It is also implicit that light ray trajectories are associated with null-like geodesic curves.) An independent postulate of motion then requires that *free* particles define only (time-like) curves satisfying a certain generally covariant differential equation involving derivatives with respect to this (proper) clock time: these curves are time-like geodesics.

'Inertial' coordinate systems are then defined as those in which the components of the affine connection vanish and the metric has the coordinate form diag$(1, -1, -1, -1)$. In such coordinate systems it follows from the geodesic assumption above that free particles satisfy the law of inertia, as desired.

Finally, Friedman investigates the coordinate representation in an inertial system of the 'symmetry group' which leaves the 'absolute' objects – the affine connection and metric – invariant. Invariance of the connection implies linearity of the coordinate transformations (see above) and that of the metric (under a continuous group) results in a 10-parameter Lie group containing the generators of translations, rotations and the Lorentz transformations. It is then easy to see that members of the full 'Lorentz' group take inertial into inertial coordinate systems, preserving the law of inertia. And any two such systems which define a uniform relative velocity in the standard configuration are related by the Lorentz transformations.

## VI. A CRITIQUE OF FRIEDMAN'S DERIVATION

It must be clarified first that Friedman has not done the impossible and derived the Lorentz transformations without picking out a clock synchrony procedure! Although he does not explicitly address the issue in his derivation, it follows from the definition of inertial coordinate systems above that null-cones at each point in the manifold are symmetrically oriented: the invariant one-way light-speed is isotropic in such systems. Thus the synchrony convention implicit in the derivation is of course Einstein's.

But what is noteworthy about Friedman's derivation is that it is ostensibly independent of any dynamical considerations. In particular it does not appear to depend on the principle of relativity PR. Friedman's space-time is not (yet) equipped with dynamical laws governing fields or other objects that couple with dynamical parameters associated with the given particles to cause non-inertial motion. To what would the principle of relativity refer? (We return to this question in Section VIII. Notice that the formulation of PR in Friedman's book appears after his treatment of relativistic kinematics.) The existence of the symmetry group is a consequence of the basic geometric structure of the manifold alone.

Yet it is precisely the remarkable dynamics-independent nature of Friedman's argument that is its weakness.

There are two principal, related queries one might raise concerning Friedman's derivation. Why *do* clocks, of arbitrary construction, act as hodometers of Minkowski space-time? And why *must* physical kinematics, i.e. the behaviour of physical rods and clocks when viewed in motion, be related to the symmetry group of the manifold? Friedman's treatment of this latter connection in his 1983 book is, I think, not consistent. As I shall discuss this problem in detail in the next section, I will put the second query aside for the moment. The reasons for the first query related specifically to the behaviour of clocks will be given shortly. First, I would like to make a couple of preliminary remarks.

It is arguable that without recourse to dynamical considerations, the highly non-trivial assumption in Friedman's derivation that particles – even those undergoing non-inertial motion – define only time-like curves in the space-time manifold lacks any physical motivation. The fact that acceleration of a massive particle to light-speed requires, in standard relativistic dynamics (which are not a consequence of relativistic kinematics alone) an infinite source of energy, explains at least why time-like trajectories for such particles stay that way.

Another aspect of the dynamics-free approach to space-time geometry was already pointed out by Sklar (1977) in 1977. He noted that in general relativity (GR), free particles only define time-like geodesics if they are spherically symmetric and spin-free[29]. Since Friedman deals only with SR in the part of his book we are concerned with, we can ignore this complication. But Sklar was concerned with a more basic problem – that of ascertaining what particles count as 'free' in the first place in order to define the geodesic assumption. As Sklar says, "we know what a free particle is . . . given our vast array of background theory". Such background theory is dynamical – it tells us when resultant forces on particles are either negligible or susceptible to compensation so as to render them effectively 'free'. (This is precisely how inertial frames are defined in the non-geometrical derivation of the Lorentz transformations[30]. ) The conclusion seems inevitable that either a significant dynamical element is implicit in Friedman's argument, or the means of *specifying* free particles is obscure.

I now return to the first query above, concerned with the status of Friedman's chronometric assumption. This issue requires some care; let me start with some remarks made by Pauli in his famous 1921 review article on relativity. After comparing Einstein's and Lorentz's points of view regarding the origin of the Lorentz transformations, Pauli recognises the "great value" of Einstein's derivation in so far as it makes no constructive assumptions about the constitution of matter, and he proceeds to write:

> Should one, then, . . . completely abandon any attempt to explain the Lorentz contraction atomistically? We think the answer to this question should be No. The contraction of a measuring rod is not an elementary but a very complicated process. It would not take place except for the covariance with respect to the Lorentz group of the basic equations of electron theory, as well as of those laws, as yet unknown to us, which determine the cohesion of the electron itself. We can only postulate that this is so, knowing that then the theory will be capable of explaining atomistically the behaviour of moving measuring rods and clocks. However, the equivalence of the two coordinate systems in relative motion will always have to be kept in mind. (Pauli 1981, p. 15)

(Related remarks made by Einstein in 1949 are cited below in Section IX.) Pauli is reminding us of a point concerning which Lorentz was painfully aware: that for a *complete*, *consistent* account of the length contraction of 'rigid' rods of arbitrary construction, the complex internal dynamics of the rod cannot be ignored. (Pauli clearly intends the point to apply to time dilation for clocks as well, but prior to Einstein's work Lorentz had no coherent view on time.) Indeed, recognition of the fact that a material rod cannot be held together in equilibrium by just electromagnetic forces was the main factor in the increasingly *ad hoc* status of Lorentz's attempts to reconcile his electron theory with experiments such as that of Michelson and Morley. Appeal had to be made also to other forces whose precise nature was unknown at the turn of the century, yet whose transformation properties somehow had to be correlated to those of the electromagnetic forces. More recently, in his attempt to resuscitate the dynamical explanation of length contraction in terms of Maxwell-Lorentz electrodynamics, Bell (1976) ran up against the same issue: in accounting for the apparent behaviour of solid bodies when viewed to be in motion he realised that one must

> . . . assume with Lorentz that the *complete theory* [of matter accounting not only for atomic stability but also for nuclear structure] is Lorentz invariant . . .

While I applaud Bell's appreciation of the dynamical element of so-called 'kinematical effects' in SR, it is arguable that he failed, unlike Pauli, to grasp that Einstein's 1905 derivation of the Lorentz transformations is itself dynamical in just the right way.

The principle of relativity PR has, as we saw in Section IV, two applications in Einstein's derivation: first to secure light-speed invariance given his light postulate, and secondly to constrain the phenomena of length contraction and time dilation. It is the second application I wish to examine now. I understand PR here to apply to *all* the dynamical laws governing the behaviour of the constituent parts of ideal rods and clocks of whatever construction – thus ensuring that the formulae for contraction and dilation (if any[31]) as a function of velocity are the same in all inertial frames. Under this view the *prima facie* oddity of a 'kinematical' effect being constrained by a *dynamical* principle is dispelled: PR is simply taking care of the Lorentz-Pauli-Bell problem at a stroke. (Whether Einstein quite saw the argument in this way as early as 1905 is debatable; but see Section IX below).

Notice that in order to predict the behaviour of *accelerating* clocks in this approach, it is necessary, as is well known, to introduce a further hypothesis. This is that the effect of a period of acceleration is nothing more than the effect of the Lorentz-Einstein dilations associated with the instantaneous velocities 'traversed' in the that period. (A related assumption is also necessary of course for accelerating rods.) This assumption is non-trivial; it does not follow from Einstein's postulates.

Let us now, by way of comparison, consider the role of clocks in Friedman's derivation and the status of the chronometric assumption, which states that these clever bodies are hodometers of the metric in space-time and hence that they dilate (in accordance with the Lorentz factor) in uniform motion and satisfy the additional clock hypothesis for accelerations. In reality of course clocks have internal structure. But since no constraints on dynamical laws are specified in Friedman's account (so far), no detailed notion of their structure and behaviour can be found. How then do they do what Friedman says they do?

In addressing this question, it might be said that it is simply in the nature of a clock that it measures the length of its own *time*-like trajectory in Minkowski space-time. But the chronometric assumption is *not* a consequence of the assumption that trajectories of clocks (and particles in general) define only 'time-like' curves in the manifold. However one chooses to characterise *a priori* the nature of an ideal, localised clock (for instance, one might say that its orbits in phase space are closed), it is hard to explain – without appeal to dynamical considerations – why such a body can read part of the postulated pre-existing Riemannian metrical structure of space-time.

Why for example does the (coordinate independent) number of ticks recorded by a clock between two events on its world-line depend not only on the events but also on the said world-line joining them? (In the next section, mention will be made of what Friedman calls 'Galilean space-time', in which clocks – again of any kind – intrinsically read a different postulated geometrical structure. Another version of the question is: *how do clocks know which space-time structure they are in?*)

In adopting Friedman's approach it is hard, then, to resist the view that such behaviour must be the result of an obscure *interaction* between that absolute structure and clocks – one unlike any other process of measurement in physics in the sense that there can be no interaction Hamiltonian for it.

Now not all students of space-time geometry necessarily see the metric structure as existing 'prior' to that of clocks, whose job it is to read it, as Friedman appears to believe. Nonetheless, even if the postulated metric is *defined* in terms of (and hence is reducible to) the behaviour of clocks, a good dose of mystery remains. Recall the point made by Ehlers, Pirani and Schild (1972) in their well-known 1972 study of the geometry of (not necessarily flat) space-time:

> It is difficult to derive *from the behaviour of clocks alone* . . . the Riemannian form for the separation [of two infinitesimally displaced points on the clock's world-line] . . . rather than some other, first-degree homogeneous, functional form in the [infinitesimal coordinate displacements] . . . Postulating this form axiomatically, one forgoes the the possibility of understanding the reason for its validity. (See also the discussion in Synge 1965, Sections I. 10 and I. 11)

The authors also question the *a priori* plausibility of the postulated connection between the chronometric assumption and the law of geodesic motion for particles and light:

> The second difficulty is that if the [components of the metric tensor] are defined by means of the chronometric hypothesis, it seems not at all compelling . . . that these chronometric coefficients should determine the behaviour of freely falling particles and light rays too. Thus the geodesic hypotheses, which are introduced as additional axioms in the chronometric approach, are hardly intelligible; they fall from heaven like [the Riemannian form for infinitesimal separations].

Ehlers *et al.* prefer in their work to *construct* the Riemannian nature of space-time on the basis of certain basic assumptions about light propagation and particles in free fall, so that neither the chronometric nor geodesic hypotheses are introduced *ab initio*.

Now it had earlier been shown in the work of Marzke (1964) that in both special and general relativity, *given* the geodesic hypotheses it is possible to construct clocks by means of free particles exchanging light rays, so that the chronometric assumption for *such* clocks is unnecessary. However, this would only partially solve Friedman's problem were he to adopt such a stratagem in SR. For how is one to account for the fact, so clearly a consequence of Einstein's approach given the 'clock hypothesis' for accelerative motions, that clocks behave in SR independently of their constitution? Why, in other words, are pendula, caesium clocks and certain biological systems all hodometers (of varying precision) of space-time *as well as* the mentioned light-particle clocks? Why *do* they dilate in the same way? (As Sklar pointed out in 1977 (*op. cit.*), this is a problem also facing the more constructive approach of Ehlers *et al.*, together with the difficulty alluded to earlier concerning the specification of 'free' particles.)

## VII. THE SPACE-TIME SYMMETRY GROUP AND KINEMATICS

Friedman is of course well aware that historically the empirical discovery that the field equations of Maxwell-Lorentz electrodynamics hold in all inertial frames was the spur to the discovery of SR. Earlier in his book he wrote that

> . . . the correct theory of electrodynamics must employ a space-time structure completely different from the space-time of Newtonian physics. (Friedman 1983, p. 108)

This followed his discussion of 'classical electrodynamics': an attempt to embed Maxwell's equations in Newtonian, or rather 'Galilean' space-time. But in this earlier discussion, although (or perhaps because?) it demonstrates Friedman's sensitivity to dynamical considerations and their relation to the geometry of space-time, the connection postulated above between physical kinematics and the symmetry group of the background 'absolute' geometrical structure of the space-time manifold is broken. In this section, I shall first examine the tension between Friedman's discussion of 'classical electrodynamics' and his above derivation of the Lorentz transformations, and then address the separate question as to how 'classical electrodynamics' is to be understood.

I shall not enter into the minutiae of Friedman's geometrical construction of Galilean space-time (Sections III. 1–2), except to say that it is equipped, like the special relativistic manifold above, with a flat affine connection, with the metric tensor above now replaced by other geometric objects which imply the absolute nature of time and the Euclidean nature of the relative spaces associated with inertial frames. (Galilean space-time is not a metric space.) Again, particles are assumed to define time-like curves and free particles define geodesics. Inertial frames are defined analogously to those in relativistic space-time. What is important for our purposes is that (the coordinate representation of) the continuous symmetry group which preserves the absolute objects is the extended Galilean group, containing as subgroups the orthogonal 'Euclidean' group (spatial rotations and translations), the group of time-translations, and the group of Galilean transformations. These last transformations play the same role *qua* boosts as the Lorentz transformations in special relativistic space-time[32].

Now imagine an inertial coordinate system to exist in which the Maxwell-Lorentz field equations (with sources) hold, along with the velocity-dependent Lorentz force law and velocity-independent charges. Friedman immediately infers from the Galilean transformations and from the fact that the field equations imply constancy (source-speed independence and isotropy) of the speed of propagating electromagnetic waves that the field equations can only hold in a set of inertial coordinate systems that are all in relative *rest*. He then shows that further geometrical objects can be introduced into Galilean space-time, along with a generally covariant equation of motion for charged particles, with the effect that this condition is satisfied: the familiar field equations of electrodynamics are valid in only 'rest' systems (or in a single 'inertial frame' in common parlance). This is 'classical electrodynamics' (Section III. 5).

I shall turn shortly to the physical status of this construction; my concern now is rather with a particular aspect of Friedman's interpretation of it. Notice that because of the nature of the new geometrical objects in classical electrodynamics, the symmetry group of the manifold is now no longer the extended Galilean group. It has collapsed down to the Euclidean transformations plus time translations ($O^3 \times T$). And yet the Galilean transformations are clearly still assumed by Friedman to represent boosts in the theory: despite not comprising a sub-group of the symmetry group they are still taken to account for physical kinematics. (In fact, on p. 108 Friedman refers to the experimental discovery that the field equations hold in *all* inertial frames, in violation of 'classical electrodynamics', in order to argue that the Galilean transformations are "incorrect".)

Actually, it is arguable that the connection between kinematics and the symmetry group of the manifold had already been broken implicitly in Friedman's formulation of 'Newtonian' space time, which appears in his book prior to the introduction of 'classical electrodynamics'. Newtonian space-time is Galilean space-time with an additional absolute geometrical object field, whose role is to pick out a privileged inertial frame, the 'relative' space of which is Newton's 'absolute space'. The symmetry group of this manifold is just that of 'classical electrodynamics': $O^3 \times T$. Now Friedman does not do much physics in Newtonian space-time, but he does say that the generally covariant theory of Newtonian gravity he formulates in flat Galilean space-time could also be given in it (Friedman *op. cit.*, Section III. 3). (It must have *some* such application, otherwise its introduction is pointless.) But in that case the symmetry group of the classical gravitational theory would not contain the subgroup of Galilean transformations. And what are the kinematics involved in the theory if not given by these transformations?

Finally in this connection the case of general relativity (GR) itself should perhaps be mentioned. As Friedman notes (p. 183), the symmetry group in this theory is just the manifold itself; its coordinate representation is the trivial group of all non-singular differentiable transformations. Yet for sufficiently flat regions of the manifold, the Lorentz transformations still of course play an important approximate role as boosts between 'local inertial' frames. Thus the connection between 'kinematics' (where applicable) and the symmetry group of the manifold is again non-existent – any such connection must now involve the tangent space structure, if it exists at all.

Where does all this leave Friedman's geometric derivation of the Lorentz transformations above? Following the example of 'classical electrodynamics', one is tempted to populate Friedman's special relativistic space-time above with some kind of dynamical object(s), whose generally covariant equations of motion violate PR: the effect is likewise to introduce a new *absolute* geometrical object into space-time which serves to pick out a privileged inertial frame. This would inevitably have the effect of reducing the symmetry group of spacetime to a subgroup of the extended Lorentz group which excludes the boost transformations. (Such a theory might be the generally covariant form of the 1905 Langevin-Bucherer theory of the deformable electron, which was consistent with light-speed invariance but in which $k$ – as in eq. (6) above – is $\gamma^{-1/3}$ and hence inconsistent with PR[33].) The "correct" coordinate transformations would have no connection with the underlying geometrical symmetry group[34]. The only way to avoid this mis-match is presumably to bar the introduction of such a dynamical theory into special relativistic space-time. But this suggests that dynamical Lorentz covariance must after all be presupposed in Friedman's derivation of the Lorentz transformations, in which case its claim to be a *bona fide* 'derivation' is threatened.

So far I have been treating Friedman's 'classical electrodynamics' (and an analogue of it in relativistic space-time) as a well-defined theoretical structure. But this is questionable: the theory as posed suffers from conceptual complications that arise precisely because it puts geometry/kinematics before dynamics. Recall first Bell's recent reminder that length contraction for solid bodies is in part – but not wholly – a consequence of the Maxwell-Lorentz field equations (plus a rudimentary atomic model). This work serves to remind us that 'rigid' material rods are not likely to retain their rest length in the frame in which the field equations are valid (let us call it $S_0$) when they undergo rapid motion – even if accelerated 'gently' to that state of motion. (Electromagnetic forces alone cannot account for the stability of matter, and other forces must be operative too in accounting fully for the structure of arbitrary rods and clocks, as we have seen earlier. But such forces do not appear in 'classical electrodynamics', so it gives no grounds for thinking that any length contraction in a moving material rod due to electromagnetic effects is wholly compensated by other internal effects.) Consider also the 'Langevin' clock composed of a light-pulse bouncing back and forth between two parallel mirrors fixed at the ends of a rigid rod, or the composite Marzke particle-light clock mentioned in the last section. Suppose each of these composite clocks starts in some state of inertial motion – say rest relative to $S_0$ – and is then boosted into another state of inertial motion. Then it is hard to see how an observer in $S_0$ can reconcile the field equations (which of course imply the constancy of the light-speed in $S_0$) with the absence of *any* time dilation for each of the composite clocks measured in that frame.

Notice that such considerations concerning specific cases of contraction and dilation due to motion relative to $S_0$ arise *without the need to know what the coordinate transformations are to moving frames*: they follow merely from the physical equations in $S_0$, and concern measurements performed by the '$S_0$-observer'. They therefore cast serious doubt on the 'correctness' of the Galilean transformations in 'classical electrodynamics' or *any* theory containing such a frame $S_0$, *if it is assumed as usual that all standard rods and clocks remain so under boosts*, that is, if it is assumed that an inertial laboratory in undergoing a boost need not recalibrate, or change the nature of, its measuring instruments. (Recall from Section IV that this assumption appears in both Keinstein's derivation of the Galilean transformations and Einstein's derivation of the Lorentz transformations, the latter transformations reducing of course to the former in the appropriate regime of low velocities, etc.)

Another route to essentially the same point (which was clearly recognised by Lorentz) comes by considering once again the Fresnel 'drag' coefficient in electromagnetism[35], first confirmed directly (to first order) in the famous non-null interference experiments by Fizeau in 1851 and Michelson and Morley in 1886. We have already seen in Section II that for Lorentz, Fresnel's factor (involving a correction to the refractive index of a transparent body due to its motion) results from the dynamical interaction between radiation and the structure of the moving transparent body. Indeed, the effect (to first order) can be shown to be a consequence of Maxwell's equations in $S_0$ for a *slowly moving* medium, and is due to reradiation from the moving secondary radiators in the medium (see for example Panofsky and Phillips 1962, Section 11–5). The result is

$$\tilde{c} \approx c/n + (1 - 1/n^2)\text{v} \tag{7}$$

where $\tilde{c}$ and $c$ are the speeds measured in $S_0$ of light inside the moving medium and *in vacuo* respectively, v is the velocity of the medium in $S_0$, and $n$ is the refractive index of the medium measured when *at rest* in $S_0$. I take it that an *exact* dynamical derivation of $\tilde{c}$ for arbitrary velocities of the medium is not available – even in principle – without knowledge of how the other (non-electromagnetic) forces at work in the medium are affected by motion relative to $S_0$.

But recall now the point made first by Laub and von Laue in 1907, that the Fresnel factor is intimately linked to the Lorentz transformations, or rather of the velocity transformation law which they imply. Relativistic kinematics alone imply the exact result for all $\text{v} < c$:

$$\tilde{c} = \frac{\{c/n' + \text{v}\}}{\{1 + (\text{v}/n'c)\}} \tag{8}$$

where $n'$ is now the refractive index of the medium *measured in the rest frame of the moving medium.* (I am considering here the ideal case where the medium is nondispersive, so that a correction due to the relativistic Doppler effect does not arise (see Jackson 1967, pp. 362–363).) In this relativistic approach, appeal can be made now to PR to equate n′ in (8) with $n$, as defined above. To first order, one recovers (7).

On the other hand, the Galilean law of addition of velocities in 'classical electrodynamics' implies exactly for all v

$$\tilde{c} = c/n' + (1 - 1/n')\nu \tag{9}$$

Now internal consistency in the theory requires that (9) be consistent with (7) *for all* $\nu \ll c$. (The electrodynamic predictions in the theory must be consistent with the kinematics.) It is not clear to me how this can be plausibly achieved for $n \neq 1$ (i.e. when the medium has a refractive index other than that of the vacuum) if, as in the relativistic derivation, the measuring rods and clocks in the rest frame of an arbitrary nondispersive medium are the same type as (or are boosted versions of) those involved in $S_0$. The defender of 'classical electrodynamics' might argue that PR fails in the theory, so that $n \neq n'$ but from the point of view of the physics given in $S_0$, it is unclear why this should be so. Recall the lesson that Bell (*op. cit.*) attributed to Lorentz – which has nothing to do with the existence of a *privileged* frame – *viz.* that

> . . . the [totality of] laws of physics in any one reference frame account for all physical phenomena, including the observations of moving observers.

In 'classical electrodynamics', all we are explicitly given in $S_0$ are the (standard) laws of electromagnetism, which, to repeat, is hardly the totality of physics. But it is still pertinent to ask: what grounds can these laws provide for surmising that $n \neq n'$? Any such grounds should apply equally well to standard electrodynamics in which $n = n'$.

Friedman does not seem to be aware of the basic dynamical reasons for adopting the Lorentzian stance that in 'classical electrodynamics' the rods and clocks assigned to $S_0$ are *not* boostable as such, without the introduction of compensations related to motion-induced deformations. (This stance means of course that an implicit assumption in Newtonian mechanics, namely that such compensations are not required for arbitrary speeds, must be abandoned, for otherwise the kinematics in both theories will not coincide as desired[36].) Moreover, it is hard to reconcile this stance with Friedman's claim, referred to earlier, that 'classical electrodynamics' was historically ruled out by *experiment*. For in all the relevant experiments bearing on electromagnetic phenomena undertaken in frames moving relatively to the putative ether, such as the Michelson-Morley experiment, standard measuring rods and clocks were assumed to be boostable, and hence the peculiar compensations apparently required in order to test 'classical electrodynamics' were not taken into account. Indeed, it seems to me that compensations are in principle available (which would make the compensated rods and clocks at rest in the moving frame 'look', in switching from the Galilean to the Lorentz chart associated with that frame, like the familiar ones) which would render the theory, as it is constructed by Friedman, on a par with Lorentz' field theory in its final version – i.e. empirically indistinguishable from 'relativistic electrodynamics'. Admittedly, this hunch requires detailed justification. But I feel enough has been said in order to throw doubt on whether Friedman's discussion of the theory sufficiently accounts for the requirements of self-consistency within it.

Finally, there is another important conceptual lesson that Friedman draws (partly) from 'classical electrodynamics' that I wish to take issue with. It is clear that the theory, as it is posed, is supposed to violate the principle of relativity. Friedman argues (*op. cit.*, p. 149) that nonetheless no violation of (essentially) Einstein's formulation of PR based on the form independence of laws (Section III above) occurs, because of the *generally covariant* nature of the theory.

I do not want to discuss the issue of whether 'classical electrodynamics' does in fact violate PR (although the above considerations do show I think that the issue is not as simple as it seems). The claim I wish to question is rather that any theory expressed in generally covariant form automatically satisfies Einstein's version of PR. One can well imagine a theory, such as the 1905 Langevin-Bucherer electron theory mentioned above, which in generally covariant form would still simplify in a non-trivial way in a given single inertial frame – as 'classical electrodynamics' *appears* to in $S_0$ – and thus violate PR in Einstein's sense. (In this connection it is worth recalling Einstein's own emphasis (1969, p. 69) on the point that in constructing GR, one is searching for field equations which are not only generally covariant, but which are *simplest* when given in that form. For a generally covariant theory defined in *flat* space-time and *satisfying* PR, the component form of the laws may – and indeed do – simplify when restricted to the class of inertial coordinate systems. It is this which accounts for the privileged status of inertial frames in pre-GR physics, a feature which Einstein regretted and sought to avoid in GR.) I therefore resist Friedman's conclusion that the notion of form invariance of laws with respect to the class of inertial frames is "much too weak" to express PR. (Friedman is led to develop (1983, Section IV. 5) a formulation of PR for space-time theories that avoids the notion of 'form invariance' of laws altogether, one similar to an earlier careful account of frame equivalence due to Earman (1974). Friedman's formulation has been criticised on technical grounds elsewhere (Sypel and Brown 1992).) How else is one to make sense, moreover, of the fact that Einstein successfully used PR in the very sense of the form invariance of the Maxwell-Hertz field equations to derive in his 1905 paper the non-trivial transformations for the field components?[37]

## VIII. 'SPACE-TIME THEORIES' AND THE RELATIVITY PRINCIPLE

Partly as a result of Anderson's seminal (1967) study of the foundations of relativistic physics, it has become common in recent decades for space-time philosophers to formulate certain theories which predate GR (Newtonian mechanics and gravity, electrodynamics, SR) in a certain geometrical mould. This involves each theory being given in a model-theoretic format incorporating a four-dimensional differentiable manifold on which is specified *first* certain 'absolute' geometrical objects and *then* further geometrical objects with dynamical content. In recasting such theories as 'space-time theories' in this precise sense, despite lingering difficulties concerning the precise definition of the notion of 'absolute' geometrical objects therein, considerable rigour and technical sophistication has been introduced into the literature on the foundations of relativity physics, as the recent books by Friedman (*op. cit.*) and Earman (1990) attest.

However, not all the theorists involved appear to concur on the significance of certain details of this approach. Earman (*ibid.*, pp. 45–47) for example, is careful to distinguish between the 'space-time symmetries' (related to the symmetry group) of a given manifold and the 'dynamical symmetries' (which appear to incorporate the boosts) of a given dynamical theory defined on that manifold. I take it that until dynamical objects and their laws are 'introduced' into a given space-time, Earman's view must be that coordinate transformations representing real boosts are simply not defined. Given this viewpoint, Friedman's geometrical derivation of the Lorentz transformations in Minkowski space-time must surely seem wrong-headed. (Earman's recognition of a connection between such coordinate transformations as they are commonly understood and dynamics is commendable, but it is of a limited kind: he does not appear to be concerned with the internal constitution of allowable rods and clocks and thus does not see the real complications inherent in 'classical electrodynamics', for example[38].)

Friedman himself states that his interest in Anderson's approach arose out of the latter's success in clarifying the 'puzzle' concerning the correct formulation of the traditional relativity principles. Friedman writes:

> These principles are standardly presented in terms of our freedom to choose any member of a certain class of coordinate systems to represent the laws of a given theory. The Newtonian principle of relativity is expressed in terms of our freedom to transform coordinate systems by a Galilean transformation, the special principle of relativity in our freedom to transform coordinate systems by a Lorentz transformation, and so on. But this way of putting the matter makes the traditional relativity principles appear trivial. Are we not always free to transform coordinate systems by any transformation whatsoever? Will not any theory, regardless of its physical content, be *generally* covariant? If so, what physical content do the traditional relativity principles express? (*op. cit.*, p. xii)

Anderson's solution was to introduce the distinction between the notions of *covariance* and *invariance/symmetry*. Thus all the pre-GR theories mentioned above are said to have the same covariance group (being formulated in *generally* covariant form), but differ in respect of their symmetry groups, which are defined in relation to the absolute objects in the theories.

I have already expressed my doubts about the success of this approach, or rather Friedman's version of it: the *a priori* connection between the symmetry group and physical kinematics is problematic, and the prioritising of absolute geometry over dynamics – an option not available in GR – leads to the difficulties we have seen in 'classical electrodynamics'. But it is not clear to me that even Friedman's *motivation* for adopting such an approach is well-founded. The puzzle concerning the relativity principles that Friedman refers to above evaporates when it is realised that the connection between the relativity principle and specific coordinate transformations is contingent, and not intrinsic, in the first place. That there exists a single relativity principle with respect to inertial motion (and not more) was argued in Section III. In Section IV it was seen (or rather recalled; the point is well known) that a specific set of coordinate transformations results only when the principle is combined with independent physical postulates (essentially Newton's second law along with force/mass velocity-independence in the 1705 case of Keinstein, the light postulate in the 1905 case of Einstein).

To emphasise just how different the view defended here is from what is now the increasingly accepted position amongst accomplished philosophers of space and time, I quote from a recent important paper on the nature of space-time by Norton, in which he writes:

> The principle of relativity of inertial motion in special relativity is the only unproblematic relativity principle of modern formulations of relativity theory. *It does not arise as a fundamental postulate of special relativity, but as an important theorem dependent on the symmetries of the Minkowski metric*. (Norton 1989, p. 1251; my emphasis[39])

Norton proceeds to give a formulation of the relativity principle, one reminiscent of Earman's discussion (1974) of frame equivalence in the language of space-time theories (and similar in spirit to the formulation found in Friedman's book). Consider a space-time theory, each model of which consists of the dynamically-allowed triple $\langle M, g_{ab}, T_{ab}\rangle$ where $M$ is a space-time manifold, $g_{ab}$ is the Minkowski metric and $T_{ab}$ a dynamical tensor field. Norton continues:

> To state the principle, we recall that that an inertial frame of reference in special relativity is a congruence of parallel timelike geodesics of the Minkowski metric. Then:
>
> *Principle of Relativity (Special Relativity)*: If $\langle M, g_{ab}, T_{ab}\rangle$ is a model of a special relativistic theory and $F$ and $F'$ are any two inertial frames, then the theory satisfies the principle of relativity only if there exists a member

$L$ of the symmetry group of the Minkowski metric $g_{ab}$ such that
(a) $L$ maps $F$ onto $F'$ and
(b) $\langle M, g_{ab}, L^*T_{ab}\rangle$ is also a model of [the special relativistic theory]. (*op. cit.*, pp. 1251–2).

($L^*T_{ab}$ here is the 'dragged' version of $T_{ab}$ under the $L$: roughly speaking the 'appearance' of $L^*T_{ab}$ in $F'$ is identical to that of $T_{ab}$ in $F$.)

Notice first that this 'statement' of PR is specific to a given space-time theory, and seems to suggest that such a theory can be 'special relativistic' and *not* satisfy the principle. Moreover, it comes with the Lorentz transformations built in. It thus has a status quite different from the principle Einstein employed in 1905: in the sense that the latter is first, universal, and secondly, used in conjunction with the light postulate in order to provide a *non-circular* derivation of the Lorentz transformations.

Furthermore, one may well ask of the two passages above whether they are mutually compatible. If PR follows, as alleged, from the symmetries of the Minkowski metric, what need is there to impose a non-trivial constraint – condition (b) above – on the dynamical laws of a 'special relativistic' space-time theory which already incorporates the metric? (Note that Earman's earlier 1974 discussion of dynamical 'frame-equivalence' in SR is based on the correct view that satisfaction of the principle for a given space-time theory incorporating the Minkowski manifold is *not* a strict consequence of the 'space-time symmetries' of the manifold.)

Norton in fact recognises that condition (b) does not strictly follow from the nature of the symmetry automorphism $L$. His point is that insofar as PR is a 'theorem' it is encapsulated in condition (a) (satisfaction of which is guaranteed for the appropriate choice of $L$) which "reminds us that the Minkowski space-time itself designates no inertial frame as preferred" – the metrical structure is preserved under change of frame[40]. Condition (b) then *independently*

> . . . stipulates that the [laws governing] additional structures defined on space-time, such as Maxwell fields or mechanical fluids, likewise do not distinguish any inertial frame as preferred. (p. 1252)

But to the extent that (b) captures some part of PR, Norton's theoremhood claim for the principle is clearly weakened by the fact that (b) is not a logical consequence of the postulated space-time structure. Moreover, I attribute to Einstein the (I think) correct view that PR has to do with the laws of physics involving fundamental dynamical objects (mass-points, forces, fields, wavefunctions, . . . ). It is by no means obvious to me that in condition (a) there is, insofar as it refers to *absolute* structure, anything relevant to PR so understood – *at least if this structure is considered ontologically prior to the 'additional structures', like fields, defined on the manifold.* In relativistic space-time that is devoid of these 'additional structures', there are no replica experiments taking place in relatively moving 'spaces' to be compared; there aren't even rods and clocks. (See the comments of Einstein cited in the next section.) PR is, in my view, *all about* condition (b). And as such, it has essentially the same *contingent* status in the context of the putative 'absolute' Minkowski space-time structure as has its three-dimensional analogue, the physical principle of spatial isotropy, in Euclidean 3-space[41].

It is worth remembering in this connection that not *all* the symmetries of the putative 'absolute' objects on the space-time manifold are even manifest in the nature of things. The symmetry group as it is standardly defined in space-time theories is arbitrarily limited to continuous symmetries; a *discontinuous* symmetry of Minkowski space-time is spatial reflection. Experiments establishing a violation of parity conservation for the weak interaction also serve to remind us of the contingent status of PR[42], bearing in mind that an argument in favour of viewing PR as a *consequence* of the symmetries of the Minkowski metric is an argument also for parity conservation. I am unaware of a reason within the space-time theories approach as to why the continuous symmetries should be given special treatment by Nature.

The approach under discussion is clearly worthy of a thorough-going analysis, which is more than can be given here. An interesting critique of Friedman's 1983 discussion of GR has already been given by Weingard and Smith (1986), who, incidentally, regret that no mention of the Kaluza-Klein, Weyl, and non-symmetric theories, which can be put in the 'space-time theory' format, is found in Friedman's book. But the problem I think is not so much that adherents of the approach do not do justice to its breadth, it is rather that they do not always do justice to its narrowness.

The point is that the whole edifice of quantum theory is essentially beyond the scope of the space-time theories approach as it is currently defined. Neither the many-body wavefunction nor the operator-valued quantum field can be adequately represented by a geometric object field defined on the point elements of the relevant space-time manifold, and serious complications due to the uncertainty relations arise in the task of finding localised quantum clocks in order to define the requisite chronometric assumptions (see Earman 1989, pp. 470–473 and Anandan 1980). (Furthermore, the appropriate geometry of quantum gauge fields lives in nontrivial fibre bundles, and not in the space-time manifold.) One searches in vain in Friedman's 1983 book for a discussion of the implications of this significant state of affairs. In particular, one wonders precisely what the relativity principle can *mean* in quantum theory (all of whose forms satisfy PR) for someone who accepts Friedman's views on the principle – and hence eschews, as we saw in the last section, the usual notion of form invariance of laws.

## IX. FINAL REMARKS

In his autobiographical notes (Einstein 1969), Einstein stressed that in SR, rods and clocks must not be considered as a different "kind of physical things" from the other matter fields that populate space-time. He wrote that in order to avoid inconsistency,

> . . . measuring rods and clocks would have to be represented as solutions of the basic equations (objects consisting of moving atomic configurations), and not, as it were, as theoretically self-sufficient entities. (p. 59)

What Einstein referred to as the 'sin' of omitting a proper dynamical analysis of rods and clocks is (still) inevitable, at least in the sense that the exact, relevant "solutions of the basic equations" are beyond us. And it is precisely the efficacy of Einstein's 1905 route to relativistic kinematics using the principle of relativity that it works even when the basic dynamical laws and/or their solutions are unknown in their totality. But this should not blind us to the fact that the behaviour of rods and clocks in motion is determined by these laws (together with the velocity-independence of certain properties such as charge). Einstein was adamant that one must avoid the view that space-time intervals

> . . . are physical entities of a special type, intrinsically different from other variables ("reducing physics to geometry", etc.). (*ibid.*, pp. 59, 61)[43]

His view, as I see it, is clear: the dichotomy between kinematics and dynamics is one of convenience, not substance.

It has been one of the objects of the present paper to support this view [44], one which finds little resonance in Friedman's 'space-time theory' approach to SR. I must stress that the objections I have raised to Friedman's approach have nothing to do with the central role of space-time geometry therein[45]. They are concerned rather with the 'absolute', dynamics-independent status the geometry acquires in fixing the kinematics of the theory, prior to consideration of the nature of the laws governing matter fields in space-time.

Where does this leave the role of geometry in SR? This is the subject of another paper, but I might here express my broad sympathy with the profound 1980 analysis of physical geometry due to Anandan (1980). In his view, a modification of Klein's Erlangen programme, geometry emerges from and is relative to the dynamical symmetry group of a given physical theory. Depending on the theory, the geometry may or may not be that of a Riemannian space; in the case of quantum theory, the ensuing geometry is defined not on a space-time manifold at all but on the symmetry group manifold, once the appropriate 'clocks' (quantum mechanical measuring instruments) are defined[46].

In short, the *prima facie* advantages of Anandan's geometrical programme are two-fold. First, it purports to accommodate both quantum theory and gauge theories. Secondly, it regards physical geometry as a convenient means of representing a certain aspect of observable physical processes, rather than (in the case of pre-quantum SR) as a theory of an unobservable pre-existing structure in space-time which is somehow connected with the symmetries of physical processes involving particles immersed in it. On the negative side, it is not clear to me whether at least in his analysis of pre-quantum theories, Anandan himself pays sufficient attention in each case to the internal structure of the physical hodometers of the relevant geometrical structure. (In the specific case of GR, however, he follows a suggestion of Penrose and argues for the quantum mechanical nature of clocks in their fundamental role as hodometers of the metrical structure of space-time. I see no reason why a version of this thesis should not apply also to the Galilean/Newtonian and Minkowski space-times.)

Be that as it may, Anandan's approach leads to a significant blurring of the usual distinction between space-time symmetries and 'internal' symmetries in quantum theory. It also seems to provide a systematic basis for the view that space-time geometry itself emerges as an aspect of the classical limit of the fundamental quantum theory, a view perhaps suggested by, *inter alia*, the experimental confirmation of the predicted change of sign of the wave-function of a fermionic particle undergoing a complete ($2\pi$ radians) spatial rotation. Let us hope that as a promising alternative to the foundations of geometry as espoused in the 'space-time theory' approach, Anandan's programme – and particularly the intricate question of its suitability to GR – gets the careful philosophical scrutiny it deserves.

*Note added in proof:* The discussion above (p. 251) of Norton's 1989 formulation of PR overlooks the fact that his condition (b) is a consequence of nothing more than the claim that the dynamical theory in question can be written in generally covariant form; this fact raises I think serious doubts about the adequacy of the formulation. Moreover, Tim Budden has pointed out (private communication) that my earlier arguments (pp. 245–248) dealing with the cogency of Friedman's 'classical electrodynamics' incorporate a misreading of the theory: they fail to recognise that the Lorentz force law in the rest frame therein (Friedman's equation III.70) is non-relativistic. Hence my suggestion that 'classical electrodynamics' is empirically indistinguishable from standard electrodynamics is incorrect, and my arguments should be evaluated in the light of this error.

ACKNOWLEDGEMENTS

Stimulating discussions on the philosophy of SR with Jeeva Anandan, Rom Harré, John Lucas, Simon Saunders, Roland Sypel, and particularly Julian Barbour and Tim Budden (who also provided me with several important references) have been of great help, even in those cases where full agreement was not reached. Instructive comments on a first draft of this paper were also gratefully received from John Norton, Roberto Torretti and Robert Weingard. My thanks go finally to Steven French and Harmke Kamminga for affording me the pleasure of contributing to this timely Festschrift. While writing the paper, I was more than usually mindful of Heinz Post's legacy as my first teacher in the philosophy of science. If I have not met his stringent demands for clarity, rigour and *good sense*, it is not for want of trying to heed his message.

*Sub-Faculty of Philosophy,*
*University of Oxford, U.K.*

## NOTES

1. The English translation I am using here is given in the Appendix of Miller (1981). Notice that this statement of the postulate already incorporates Einstein's convention for synchronising distant clocks in the resting frame. A recent, detailed analysis of the origins and logical status of the light postulate in the 1905 paper is found in Brown and Maia forthcoming 1992; much of the discussion below rests on that analysis.
2. The earliest direct evidence, based on binary star observations due to de Sitter, was published in 1913 (and questioned by Fox in 1962); for more details the reader may consult the authoritative study of the foundations of relativity theory by Torretti (1983, note 12, p. 293).
3. The reason was Planck's 1900 work on black-body radiation. Despite the fact that Einstein's earlier 1905 arguments for his light-quantum hypothesis (which implied that the Maxwell-Lorentz equations could not be valid at the micro-level) did not rely on the Planck distribution law, this was their source of inspiration. (See Einstein 1969, pp. 49–53).
4. The term 'speed' here should be interpreted in its two-way sense. The isotropy of the one-way speed of light is a consequence, of course, of Einstein's convention for synchronising distant clocks.
5. Being wavelength dependent, the partial drag coefficient seemed to imply the existence of a continuum of luminiferous ethers.
6. Mention of Einstein's analysis of the emission theory is found in Stachel 1982, and Torretti *op. cit.*, p. 50.
7. In the 1905 paper, Einstein states that the 'resting' frame can be chosen arbitrarily from the frames in which the laws of mechanics hold good, but this is clearly done with the independent relativity principle in mind, and tends to obscure the historical origins of the light postulate.
8. Nugaev (1988) has recently provided an illuminating discussion of the role of quantum theory in the genesis of SR, a topic hitherto somewhat neglected.
9. The velocity-independence of inertial and hence gravitational mass is also required in the proof. (Analogously, no proof of the Lorentz-invariance of Maxwell's equations is possible without the assumption that charge is velocity independent.) An instance of the common, erroneous claim that the (inertial) frame-independence of Newton's second law, for instance, follows merely from the fact that accelerations are invariant under the Galilean transformations, is found in Friedman 1983, p. 13.
10. Penrose has conjectured (1987) that Newton's early commitment to the relativity principle may indeed have been instrumental in his adoption of the corpuscular theory of light.
11. See the discussion after Definitions V, VI and VII, and particularly the Scholia at the ends of section XII of Book I and section V of Book II.
12. I have overlooked the fact that Galileo did not consistently restrict himself to uniform, rectilinear motion. For a careful discussion of this point, see Barbour *op. cit.*, Chapter 7.
13. This version of the relativity principle is not to be read as purely 'phenomenological': theory is required of course to specify precisely what the relevant initial conditions are in each

space, and what effects count as potential falsifiers of the principle. This is what makes Einstein's definition (see below) more satisfactory. For a recent discussion of this point, see Brown and Maia *op. cit.*, section III.

14. The important difference of course between a sound medium and the ether is that the former is in principle boostable in bulk.
15. The problem of moving bodies in 19th century ether theories was of course more intricate than I have been able to indicate. A useful review of the problem is found in Torretti 1983, section 2.2. Like Torretti (*ibid.*, p. 45), I find the inferences Lorentz drew from his theorem of corresponding states hard to follow, given his reticence to specify the operational meaning of 'local time'.
16. This was not quite all: a full account of the origins of SR would also stress Einstein's important realisation of the conventional nature of distant simultaneity, and the possibility, given a natural convention for distant clock synchrony, of the relativity of simultaneity. In this connection, Torretti's analysis (*ibid.*, sections 3.1–3.3) is exemplary.
17. One of these grounds is the phenomenal symmetry involved in the case of electromagnetic induction. Much has been written on the significance of this argument for Einstein's discovery of SR. I see its significance as more than one of several arguments for the relativity principle (which it undeniably was for Einstein), in the sense that its resolution does not follow simply from adoption of the principle and is thus not found in the 'kinematical' part of the paper. It was specifically a challenge to provide a 'constructive' frame-independent language of reality for such electrodynamic effects, as Post (1971 p. 244) has stressed. This is precisely what Einstein provided in the electrodynamic part of his paper, where he solves the problem of "currents produced by the relative motion of a magnet and a conductor".
18. This qualification quite clearly holds for Einstein, but it has been questioned, for example in Stiegler 1952 and Tzanakis and Kyritsis 1984, who defend an idiosyncratic, weakened version of PR. For more discussion of this point, see Brown and Maia *op. cit.*.
19. The reader may usefully compare equation (3) with the acceleration transformation formula in SR.
20. The claim by Rindler (1985) that reciprocity is a consequence of spatial isotropy holding in all inertial frames has been questioned recently by Brown and Maia (*op. cit.*, Appendix II).
21. For early use of the relativity principle as a postulate, see Penrose 1987 and Barbour 1989, pp. 464–7; 575.
22. I am grateful to Roberto Torretti for pointing out (private communication) that full recognition of the role of the Galilean group in underlying the relativity theory in classical mechanics occurred only this century.
23. Had Keinstein wished to countenance this invariance principle as a postulate, he could in principle have anticipated the more complicated derivation of the Galilean transformations found, say, in Berzi and Gorini (1969), who assume additionally only PR and spatial isotropy.
24. Recently, anisotropic relativistic kinematics in two dimensions were discussed in Brown 1990, where it was shown that Winnie's 1970 'equal passage times' principle is connected with isotropy, not PR. (An error elsewhere in Brown *op. cit.* was pointed out in Brown and Maia *op. cit.*, footnote 6.) Further foundational implications of anisotropic relativity were developed in Budden 1992. Neither of these authors was aware of the earlier formal treatment of the subject in Bogoslovsky 1977.
25. Note that Einstein could have used the isotropy of length contraction and time dilation to determine his clock-synchrony convention. For it follows from eq. (6) in Brown *op. cit.* that, given the (synchrony-independent) principle of round-trip light-speed invariance, such isotropy holds only if clocks are synchronised so as to render the one-way light-speed isotropic.
26. It was shown by Tzanakis and Kyritsis (*op. cit.*) in 1984 that if reciprocity is granted, the

derivation does not actually need light-speed invariance; it is enough to assume along with the group property of the transformations and spatial isotropy that the light-speed is 'constant' in each frame but not necessarily invariant. A strengthened version of this result, which does not depend on reciprocity, is found in Brown and Maia *op. cit.*.

27. Winnie (*op. cit.*) used a variable-synchrony formulation of SR to establish equivalence. Details of Eddington's earlier 1923 argument are found in Torretti *op. cit.*, pp 226–7.
28. Berzi and Gorini (*op. cit.*) provided a derivation of linearity supposedly based on Einstein's homogeneity principle, or rather a formulation of it due to Lalan. This derivation has been criticised by the ever-alert Torretti (*op. cit.*, note 8, p. 297) on precisely the grounds that it is based on the homogeneity of space-time and not space and time.
29. More recently, Weingard and Smith (1986) have discussed the difficulties posed by this fact for Friedman's 1983 discussion of GR.
30. The requisite grounds for specifying free particles could be given by knowledge of certain physical laws defined in a single, chosen frame, and do not necessarily involve knowledge of the inertial coordinate transformations nor commitment to the principle of relativity.
31. It is easy to show from eq. (6) above that the invariance of the light-speed alone does not imply either finite length contraction, or finite time-dilation, although it does imply that both cannot be zero.
32. It is noteworthy that in deriving the Galilean transformations, Friedman nowhere indicates the convention specifying how distant clocks are synchronised in each inertial frame of reference, without which the (temporal) transformations are devoid of physical meaning.
33. See the discussion of this theory, and Poincaré's treatment of it, in Miller *op. cit.*, p. 82.
34. Another form of this argument has been developed recently by Budden (1992), who introduces a dynamical theory in two-dimensional Minkowski space-time which this time is consistent with PR, but which violates spatial isotropy.
35. I am grateful to Tim Budden for help in formulating this argument.
36. Acceptance of Maxwell's electrodynamics in $S_0$ already ensures the failure of two other tenets of Newtonian mechanics: the third law of motion, and the assumption that forces are velocity independent.
37. The notion of form invariance of laws is important elsewhere in physics, an obvious example being the derivation of gauge fields by demanding that the equation of motion in quantum theory be form invariant under local gauge transformations. A recent detailed defence of the notion of form invariance in the context of Friedman's criticism is found in the Appendix in Weingard and Smith *op. cit.*.
38. Earman introduced his own version of the theory in his 1974, but only later in his 1990, pp. 51–55 does he pose difficulties for this construction. He mentions that there are two different versions of the theory which have different empirical predictions. (These versions arise from looking at the 'downstairs' and 'upstairs' versions of the Maxwell tensor, which are required to transform covariantly and contravariantly respectively under the Galilean transformations.) But there is surely a potential infinity of such odd theories, all differing according to the postulated transformation properties of the electric and magnetic field components under the Galilean transformations, which themselves, as we have seen, require a non-standard interpretation. The fact that any such theory "loses contact with historical reality" (p. 54) and the outcome of ether wind experiments etc. is hardly surprising. The relevant question is not whether the theory is right, but whether it is coherent.
39. In apparently a somewhat similar vein, Torretti has stated (1983) that whereas PR for Einstein is an axiom,

> . . . one cannot say that in the more mature formulations of Special Relativity it is left unexplained, for . . . it is a natural consequence of the structure of Minkowski space-time. (p. 87).

But this is severely qualified by a later passage on p. 90:

> The comparative ease with which [Minkowski space-time] can be defined should not

blind us to its physical significance, which is solidly grounded on the physical hypotheses that go into the definition of Lorentz charts.

40. Analogously, Friedman (*op. cit.*, p. 91) clearly regards the 'Galilean' relativity principle as a consequence of the symmetry group of 'Galilean space-time'. This would seem to imply that the principle fails in 'Newtonian space-time', a questionable outlook (see Sypel and Brown 1992) which appears to be confirmed by Friedman later in the book on p. 154.
41. The fact that we can successfully express physical laws in the 4-tensor calculus no more follows from the nature of pre-given Minkowski space-time, than the fact that we can successfully express physical laws holding in a given inertial frame in the 3-tensor calculus follows from the nature of pre-given Euclidean space.
42. Torretti recommends (*op. cit.*, p. 66) on the basis of parity violation that the "admissible" transformations in SR should be limited to the proper orthochronous Lorentz (Poincaré) group, a subgroup of the symmetry group associated with the manifold. See also the discussion in Lucas and Hodgson 1990, Chapter 7.
43. It is noteworthy that when Einstein summarises Minkowski's contribution to SR in these notes (p. 59), he praises Minkowski's discovery of the relevant four-dimensional language of invariants (4-tensor calculus), but makes no reference to geometry.
44. Perhaps things would be less confusing were one to refer to the Lorentz transformations, for example, as 'space-time dynamics', or some such phrase! Of course the intertwining of kinematics and dynamics is not restricted to post-1905 physics, as should be clear from Section IV above. Recognition of this point is also found in Torretti's recent discussion of Newtonian mechanics (1990, p. 109).
45. Nor, I might add, are the objections based on any anti-realist philosophical posture, or more specifically on an anti-'substantivalist' attitude toward space-time ontology. The reality of the space-time manifold itself may be open to question, but not I think on grounds related to the arguments presented in this paper.
46. Recent developments in Anandan's geometric approach to quantum mechanics are found in Anandan 1991.

## BIBLIOGRAPHY

Anandan, J. (1980), 'On the Hypotheses Underlying Physical Geometry', *Foundations of Physics*, 10, pp. 601–629.

Anandan, J. (1991), 'A Geometric Approach to Quantum Mechanics', *Foundations of Physics*, 21, pp. 1265–1284.

Anderson, J.L. (1967), *Principles of Relativity Physics*, Academic Press, New York.

Barbour, J.B. (1989), *Absolute or Relative Motion? Vol. 1: The Discovery of Dynamics*, Cambridge University Press, Cambridge.

Bell, J.S. (1976), 'How to Teach Special Relativity', *Progress in Scientific Culture*, 1 (2); also in J.S. Bell, *Speakable and Unspeakable in Quantum Mechanics*, Cambridge University Press, Cambridge, pp. 67–80.

Berzi, V. and Gorini, V. (1969), 'Reciprocity Principle and the Lorentz Transformations', *Journal of Mathematical Physics*, 10, pp. 1518–1524.

Bogoslovsky, G.Y. (1977), 'A Special-Relativistic Theory of the Locally Anisotropic Space-Time', *Il Nuovo Cimento*, 40B, pp. 99–115, 116–133.

Brown, H.R. (1990), 'Does the Principle of Relativity Imply Winnie's (1970) Equal Passage Times Principle?', *Philosophy of Science*, 57, pp. 313–324.

Brown, H.R. and Maia, A. (1992), 'Light-speed Constancy versus Light-Speed Invariance in the Derivation of Relativistic Kinematics', *British Journal for the Philosophy of Science*, forthcoming.

Budden, T. (1992), 'The Relativity Principle and the Isotropy of Boosts', in D. Hull, M. Forbes and K, Okruhlik (eds.), *PSA 1992*, Volume 1, Philosophy of Science Association, East Lansing, Michigan, 1992, pp. 528–541.

Earman, J. (1974), 'Covariance, Invariance, and the Equivalence of Frames', *Foundations of Physics*, 4, pp. 267–289.

Earman, J. (1989), 'Locality, Nonlocality and Action at a Distance: a Skeptical Review of Some Philosophical Dogmas', in R. Kargon and P. Achinstein (eds.), *Kelvin's Baltimore Lectures and Modern Theoretical Physics*, MIT Press, Cambridge, pp. 449–490.

Earman, J. (1990), *World Enough and Space-Time, Absolute versus Relational Theories of Space and Time*, The MIT Press, Cambridge.

Ehlers, J., Pirani, F.A.E., and Schild, A. (1972), 'The Geometry of Free Fall and Light Propagation' in L. O'Raifeartaigh (ed.), *General Relativity: Papers in Honour of J.L. Synge*, Clarendon, Oxford, pp. 63–84.

Einstein, A. (1905), 'Zur Electrodynamik bewegter Körper', *Annalen der Physik*, 17 (4), pp. 891–921.

Einstein, A (1907), 'Relativitätsprinzip und die aus demselben gezogenen Folgerungen', *Jahrbuch der Radioaktivität*, 4, pp. 411–462 ; 5, pp. 98–99 (Berichtigungen).

Einstein, A. (1969), 'Autobiographical Notes', in P.A. Schilpp (ed.), *Albert Einstein: Philosopher-Scientist*, Open Court, La Salle Illinois, 3rd ed.

Fock, V. (1959), *The Theory of Space, Time and Gravitation*, Pergamon, London.

Friedman, M. (1983), *Foundations of Space-Time Theories*, Princeton University Press, Princeton.

Hirosige, T. (1969), 'Origins of Lorentz' Theory of Electrons and the Concept of the Electromagnetic Field', *Historical Studies in the Physical Sciences*, 1, pp. 1561–269.

Hirosige, T. (1976), 'The Ether Problem, the Mechanistic Worldview, and the Origins of the Theory of Relativity', *Historical Studies in the Physical Sciences*, 7, pp. 3–82.

Hoffman, B. (1982), 'Some Einstein Anomalies', in G. Holton and Y. Elkana (eds.), *Albert Einstein; Historical and Cultural Perspectives, The Centennial Symposium in Jerusalem*, Princeton University Press, Princeton.

Jackson, J.D. (1967), *Classical Electrodynamics*, John Wiley & Sons, New York.

Lucas, J.R. and Hodgson, P.E. (1990), *Spacetime and Electromagnetism*, Clarendon Press, Oxford.

Marzke, R.F. and Wheeler, J.A. (1964), 'Gravitation as Geometry – I: The Geometry of Space-Time and the Geometrodynamical Standard Meter', in H. Y. Chiu, and W. F. Hoffman (eds.), *Gravitation and Geometry*, Benjamin, New York, pp. 40–64.

Miller, A.I. (1981), *Albert Einstein's Special Theory of Relativity*, Addison-Wesley, Mass.

Newton, I. (1934), *Sir Isaac Newton's Mathematical Principles of Natural Philosophy and his System of the World*, trans. by A. Motte 1729, revised by F. Cajori, University of California Press, Berkeley.

Norton, J. (1989), 'Coordinates and Covariance: Einstein's View of Space-Time and the Modern View', *Foundations of Physics*, 19, pp. 1215–1263.

Nugaev, R.M. (1988), 'Special Relativity as a Stage in the Development of Quantum Theory', *Historia Scientiarum*, 34, pp. 57–79.

Panofsky, W.K.H. and Phillips, M. (1962), *Classical Electricity and Magnetism*, Addison-Wesley, Reading, Mass.

Pauli, W. (1981), *Theory of Relativity*, Dover; English translation of 'Relativitätstheorie', *Encyclopädie der matematischen Wissenschaften*, Vol. V19, B.G. Teubner, Liepzig, 1921.

Penrose, R. (1987), 'Newton, Quantum theory and Reality', in S. Hawking and W. Israel (eds.), *Three Hundred Years of Gravitation*, Cambridge University Press, Cambridge, pp. 17–49.

Post, H. (1971), 'Correspondence, Invariance and Heuristics; In Praise of Conservative Induction', *Studies in the History and Philosophy of Science*, 2, pp. 213–255.

Rindler, W. (1985), *Introduction to Special Relativity*, Oxford University Press, Oxford.

Saunders, S.W. and Brown, H.R. (1991), 'Reflections on ether', in S.W. Saunders and H.R. Brown (eds.), *The Philosophy of Vacuum*, Clarendon Press, Oxford, pp. 27–63.

Sklar, L. (1977), 'Facts, Conventions, and Assumptions in the Theory of Spacetime', in J. Earman, C. Glymour, and J. Stachel (eds.), *Foundations of Space-Time Theories*, Minnesota Studies in the Philosophy of Science, Vol. 8, University of Minnesota Press, Minneapolis, pp.

206–274; also in L. Sklar, *Philosophy and Spacetime Physics*, University of California Press, 1985, Chapter 3.

Stachel, J. (1982), 'Einstein and Michelson, the Context of Discovery and the Context of Justification', *Astronomische Nachrichten*, 303, pp. 47–53.

Stiegler, K.D. (1952), 'Sur le principe de la constance de la vitesse de la lumière', *Comptes Rendus*, 234, pp. 1250–1252.

Synge, J.L. (1965), *Relativity: The Special Theory*, North Holland, Amsterdam.

Sypel, R., and Brown, H.R. (1992), 'When is a Physical Theory Relativistic?', in D. Hull, M. Forbes and K. Okruhlik (eds.), *PSA 1992*, Volume 1, Philosophy of Science Association, East Lansing, Michigan, 1992, pp. 507–514..

Taylor, E.F. and Wheeler, J.A. (1966), *Spacetime Physics*, W. H. Freeman and Company.

Torretti, R. (1983), *Relativity and Geometry*, Pergamon, Oxford.

Torretti, R. (1990), *Creative Understanding: Philosophical Reflections on Physics*, University of Chicago Press, Chicago.

Tzanakis, C. and Kyritsis, C. (1984), 'On Special Relativity's Second Postulate', *Annales de la Fondation Louis de Broglie*, 9, pp. 343–352.

Weingard, R. and Smith, G. (1986), 'Critical Notice: Michael Friedman's Foundations of Space-Time Theories', *Philosophy of Science*, 53, pp. 286–299.

Winnie, J. (1970), 'Special Relativity without One Way Velocity Assumptions', *Philosophy of Science*, 37, pp. 81–99, 223–238.

JAMES T. CUSHING

# Underdetermination, Conventionalism and Realism: The Copenhagen vs. the Bohm Interpretation of Quantum Mechanics

## I. INTRODUCTION

In an article from which the present *Festschrift* takes its title (Post 1971), Heinz Post discusses heuristic strategies with the aid of which scientists generate a successor (or more general, covering) theory *L* from a theory *S* that has been pressed beyond the limits of its validity. These "'theoretic' guide lines to new theories" (*ibid.*, p. 217) he broadly classifies as invariance principles and a *General Correspondence Principle* (*ibid.*, p. 228). Our comments here and this paper in general are related most directly to his correspondence principle, one feature of which consists in conserving in *L*, not only the successful empirical consequences of *S*, but also as many of the explanatorily useful and conceptually desirable features of *S* as possible, as well as requiring the new theory *L* to yield the old one *S* as a well-defined (mathematical and, perhaps, even conceptual) limit as some relevant parameter or physically significant quantity is varied. It is certainly true that both Bohr and Bohm[1] used such heuristic guidelines to generate quantum mechanics from classical mechanics. Post claims (and quite correctly, as *general* characteristics of theory construction) that the procedure of generating a new theory (or of generalizing and old one) "is conservative (as every good scientist is)." (*ibid.*, p. 218).

However, as the present paper indicates and as we have argued at greater length elsewhere (Cushing 1992), the 'successful', preferred 'Copenhagen' version of quantum mechanics has far fewer (conceptual *or* formal) characteristics in common with classical mechanics than does the not generally accepted, 'causal' quantum mechanics of David Bohm. That is, in several respects Bohm's quantum mechanics has greater correspondence (in Post's sense) with classical mechanics than does Bohr's. It is true that Bohm himself did use both the past success of (interim) statistical theories finally being underpinned by detailed, (classically) deterministic successor theories of microentities (Bohm 1952, p. 168) and the essentially circular and potentially self-fulling nature of 'Copenhagen's' claim to completeness and finality (*ibid.*, p. 169) as guides and motivating factors in his own formulation of a causal,



*S. French and H. Kamminga (eds.), Correspondence, Invariance and Heuristics*, 261–278.

deterministic, realistic interpretation of quantum mechanics[2]. So, the use by both protagonists of general heuristic guidelines does lend support to Post's broad claims. However, the competitor that adhered more strongly to Post's correspondence principle has lost out – surely an undesirable result for the efficacy of such guidelines!

Furthermore, Post accepts as apparently unproblematic the 'fact' that " . . . at any one time there is usually only one dominant general theory . . . covering a particular field" (Post *op. cit.*, p. 220; see also fn. 58, p. 239) without asking *why* this should be so[3]. This unnecessary (and, perhaps, even unwarranted) restriction to just the 'Copenhagen' formulation of quantum mechanics even generates an exception to Post's own scheme:

> Paradoxically, the only counterexample we have been able to find to the General Correspondence Principle is the paradigm example of the relation of quantum mechanics to classical mechanics. Contrary to the impression that may be given in some textbooks, it is not possible to reduce quantum mechanics to classical mechanics except 'locally', *i.e.*, with respect to certain sub-theories such as some of those involving angular momentum. Ehrenfest's theorem establishes a correspondence between the motion of the centre of a quantum mechanical distribution and the motion of the corresponding classical mass point. No correspondence with respect to higher momenta [*sic*] of the distribution has been found.
>
> This failure to establish a general correspondence between quantum mechanics and classical physics (in the limit where $\Delta P \, \Delta Q \gg h$ ) should in our view be regarded as a shortcoming of quantum mechanics (at least, of the theory of 'first quantization') in its claim to the status of *L*-theory, rather than as a breakdown of the General Correspondence Principle. This failure of quantum mechanics, together with its incompleteness, indicates clearly the need for a new, better *L*-theory. (Post 1971, pp. 233–234)

That is, one could attempt to argue that Post's own *General Correspondence Principle* could be used to select Bohm's causal quantum mechanics over the 'Copenhagen' one as the proper successor to classical mechanics! We hope the reader will be convinced by the present paper, though, that such is not a robust enough move to counter the pervasive underdetermination that we show to exist between these two versions of quantum mechanics. (Even the additional criterion of heuristic scope or fertility does *not* resolve this particular case of underdetermination.) Our analysis is, we believe, consonant with Post's own view of the importance of looking at *actual* scientific practice and supportive of his disdain for mere "methodology of *post-hoc* criticism." (*ibid.*, p. 215). Now, let us turn to the details of our case study and argument.

In her recent article, Yemima Ben-Menahem (1990) has nicely summarized the problem that the Duhem-Quine underdetermination thesis poses for the scientific realist. Succinctly put, if there are two equally empirically adequate successful scientific theories (agreeing on all *possible* empirical tests and, therefore, being observationally indistinguishable) that support radically different (and, in fact, incompatible) ontologies, then such a situation must frustrate the scientific realist in his (or her) search for *the* correct scientific theory that gives a true picture of the world (even with limits set by reasonable caveats). As long as the two theories under consideration differ in relatively minor respects as regards their ontologies about the furniture of the world, one can simply decide to bracket these as inessential matters. The case we consider involves one theory that represents the fundamental physical processes in the world as being inherently and *irreducibly* indeterministic and another theory that is based on an *absolutely* deterministic behavior of the physical universe. This would not appear to be a minor or irrelevant difference.

It is perhaps worthwhile here to enter a disclaimer or two. We are *not* centrally concerned with the question of the *refutation* of scientific theories. That is, in terms of the title of a well-known collection of essays on the Duhem-Quine thesis, *Can Theories Be Refuted?* (Harding 1976), we would be willing to give away a 'yes' answer in the following sense. Even if one grants that there are many theories that can reasonably be rejected on evidential grounds (those are the *easy* cases!), there still remain other cases, we claim, in which viable, fertile theories have been rejected. Our position is not a radical one that claims that practical underdetermination always exists in all cases[4], but rather that there are some cases (at least *one* important instance) in which genuine underdetermination does exist. The case discussed in this paper does not concern just the (mere) compatibility of two essentially different theories with the presently *available* data, but involves a much more deep-seated indistinguishability. While a choice can be, and has been, made on the basis of non-evidential criteria, the question must then be faced of the basis for such criteria and of the role historically contingent factors have played in fashioning them. One must resist an urge to seek resolution of the underdetermination problem in terms of future developments that *may* take place in science. That alone would be more a declaration of belief than an argument. It is also true, though, that, even if one grants validity to the case presented in this paper, it would remain just *one* example from theoretical physics (although another such candidate has recently been developed in detail (Cushing 1990)) and may be peculiar to that area, having little relevance to the philosophy of science in general.

The basic issue here is a belief in the (at least effective) uniqueness of a correct scientific theory, with the selection process being 'objective' and not involving in any ineliminable fashion 'subjective' criteria such as coherence ('beauty'), simplicity or minimum mutilation (Ben-Menahem *op. cit.*, p. 267). By coherence we most specifically do *not* mean just lack of logical contradiction, since both theories we discuss in the next section are logically consistent and neither is pejoratively *ad hoc* in nature. Scientists typically take for granted the practical uniqueness (in any given era) of successful scientific theories. Thus, Einstein in an address delivered before the Berlin Physical Society in 1918 on the occasion of Max Planck's sixtieth birthday tells us that:

> The supreme task of the physicist is to arrive at these universal elementary laws from which the cosmos can be built up by pure deduction. There is no logical path to these laws; only intuition, resting on sympathetic understanding of experience, can reach them. In this methodological uncertainty, one might suppose that there were any number of possible systems of theoretical physics all equally justified; and this opinion is no doubt correct, theoretically. But the development of physics has shown that at any given moment, out of all conceivable constructions, a single one has always proved itself decidedly superior to all the rest. Nobody who has really gone deeply into the matter will deny that in practice the world of phenomena uniquely determines the theoretical system, in spite of the fact that there is no logical bridge between phenomena and their theoretical principles. . . . (Einstein 1954, pp. 221–222)

Notice that, curiously enough, Einstein allows the *theoretical* (i.e., logical) possibility of more than one empirically adequate theory, but then goes on to make the (rather startling) declaration of faith that at any given time the "world of phenomena" (which sounds pretty objective) uniquely determines one theory as superior to all others. Similarly, Heisenberg, in his retrospective reconstruction of what was going on in Copenhagen in 1926, states:

> I wanted to start from the fact that quantum mechanics as we then knew it already imposed a unique physical interpretation of some magnitude occurring in it . . . so that it looked very much as if we no longer had any freedom with respect to that interpretation. Instead, we would have to try to derive the correct general interpretation by strict logic from the ready-to-hand, more special interpretation. (Heisenberg 1971, p. 76)

It is just this (even practical or effective, as opposed to merely conventional) uniqueness that we refute, with a specific example, in the next section.

Also, we point out a second difficulty for the scientific realist regarding the value of rational reconstruction as a means of evaluating a successful scientific theory. That is, even though a philosopher can construct an abstract and coherent representation of an accepted scientific theory in a fashion that is logically independent of the actual historical (and contingent) process of construction and selection of that theory, we argue that (at least for the important example of this paper – namely, quantum mechanics) such an exercise (while not wholly unimportant) is at best a consistency check, and one that is not actually wholly decoupled from the contingent (but not uniquely logically compelled) choices actually made by the creators of the theory being reconstructed for evaluation.

Let us now examine the specifics of our example.

## II. TWO EMPIRICALLY INDISTINGUISHABLE THEORIES

For the purpose of this paper, we make a simple distinction between two components of a scientific theory: its formalism and its interpretation. The claim is that these are conceptually separable, even if they are often entangled in practice. To simplify matters, we also restrict our remarks to theories in modern physics, since that will be sufficient for our purposes here. Very informally, what we mean by a formalism is a set of equations and a set of calculational rules for making predictions that can be compared with experiment (i.e., "getting the numbers right"). It is important to stress that we need only the fact that both standard quantum mechanics and Bohm's version use exactly the same set of rules for predicting the values of quantities that can actually be *observed* . There may be additional quantities that have in-principle well defined values for Bohm (but not for 'Copenhagen'), but these are not accessible to observation (certainly not now and, perhaps, never). The (physical) interpretation refers to what the theory tells us about the underlying structure of these phenomena (i.e., the corresponding picture story about the furniture of the world – an ontology). Hence, *one* formalism with *two* different interpretations counts as *two* different theories.

The specific example we consider is (nonrelativistic) quantum mechanics, which is surely considered to be a fundamental theory of modern physics. (Optimistically, it is *the* basic theory of *all* physical phenomena.) Since this is not the place to go into great mathematical detail (nor is it necessary to make our point), we can somewhat elliptically sketch the *formalism* of quantum mechanics as consisting of a vector $\psi$ representing (at the very least) our knowledge of the state of the physical system under consideration, a dynamical equation (e.g., the Schrödinger equation) that gives the time evolution of $\psi$, a set of operators $A$ (representing physical observables having possible values $a_j$), and rules for calculating average values of a sequence of observations (in terms of $A$ and of $\psi$). In brief quantitative outline, the rules and postulates that are usually employed in making quantum-mechanical calculations are the following:

i. a state vector (e.g., $\psi$ ) – a vector, in a Hilbert space $\mathfrak{H}$, representing the state of the physical system;
ii. a dynamical equation (e.g., the Schrödinger equation)

$$H\psi = i\hbar \frac{\partial \psi}{\partial t} \tag{1}$$

giving the time evolution of the state vector $\psi$ under the influence of the Hamiltonian $H$ for the physical system;
iii. a correspondence between (hermitian) operators $A$ in $\mathfrak{H}$ and physical observables $a$. These physical observables $a$ can take only the eigenvalues $a_j$ where

$$A\ \psi_j = a_j\psi_j; \tag{2}$$

iv. ensemble averages for a series of observations of $a$ are given as $\langle \psi \, |A| \, \psi \rangle$;

v. a projection postulate (either explicitly or effectively assumed) upon measurement

$$\psi = \sum_k a_k \psi_k \rightarrow \psi_j \tag{3}$$

From this formalism follows the Heisenberg uncertainty, or indeterminacy, relation. This is related to Bohr's complementarity principle. For our purposes here, a special case of complementarity that will serve as an illustration is the well-known wave-particle duality according to which a physical system (e.g., an electron or a photon) behaves *either* as a wave *or* as a particle, depending upon the context or environment.

This formalism of quantum mechanics has been so stunningly successful (beyond argument *the* most successful theory in the history of physics) that we simply accept it (here, in our argument) as being completely empirically adequate for any set of physical phenomena to which it has ever been applied in the sixty-five years since its inception. The point at issue will not be the formalism, but rather its interpretation.

Now the standard (almost universally accepted by practicing physicists) 'Copenhagen' interpretation of quantum mechanics requires a discontinuous (and inherently uncontrollable) change ('jump' or 'collapse') of the state vector $\psi$ upon observation (or 'measurement') of the physical system under consideration. This is the notorious measurement problem of quantum mechanics. An implication of this is that there is an *essential* (ineliminable) indeterminacy at the most fundamental level of physical processes. If taken as telling us something about the nature of our physical world, the Copenhagen interpretation of quantum mechanics requires an ontology that is irreducibly indeterministic so that we cannot speak even *in principle* of the continuous evolution (or about the trajectory in a space-time background) of a microsystem.

On the other hand, there is an interpretation, due to David Bohm (1952), of precisely the *same* formalism of quantum mechanics in which microsystems evolve completely deterministically and follow well-defined, continuous trajectories in a space-time background. By means of a mathematical transformation alone, the dynamical equations of quantum mechanics (e.g., the Schrödinger equation) can be put into the form

$$\frac{dp}{dt} = F = -\nabla(V + U) \tag{4}$$

where $V$ is the usual classical potential and $U$ is the so-called quantum potential defined as

$$U = -\frac{\hbar^2}{2m}\frac{\nabla^2 R}{R}$$

$$R = |\psi| \tag{5}$$

Here $\psi$ is the wave function and is the solution to the Schrödinger equation. In this interpretation, there are both a particle (with the usual state variables $r$ and $v$) and an associated wave ($\psi$). The particle follows definite (if, at times, highly irregular) trajectories (in a space-time continuum), but there are (instantaneous) nonlocal influences produced by the quantum potential. Here $\psi$ represents the effect of the environment on the microsystem (or particle). There is no collapse of the wave function upon observation, but we (effectively) discover where the particle is. We recover the Heisenberg uncertainty relations (as limitations on the accuracy of our measurements due to the effects of the quantum potential) and all of the statistical predictions of standard quantum mechanics. On this interpretation, a microsystem behaves as a classical chaotic system (of the type now so much discussed in nonlinear dynamics).

In particular, the Bohm theory is identical in its predictions to the standard one provided:

i. $\psi = Re^{iS/\hbar}$ is a solution to the Schrödinger equation

ii. $v = \frac{1}{m} \nabla S$ is the velocity of the particle

iii. the precise location of a particle is not predicted or controlled, but has a statistical (ensemble) distribution according to the probability density $P = |\psi|^2$

These three assumptions are logically independent.

Let us make a few comments on these three postulates. Here (i) and (ii) can be lumped together under 'dynamics', while (iii) is a special postulate needed if Bohm's scheme is to remain absolutely observationally indistinguishable from the Copenhagen version. Specifically, Bohm himself says of (iii):

> That we do not predict or control the precise location of the particle, but have, in practice, a statistical ensemble with probability density $P(x) = |\psi(x)|^2$. The use of statistics is, however, not inherent in the conceptual structure, but merely a consequence of our ignorance of the precise initial conditions of the particle. (1952, p. 171)

What is to be emphasized here is that this ought not be taken as a nakedly and pejoratively *ad hoc* assumption. The state vector plays two roles in Bohm's theory: its primary conceptual function is to represent the influence of the environment (through the quantum potential $U$) and, secondarily, $\psi$ gives the actual probability distribution $P$. Since these are logically distinct concepts, Bohm offers a rather detailed argument (1953) to show that the 'dynamics' – (i) and (ii) above – plus random interactions will drive *any* initial probability distribution $P$ to its equilibrium value of $|\psi|^2$. The purpose in making these remarks is to urge that (iii) need not (ought not?) be viewed as an *ad hoc* assumption made just to produce agreement with the Copenhagen version of quantum mechanics.

A comment or two on the relevance of Bell's theorem may be appropriate here. The details of Bell vs. quantum mechanics are actually fairly involved and subtle, but the upshot of it all is this (Cushing and McMullin 1989). The Bell inequalities are derived for what are usually termed local hidden-variables theories and *some* of the predictions of the *formalism* of quantum mechanics ['Copenhagen', Bohm's, (almost) anyone's] violate (at least some of) these inequalities. Experimental results also violate (some of) the Bell inequalities, but this comparison can be made *only* with the aid of certain auxiliary assumptions, beyond those of quantum mechanics or of Bell. No direct, auxiliary-free experimental violations of Bell's inequalities or of quantum mechanics are known. But here the actual experimental situation is not relevant. What is true is that no *local* hidden-variables model can reproduce *all* of the predictions of (the formalism of) quantum mechanics. Since Bohm's theory *does* make exactly the same predictions as quantum mechanics, it follows that Bohm's hidden-variables theory must not be a *local* hidden-variables theory, and, indeed, such is the case. (In fact, it was just this possibility that got John Bell going on his 'theorem' decades ago.) Hence, the present case is one of observational equivalence and Bell has nothing to say there. It is definitely *not* a case in which two theories looked equivalent but ultimately were found to be empirically discriminable. This equivalence has *not* been historically circumscribed, except by *fiat* (as we argue below).

In Bohm's theory, the position variable plays a distinguished role, unlike the situation in standard quantum mechanics where all variables (or representations) enter on the same footing. As for this privileged status that the position variable (or space-time) plays in the Bohm theory, one could claim that this is a *virtue*. After all, we *do live* in position space (not, for example, in momentum space or the like). A theory's recognition of this special status, as a reflection of the actual physical world, need not be taken as a drawback of the theory. The general representational symmetry of the standard view may be a *calculational* convenience, but this *formal* convenience is available for Bohm too. Bohm's theory has a relational holism about it too, since what specific value for a property is actually observed does depend upon what one chooses to measure (i.e., upon the environment as represented by the wave function $\psi$ via the quantum potential).

It is also true that Bohm is committed to genuine nonlocality in nature (i.e., a not-directly-observable or controllable action-at-a distance). It is this mechanism that, ultimately, *produces* the long-range quantum correlations. On some representations, Copenhagen quantum mechanics simply postulates, or accepts, quantum correlations as a primitive, or brute, fact of nature (see, for example, van Fraassen 1985 or Fine 1989). Thus, one could ask, is it so that, in the Copenhagen version, the quantum violations of the Bell inequalities can be *understood* as resulting from distant correlations alone? The issue truly is one of *understanding*. Of course, in any discussion or explanation, there are always certain irreducible primitives that must be accepted as not further explainable. But, *is* one set of primitives as good as, or as understandable as, another? That's another, longer story and we have, for this case, discussed it elsewhere (Cushing 1991). Finally, it comes down to a choice involving not only empirical evidence, but predilections as well.

There is incidentally, a middle-road interpretation – the so-called sum-over-histories one – in which one has actual physical paths yet no nonlocality (Sinha and Sorkin 1991). But that's yet more underdetermination.

Now, at first sight this nonlocality might suggest a conflict with the first signal principle of special relativity. But this is not so, since these long-range correlations cannot be used for signalling (Cushing and McMullin *op. cit.*). So, the Bohm interpretation does demote special relativity from the status of an 'absolutely' universal foundational theory and demands relativistic invariance only of the *observational* content of a physical theory. Why is this so bad? After all, Einstein's postulates for special relativity (i.e., form invariance of the laws of physics – no possible *detection* of an absolute velocity – and the constancy of the speed of light for all inertial observers) were themselves based on observational consequences.

Bohm does *allow* us to be realistic about actual space-time trajectories, but at a price. *All* interpretations of quantum mechanics have price tags attached. Again, it's a buyer's market. In terminology Arthur Fine has used in another context (1986, p. 87), these interpretations, or worldviews, get *entheorized* into a general network of concepts and commitments and one must arrange the trade-off most acceptable to oneself (*my* characterization here, *not* Fine's). Let us recall, though, that the *purpose* of Bohm's 1952 papers was to show that trajectories are possible and that, therefore, part of the Copenhagen *dogma* (i.e., the alleged impossibility of such trajectories) is false. Since Bohm's theory has the same formalism as does the Copenhagen interpretation, it is not surprising (and, it is *true* (Bohm 1952; Cushing 1990, Sec. 9.5)) that this theory produces exactly the same predictions (in *all* conceivable situations) as does the standard Copenhagen version of quantum mechanics. This claim is not controversial and is generally accepted by experts working on foundational problems in quantum mechanics (Lahti and Mittelstaedt 1991).

Our use of 'empirical equivalence' here is essentially synonymous with 'observational equivalence'. Philosophers may find it useful to distinguish between these two terms[5], but we do not here, and basically for the following reason. The *Random House Dictionary* has the following entry:

> *empirical* – derived from or guided by experience or experiment; depending upon experience or observation alone; provable or verifiable by experience or experiment.

Thus, even though a dictionary of American English is hardly the court of final appeal in a philosophical discussion, use of the term 'empirical', *prima facie*, might not seem indefensible for what we have in mind. We really are interested in absolute observational equivalence. Let us suggest cashing out the finer distinction 'empirical' may have over 'observational' in terms of different *ontologies*, rather than different empirical contents, for 'Copenhagen' vs. 'Bohm'. That is, each interpretation makes different *ontological* statements about the world – for instance, the actual location of a pointer before observation – but both make the same *observational* claim. Is this an *empirical* difference? There is an in-principle ontological difference, but *de facto* practical equivalence. Perhaps this *is* an issue for philosophers to decide.

So, here we have two *actual* (not just fancifully concocted for argument's sake) absolutely empirically indistinguishable scientific theories that have diametrically opposed ontologies (i.e., indeterministic vs. deterministic). We do not wish to deny that the formalism of quantum mechanics may change (as *all* previous formalisms have eventually), but, should that be required, 'Copenhagen' and 'Bohm' may both have to go! Much turns on how important one takes an *actual*, long-standing case of underdetermination to be.

## III. A DILEMMA FOR THE REALIST

Now this situation represents, in a sense, a double threat to the scientific realist. To begin with, the almost universally accepted Copenhagen interpretation has traditionally been a serious (perhaps, even arguably, an insurmountable) challenge to a realistic construal of quantum mechanics. The core of the difficulty is the measurement problem, one entailment of which is that a physical system cannot (*in principle*) possess definite (but merely unknown to us) values of all physically observable attributes (such as position and velocity). Taking such a theory seriously as an actual representation of the physical world (at the level of *individual* microentities) requires that we accept a rather bizarre ontology, one that may not even be conceptually coherent. The measurement problem has been around now for sixty-five years and has resolutely defied any successful, generally-accepted solution. This has provided effective ammunition for the antirealist who begins at the level of microphenomena, accepts ('Copenhagen') quantum mechanics as *the* fundamental and exact theory of all physical processes, and then throws down the challenge to the realist to construct a coherent (*realistic*) ontology consistent with the demands of quantum theory (van Fraassen 1980). That is, such an antirealist begins his (or her) argument in the micro-realm, extrapolates to the macro-realm of everyday experience and leaves the ensuing conundrum at the doorstep of the realist. In fairness, though, we must point out that the realist has a relatively easy going in the domain of macro-phenomena (e.g., everyday objects, bacteria, dinosaurs, etc. (McMullin 1984)), but then encounters difficulties in carrying these explanatory resources down to the domain of micro-phenomena.

At first sight, the Bohm interpretation of quantum mechanics would seem to offer consolation and a potentially powerful means of rebuttal (of antirealism) to the realist. That is, this interpretation, which represents a microentity as a wave *and* a particle (not as a wave *or* a particle as does 'Copenhagen'), lends itself readily to a realist construal of even fundamental physical processes that develop completely deterministically in a continuous space-time background. (True, there are some highly nonlocal, nonclassical effects present, but this is equally true for the Copenhagen interpretation.) However, while it is true that this Bohm interpretation is consonant with (and even conducive to) a realist position, it is empirically indistinguishable from an ontologically incompatible interpretation ('Copenhagen'). So, the realist has no grounds for requiring a realistic interpretation other than predilection or fiat. Realism is in double jeopardy here: 'Copenhagen' is anathema to realism, while 'Bohm', which provides a consistent realistic interpretation, presents an underdetermination dilemma and thus blocks the realist from achieving the desired goal. That is, *if* one can erect mutually incompatible ontologies on a given formalism, then that does (or may) pose a genuine problem for the realist since an opponent would have to prove or argue strongly for a claim that once genuinely different ontologies are proposed it will be possible to extend the formalism along different lines, *because distinct ontologies involve distinct physical magnitudes*. This *may* happen, but *need* it? That seems more a declaration of belief than an argument for a position.

Of course, there is one obvious move still open to the realist at this point. He can claim that any two theories that are empirically indistinguishable can (by *definition*) differ only in inessentials. To set the stage for this, let us recall that Erwin Schrödinger, as early as 1922, expressed the belief that it is impossible to decide on the basis of observation whether the world is basically deterministic or indeterministic. Which description one uses is a purely pragmatic matter, dictated by convenience. Thus, in his 1922 inaugural address before the Prussian Academy of Sciences, Schrödinger claimed:

> The most that can be decided [on this issue] is whether the one or the other concept leads to the simpler and clearer survey of all the observed facts. (Hanle 1979, p. 268)

So, the scientific realist *could* write off the difference between indeterminism and determinism in the ontology of the world as an insignificance[6], but that would indeed be strange for one concerned with a reliable and meaningfully complete picture of the world. This radical conceptual difference between inherent indeterminism and absolute determinism also makes it virtually impossible to conceive of a 'dictionary' that would map the language of one of these theories onto the other (i.e., to map a concept onto its negation).

## IV. A FURTHER PROBLEM FOR THE REALIST

Of course, it is not absolute endgame for the realist who still wants to have *a* theory that will yield *the* correct and actual ontology of our world. Since empirical (in)adequacy and logical (in)consistency together do not alone provide sufficient criteria to choose between the two theories presented above, one can enlarge these criteria to include factors such as fertility, beauty, coherence, naturalness and the like. Certainly, it is the case that the scientific community quite early on did make a decisive selection in favor of the Copenhagen interpretation. The rival interpretations present at that time (*circa* 1927) were Erwin Schrödinger's wave picture and Louis de Broglie's pilot-wave model (which was a precursor to Bohm's conceptually similar interpretation some twenty-five years later). That is, much as Einstein claimed in the quotation at the beginning of this paper, *one* theory was in fact chosen. Could the actual criteria used have been objective, or at least atemporal in the sense that they are not, in an essential way, (unstable) products of historical contingency or accident? We indicate below that either of these theories passes equally well a test for fertility, in the sense of possessing the internal resources to cope with anomaly and new empirical developments that actually occurred, as well as for suggesting new avenues for research and generalization. And, we argue, an examination of the actual historical record shows that key motivating factors, for certain crucial assumptions about the features that an 'acceptable' theory must have, were based upon the philosophical predilections of the creators of the Copenhagen version of quantum mechanics and upon highly contingent historical circumstances that could easily have been otherwise. Below we sketch this 'history' of the origin and acceptance of the Copenhagen view, since this was the victorious program and we want to focus on the contingent character of that victory. Of course, a similar investigation can be carried out (and, in fact, it has been; see Cushing 1992) for the opposing 'continuity' view (involving Einstein, de Broglie and Schrödinger). The results are similar, but there is no space here for that part of the story.

Our intention in claiming equal fertility for this historically counterfactual scenario is to stress that Bohm's interpretation was not an *ad hoc*, stillborn creation that could have matched 'Copenhagen' only for the simplest cases (say, *circa* 1925), *but then would not have been able to grow and mature with its own internal resources* and without recourse to *ad hoc* moves at each turn. That requires a detailed historical and technical analysis that can be, and *is* being, done, but to which we do not even allude here. On the other hand, we do *not* claim that versions of Bohm's theory (or extensions thereof) have yet been generated that have matched *all* of the successes of the standard approach. Look at how *few people* have worked on this alternative program. Nor are we claiming more people *should* work on it. Our purpose here is to argue for a genuine underdetermination of the majority worldview, not to argue in favor of Bohm's. And, if there had been a different majority view, there is little reason to suppose that we would necessarily have arrived at the same place (i.e., set of questions and experiments) we are at now. In part, this would also be our response to some of the conceptual (but not formal) difficulties in producing Bohm-style quantum field theories for fermions. The Lord – malicious or not – knows well enough the (mathematical and possibly even conceptual) problems that plague conventional quantum field theory!

For brevity let us represent the historical situation in the first quarter of the present century as follows. There were two schools of thought and of workers, one of which (Bohr, Heisenberg, Pauli, Born, Jordan) led to the matrix-mechanics formulation of quantum mechanics, the other (Einstein, de Broglie, Schrödinger) to the wave-mechanics version. The basic issue separating these two groups was, respectively, discontinuity versus continuity, and this eventually evolved to encompass an indeterminism-determinism opposition. Each school began with a different set of philosophical predilections/commitments and each focused on a different class of physical phenomena (discrete spectral lines versus electromagnetic radiation, respectively) as presenting the paradigm puzzle to be solved. The sources of these different philosophical influences upon the major protagonists have been extensively documented by historians of science (Beller 1983a; 1983b; Cassidy 1976; Jammer 1974; 1989) and this need not detain us here. The point relevant to the present discussion is that this discontinuity-continuity dichotomy was contingently rooted in philosophical commitments and in specific classes of physical phenomena. Neither such commitments alone nor the phenomena alone themselves fix the course of development of the theory-selection process in science. Rather, it is the intersection of general predilections like these with the puzzles presented by physical phenomena that results in a definite theory or research program.

The small number of central players involved in laying the foundations of the matrix-mechanics program suggests that this was a closed group. This becomes all the more plausible when we realize that Pauli and Heisenberg were both Ph.D. students with Sommerfeld at Munich, each in succession were then Born's assistants at Göttingen and later worked with Bohr at Copenhagen. Each of these young students was greatly impressed by Bohr's 1922 Göttingen lecture. Jordan was also a student at Göttingen at this time. Philosophical factors in Bohr's own background inclined him toward, or at the very least made him receptive to, a discontinuous structure in nature at the most fundamental level and, eventually, to a doctrine of complementarity between opposites. This element of discontinuous transitions is a central feature in his 1913 'semiclassical' model for the hydrogen atom. This was certainly the current language for discussing atomic phenomena in Sommerfeld's school. So, *prima facie*, there is a case that Pauli and Heisenberg as young students were impressionable and naturally accepted this central tenet of atomic theory. However, this *inclination* toward credence is scarcely sufficient to account for the prevailing strength of this conviction on the discontinuity vs. continuity issue.

Largely due to the *failure* of certain classical approaches, the main players took up various *philosophical* positions on what was and was not *possible* in principle. That is, these were not logical or in-principle refutations, but strong, practical beliefs that became dogma. Thus, Bohr's own Ph.D. dissertation argued that the failure of the classical electron theory of metals was attributable to a fundamental insufficiency of the classical principles themselves. Both Pauli and Heisenberg had been involved in Born's program of attempting to apply the old quantum theory, with its orbitals, to molecular systems and the utter failure of this approach convinced them that electron orbitals were meaningless. The Stern-Gerlach experiments (1921/22) (notice the name – *space quantization* – originally used for this effect!) supported quantum predictions, rather than a *continuous* (classical) magnetic moment. Pauli's success with the Zeeman effect (1924) in terms of a classically non-describable two-valuedness in the quantum-theoretical properties of the electron further strengthened this belief in nonvisualizability. He was convinced as well that the exclusion principle could not follow from classical mechanics or from the old quantum rules. Heisenberg began to be converted to Bohr's and Pauli's views on the failure of mechanics. The failure of the Bohr-Kramers-Slater theory (in which energy conservation had been given up) in 1925 indicated to Bohr that " . . . a complete renunciation of the usual space-time methods of visualization of the physical phenomena . . . " (Dresden 1987, p. 212)[7] would be necessary for further progress. Heisenberg's matrix mechanics provided this.

It was the 'collision' between two forms of quantum mechanics that provided the impetus for the formulation of a consistent interpretation of quantum mechanics (Beller *op. cit.*). It is not uncommon for scientists to believe that there is just one (*unique*) law or theory. Such a belief would justify one in looking for, or attempting to formulate, *the* correct version of quantum mechanics. Heisenberg's faith in the finality of quantum mechanics was essential for his struggle to fashion the 'Copenhagen' interpretation via his uncertainty relations.

Born believed that microscopic coordinates were unmeasurable and, therefore, irrelevant. Through his analysis of scattering processes with Schrödinger's formalism, Born came to the opinion that even *perfect* initial information still led to uncertainty in the result and this implied, for him, a lack of causality. A propos of these issues, Heilbron (1988) asks what was the warrant for the complementarity principle as being complete and the final word in forbidding even the *possibility* of a description of microphenomena that is both causal and pictured in a continuous space-time. His response is that (*thus far*) experience has shown the validity of complementary pairs of descriptions and that belief in the ultimate *necessity* of complementarity rests on the *subjective* epistemological criterion of the need for classical concepts and on the indivisibility of atomic phenomena (i.e., Bohr's act of faith). In Heilbron's representation, the Copenhagen interpretation defined itself as true and strengthened its hold on physics, rewriting history so that Einstein, de Broglie and Schrödinger largely fade from view, thus leaving 'Copenhagen' as the only intelligible version of quantum mechanics. Heilbron also suggests some of the 'external' factors that were present when the 'Copenhagen' interpretation was being formulated:

> [T]he Pauli letters testify to the tremendous psychological *Zwang* ('constraint', another term of the Copenhagen art) under which the quantum physicists around Born and Bohr labored. Born himself oscillated between modest self-confidence and deep self-doubt; Bohr was frequently ill; Pauli hovered on the verge of breakdown; while Heisenberg, maintaining a healthy flippancy, succeeded in rescuing them all. Their correspondence is filled with strong words expressing despair, misery, and resignation, or joy, hope, and elation. . . . [T]hese romantic expressions . . . suggest the frame of mind that soon generated a religion of complementarity. . . . (Heilbron 1985, p. 391)

Beller (*op. cit.*) has also indicated personal and sociological factors that lent urgency to the Copenhagen school's establishing the hegemony of its interpretation.

A crucial encounter between the proponents of these two interpretations of quantum mechanics occurred at the Fifth Solvay Congress in 1927. There Louis de Broglie presented a theory in many ways similar to the Bohm interpretation that we outlined in a previous section of this paper. Pauli presented an example that appeared to show de Broglie's theory to be incoherent and de Broglie responded poorly (although subsequent analysis has shown Pauli's objections to be specious). While de Broglie's views received no real support, the Copenhagen adherents maintained a unified front and carried the day (both at this Solvay Congress and elsewhere). However, an examination of subsequent results in the physics literature on work in the causal (i.e., de Broglie-Bohm) quantum theory shows that it is a logically coherent interpretation of quantum phenomena and that micro-physics could have gone the (in-principle) deterministic route, rather than the now generally-accepted indeterministic one.

So, what is the relevance of this claim for the scientific realist? Well, if the historical record indicates a rearrangement of highly contingent factors could plausibly have led to a radically different scientific theory and world-view being accepted as correctly and uniquely representing the physical world, then one can reasonably question the value of a philosopher's rational reconstruction to pass judgement on a scientific theory. Philosophers typically study the successful theory already accepted by the scientific community. They doubtless have the ability to reconstruct rationally, and hence to legitimate, *any* theory that has already survived the scrutiny of the scientific community. But for historical contingency, though, they might find themselves doing just as well reconstructing and justifying an *essentially* different, equally successful and widely accepted theory. So, what is the value of the exercise, except as a check for noncontradiction? Each of these reconstructions could be equally *rational*, but there would not necessarily be anything rational to choose between them.

## V. CONCLUSIONS

The purpose of this paper has been to consider an underdetermination thesis, not just as some in-principle or abstract logical possibility, but as a real and practical problem that should be faced in our most successful scientific theory to date – quantum mechanics. If we do not 'see' this feature as an obvious fact, it is only because we are not sufficiently sensitive to the essential role played by highly-contingent historical factors at certain crucial junctures of theory construction and selection. I have argued that, with a different but equally plausible set of contingent factors, our rationally reconstructed and justified view of the physical world could be diametrically opposed to what it now is. However, most emphatically, my aim is *not* to argue in favor of 'Bohm' over 'Copenhagen' (nor *vice versa*). Here, I remain agnostic on that choice. And, of course, this underdetermination is only modulo no change in the formalism.

## ACKNOWLEDGEMENTS

This research was supported in part by the National Science Foundation under grant number DIR 89 08497.

I want to thank Larry Laudan and Jarrett Leplin for correspondence on (but not agreement over the content of) an earlier draft of this paper and, most especially, Arthur Fine for extended and helpful comments and criticisms on my paper on this topic given at the Central Division Meeting of the American Philosophical Association in Chicago in April, 1991. Finally, it is a great pleasure to be able to dedicate this paper to Heinz Post who, along with Michael Redhead, was so very encouraging to me when I first became interested in the philosophical implications of modern physics.

*Departments of Physics and of Philosophy*
*University of Notre Dame, Notre Dame*

## NOTES

1. Here the terms 'Bohr' and 'Bohm' are used to stand for entire casts of actual historical characters belonging to two schools, one of which ('Bohr') formulated the now almost universally accepted 'Copenhagen' version of quantum mechanics and the other the now largely ignored 'causal' version.
2. Post (1971, pp. 219–220) refers only to the 'Copenhagen' version of quantum mechanics and does not even acknowledge the existence of viable alternatives.
3. This is discussed at length in Cushing 1990 and 1992.
4. Laudan and Leplin (1991) argue against that.
5. I thank Arthur Fine for pointing this out to me and Keith Abney for discussion about this distinction.

6. Harmke Kamminga has kindly informed me that Russell (1917, Chapter IX) also argued for the epistemological (and empirical) indistinguishability between determinism and indeterminism and used this indistinguishability to argue that Laplacean determinism is vacuous (see especially pp. 199–208). I thank her for this interesting and relevant reference.
7. This is a quotation of Dresden's summary (not a direct quote of Bohr himself).

## BIBLIOGRAPHY


Beller, M. (1983a), *The Genesis of Interpretations of Quantum Physics, 1925–1927*, unpublished Ph.D. dissertation, University of Maryland.

Beller, M. (1983b), 'Matrix Theory Before Schrödinger', *Isis*, 74, pp. 469–491.

Ben-Menahem, Y. (1990), 'Equivalent Descriptions', *British Journal for the Philosophy of Science*, 41, pp. 261–279.

Bohm, D. (1952), 'A Suggested Interpretation of the Quantum Theory in Terms of 'Hidden' Variables, I and II', *Physical Review*, 85, pp. 166–193.

Bohm, D. (1953), 'Proof That Probability Density Approaches $|\psi|^2$ in Causal Interpretation of the Quantum Theory', *Physical Review*, 89, pp. 458–466.

Cassidy, D.C. (1976), *Werner Heisenberg and the Crisis in Quantum Theory, 1920–1925*, unpublished Ph.D. dissertation, Purdue University.

Cushing, J.T. (1990), *Theory Construction and Selection in Modern Physics: The S Matrix*, Cambridge University Press, Cambridge.

Cushing, J.T. (1991), 'Quantum Theory and Explanatory Discourse: Endgame for Understanding?', *Philosophy of Science*, 58, pp. 337–358.

Cushing, J.T. (1992), 'Causal Quantum Theory: Why a Nonstarter?', in (Selleri 1992, pp. 39–68).

Cushing, J.T., and McMullin, E. (eds.) (1989), *Philosophical Consequences of Quantum Theory: Reflections on Bell's Theorem*, University of Notre Dame Press, Notre Dame.

Dresden, M. (1987), *H.A. Kramers: Between Tradition and Revolution*, Springer-Verlag, New York.

Einstein, A. (1954), 'Principles of Research' in *Ideas and Opinions*, Dell Publishing Co, New York.

Fine, A. (1986), *The Shaky Game: Einstein, Realism and the Quantum Theory*, The University of Chicago Press, Chicago.

Fine, A. (1989), 'Do Correlations Need to Be Explained?' in (Cushing and McMullin 1989, pp. 175–194).

Hanle, P.A. (1979), 'Indeterminacy Before Heisenberg: The Case of Franz Exner and Erwin Schrödinger', *Historical Studies in the Physical Sciences*, 10, pp. 225–269.

Harding, S.G. (ed.) (1976), *Can Theories Be Refuted?*, Reidel, Dordrecht.

Heilbron, J.L. (1985), 'Review of Jagdish Merha and Helmut Rechenberg's *The Historical Development of Quantum Theory*', *Isis*, 76, pp. 388–393.

Heilbron, J.L. (1988), 'The Earliest Missionaries of the Copenhagen Spirit' in E. Ullmann-Margalit (ed.), *Science in Reflection*, Kluwer, Dordrecht, pp. 201–233.

Heisenberg, W. (1971), *Physics and Beyond*, Harper & Row, New York.

Jammer, M. (1974), *The Philosophy of Quantum Mechanics*, John Wiley & Sons, New York.

Jammer, M. (1989), *The Conceptual Development of Quantum Mechanics*, 2nd ed., Tomash, New York.

Lahti, P., and Mittelstaedt, P. (eds.) (1985), *Symposium on the Foundations of Modern Physics*, World Scientific Publishing Co, Singapore.

Lahti, P., and Mittelstaedt, P. (eds.) (1991), *Symposium on the Foundations of Modern Physics 1990*, World Scientific Publishing Co, Singapore.

Laudan, L., and Leplin, J. (1991), 'Empirical Equivalence and Underdetermination', *The Journal of Philosophy*, 88, pp. 449–472.

McMullin, E. (1984), 'A Case for Scientific Realism', in J. Leplin (ed.), *Scientific Realism*, University of California Press, Berkeley, pp. 8–40.
Post, H.R. (1971), 'Correspondence, Invariance and Heuristics', *Studies in History and Philosophy of Science*, 2, pp. 213–255.
Russell, B. (1917), *Mysticism and Logic*, George Allen & Unwin Ltd, London.
Selleri, F. (ed.) (1992), *The Wave-Particle Duality*, Plenum Publishing Co., London.
Sinka, S., and Sorkin, R.D. (1991), 'A Sum-Over-Histories Account of an EPR (B) Experiment', *Foundations of Physics Letters*, 4, pp. 303–335.
van Fraassen, B.C. (1980), *The Scientific Image*, Oxford University Press, Oxford.
van Fraassen, B.C. (1985), 'EPR: When is a Correlation Not a Mystery?' in (Lahti and Mittelstaedt 1985, pp. 113–128).

ARTHUR FINE

# Measurement and Quantum Silence

*Nowadays, we have stripped Maxwell of his philosophy and retained only his equations. Perhaps we should do a similar job on quantum mechanics.*

(H.R. Post 1974, p. 14.).

## I. THE PROBLEM

The central problem in the interpretation of the quantum theory is how to understand the superposition of the eigenstates of an observable. To a considerable extent scientific practice here, especially as codified in versions of Bohr's Copenhagen interpretation, follows an interpretive principle that I have elsewhere called the *Rule of Silence* (Fine 1987). That rule admonishes us not to talk about the values of an observable unless the state of the system is an eigenstate, or a mixture of eigenstates, of the observable in question. With regard to the Rule of Silence, as in other matters bearing on the interpretation of the quantum theory, Einstein was one of the first to realize that there can be difficulties. They appear as soon as we look at something like an explosion; i.e., the interaction between a micro- and a macrosystem that involves the amplification of a microphenomenon to macroscopic scale (Fine 1988 Chap. 5, esp. p. 78ff.). John Bell describes the difficulty over the Rule of Silence this way.

> The 'Problem' then is this: how exactly is the world to be divided into speakable apparatus . . . that we can talk about . . . and unspeakable quantum system that we cannot talk about? (Bell 1987, p. 171)

The "Problem", of course, is the quantum measurement problem. It is set by a series of results that make up the *insolubility theorem* (Wigner 1963; Fine 1970; Shimony 1974). According to that theorem no unitary evolution of states corresponding to a measurement yields a mixed object-apparatus state in which the indicator variable on the apparatus shows definite results, even under minimal restrictions on what counts as a measurement interaction.



*S. French and H. Kamminga (eds.), Correspondence, Invariance and Heuristics*, 279–294.

The measurement problem poses an obstacle to what some regard as a necessary condition for an acceptable physical theory; namely, that it stand in a correspondence relation to its predecessors.

> Roughly speaking, this is the requirement that any acceptable new theory *L* should account for the success of its predecessor *S* by 'degenerating' into that theory under those conditions under which *S* has been well confirmed by tests. (Post 1971, p. 228)

Such a correspondence with classical physics was one of the touchstones that Einstein employed in constructing relativity, and in judging the plausibility of various proposals for new physical theories (Fine 1988, Chap. 2). Einstein's rejection of Bohm's (1952) hidden variables approach to the quantum theory, for example, was based in part on his contention that the Bohm theory did not enable one to retrieve the classical and well-confirmed account of a ball rebounding elastically between two walls. According to Einstein this violated

> . . . the well-founded requirement that in the case of a macro-system the motion should agree, approximately, with the motion following from classical mechanics. (Born 1953, p.39)

In his response to the criticism Bohm rejected the methodology of correspondence principles, allowing it some value in guiding the search for new theories, but urging that no such general considerations can provide a good basis for rejecting an existing and well-confirmed theory (*op. cit.*, pp. 18–19). In their correspondence over this issue in 1951, Bohm reminded Einstein that the quantum theory never issues in accounts of how objects are likely to behave, but rather only in accounts of what we are likely to *observe* regarding their behavior. In the Bohm theory, moreover, objects have initial values, and measurements of those objects, while they may disturb the initial values, always issue in results. In fact the Bohm theory actually satisfies a modified correspondence principle: where the classical account itself is well-confirmed, the Bohm theory 'degenerates' into the classical account of *what we are expected to observe* under well-defined conditions of observation. Given the fundamental role of measurement in the quantum theory, this 'observational' principle would seem to be the proper version of correspondence there. Unless we simply ignore the measurement problem, however, the quantum theory does not satisfy even this modified correspondence principle. For the insolubility theorem makes highly problematic indeed just "what we are expected to observe". According to the Rule of Silence, it may be nothing.

Despite the failure of general correspondence between classical mechanics and the quantum theory, Heinz Post does not want to regard the development of the quantum theory as running counter to his correspondence-driven heuristic. Instead, regardless of Bohm's advice, he would blame the failure of correspondence on the quantum theory itself, which (like Einstein) he finds unacceptable on roughly realist grounds. Post looks forward to a more satisfactory and realist theory that would yield a general correspondence with classical mechanics. In the sequel I explore a more pragmatic and less visionary goal; namely, the prospects for reconciling the existing and well-confirmed quantum theory with what I referred to above as "observational correspondence". This requires a constructive response to the insolubility theorem.

Responses to the insolubility theorem constitute so-called 'solutions' to the measurement problem. Generally these responses sacrifice the Rule of Silence by allowing talk of definite values in certain special, superposed states. In giving voice to the unspeakable these responses constitute hidden variables theories. Among them are the radical de Broglie-Bohm pilot wave theory (de Broglie 1956; Bohm 1952), as well as more conservative solutions that 'approximate' the final superposed state in a measurement by an appropriate mixture, often achieved only in the limit. Other responses seek to respect the Rule of Silence by sacrificing the unitary dynamics instead. Below I will look briefly at both kinds of solutions. First I want to discuss the problem itself, in order to frame it in just the right way. I hope that way will prepare us for a rather different kind of approach. It is an approach that satisfies the desire that "fundamental theory permit exact mathematical formulation" (Bell *op. cit.*, p. 171). Bell recommends this objective as an antidote to the loose pragmatism of the quantum theory, and if one's reservations are actually about looseness, then my approach may help. If the reservations are about pragmatism, however, then because my approach is also pragmatic it may not help enough. In that case we will still have a problem, although I am not sure whether this problem would concern physics or the philosophy of physics. Perhaps it does not matter.

I frame my discussion in terms of the most familiar example, that of a Stern-Gerlach measurement of a component of spin.

## II. THE REFORMULATION

In a Stern-Gerlach measurement a spin-1/2 particle passes through a magnetic field inhomogeneous, say, in the $z$-direction. The action of the field correlates the microscopic position of the particle with the spin in the $z$-direction, spatially separating the state into spin up and spin down components that move toward two separate luminescent screens; say, $U$ and $D$. When a particle strikes a screen ($U$ or $D$) it puts electrons in the screen into an excited state. The electrons quickly wind down, and as they return to their ground state they emit photons. This produces a flash of light that marks a visible spot on the screen. The relative frequency with which flashes occur in the $U$ and $D$ screens, for a beam of particles prepared in the same initial state $\psi$, yields the probability in $\psi$ for spin-up and spin-down (respectively) in the $z$-direction.

A single spin measurement produces a flash on either the $U$ or the $D$ screen as its result. That result conveys virtually no information about the spin of the particle in the initial state. All we can conclude if, say, $U$ flashes is that the initial state was not a $z$-spin-down eigenstate. In particular, nothing follows from the $U$ flash about initial spin values. Instead of 'revealing' spin values, the measurement transfers the whole initial probability distribution for spin up or down in the $z$-direction to the probability for flashes in $U$ or $D$. It achieves this transfer by way of an amplification that leaves a thermodynamically irreversible and accessible record. If we write the initial spin state $\psi$ as

$$\psi = a\,|\uparrow\,> + \,b|\downarrow\,> \tag{1}$$

then the relative frequency of the flashes determines $|a|^2$ and $|b|^2$. The flashes do not distinguish between a beam of particles initially in the pure state $\psi$ and a beam initially in the corresponding mixed state $\rho$ given by

$$\rho = |a|^2\,|\uparrow\,><\,\uparrow| + |b|^2\,|\downarrow><\,\downarrow| \tag{2}$$

Because measurements produce results that are macroscopically accessible, the inability to distinguish between an initial pure state and the corresponding mixture is a characteristic feature of quantum measurement procedures. On the macroscopic scale there seems to be no distinction between pure states and mixtures, and hence no way of using the results of macroscopic measurements (of a single variable) to tell the difference.

One can turn this characteristic feature of quantum measurements around. That is, we can ask whether the end product of a typical measurement interaction would differ, depending on whether the initial object state were the pure state $\psi$ or the mixture $\rho$. In the Stern-Gerlach experiment sketched above the end product is a series of flashes and, as we have noted, the statistics do not differ for these pure and mixed starting states. If we describe the measurement quantum mechanically, however, then there ought to be a difference. In the case where we start with $\rho$, the linear Schrödinger evolution produces a transition to a final mixed object-apparatus state:

$$\rho \otimes \alpha \Rightarrow \rho_F = |a|^2\,|\uparrow\,><\,\uparrow| \otimes |u><u| + |b|^2\,|\downarrow\,><\,\downarrow| \otimes |d><d| \tag{3}$$

where $|u>$ and $|d>$ represent states of the $U$ and $D$ screens in which a suitable number of the electrons in the screen glow, and $\alpha$ is the density operator for the specially tuned starting state of the whole measurement apparatus. In case $\psi$ were the starting state, there would be a transition to a pure object-apparatus state

$$|\psi><\psi| \otimes \alpha \Rightarrow \rho_F + \rho_I \tag{4}$$

where

$$\rho_I = ab^*\,|\downarrow\,><\,\uparrow| \otimes |d><u| + a^*b\,|\uparrow\,><\,\downarrow| \otimes |u><d| \tag{5}$$

The term $\rho_I$ arises from the 'interference' between the up and the down terms present in the initial pure state and absent in the mixed one. As we have seen, it does not show up in the Stern-Gerlach measurement, nor would we expect it in any interaction that produces macroscopically accessible results. That is, whether the starting state is pure or not, what we observe in practice is the transition to the mixed state $\rho_F$. This is exactly what we would expect to observe from a physical point of view. This expectation is not satisfied by the quantum theory, however, which is one way of reformulating the quantum measurement problem.

In its usual formulation the measurement problem asks us to account for the fact that measurements have definite results. This suggests that something is missing from the usual story; namely, the actual registration of a result. The 'solution' then seems to require some addition to the theory, an addition that (somehow) puts results in. The reformulation above emphasizes just the opposite. Instead of suggesting that something is missing from the measurement story, that formulation emphasizes that the usual story is actually too full. If we want to accomplish a transition from an initial pure case to the right mixture, it seems that we have to lose something, not gain it. What we need to lose is the possibility of distinguishing between an initial pure case, like $\psi$, and the corresponding mixture, like $\rho$. Measurements proceed just *as if* the interactions always start out from an initial mixture, regardless of whether the initial state is pure.

## III. TWO PROPOSALS

Proposed solutions to the measurement problem can be graded on how well they succeed in explaining this *as if* feature of measurements. Why do measurements proceed *as if* the initial state of the measured system were mixed? There are two kinds of proposals that seem popular in the recent literature: replacement solutions, which modify the Rule of Silence while retaining the Schrödinger evolution, and collapse ones that change the unitary dynamics.

Replacement proposals (e.g., Machida and Namiki 1988; Fukuda 1987a, 1987b; Kobayashi and Ohmomo 1990) work roughly like this. They replace (4) with

$$|\psi><\psi| \otimes \alpha \Rightarrow \rho_F + \tilde{\rho}_I \qquad \text{(4R)}$$

where $\tilde{\rho}_I \to 0$ in the infinite limit of some suitable parameter (like relative time, size, or degrees of freedom – or some combination of these). The adjusted interference term $\tilde{\rho}_I$ usually results from eliminating apparatus variables that seem to do no work as indicators of the quantity being measured. (Here I follow Kobayashi and Ohmomo *op. cit.*) For example, in the Stern-Gerlach experiment the spot on the scintillation screen results from a loss of energy by the excited electrons as they make the transition to the ground state. The particular position coordinates of the electrons in the screen are not important. So, in coordinate representation, if we take the evolved state function and simply integrate over the positions of all the electrons, we can still track the energy shift. This effects the replacement of $\rho_I$ by $\tilde{\rho}_I$. Although the argument is not usually stated this way, the fact seems to be that the neglected variables (here the position coordinates of the electrons in the screen) actually do too much work. For they couple to the interference terms in the initial pure state $\psi$ in a way that distinguishes between interactions starting from pure states and interactions starting from mixtures. These terms need to be dropped (or traced out) in order to make sure that the measurement does not carry too much information about the initial state. Thus, the very fact that needs explaining, that information distinguishing pure cases from mixtures is lost in transit, is used to adjust the penultimate state so that it comes out the right mixture in the limit. A common criticism of replacement theories is directed at this final limiting operation (see Bell *op. cit.*, p.45). My point here is more basic. From the perspective of providing an acceptable explanation, replacement theories are circular. They use the fact they are supposed to explain (that information distinguishing pure cases from mixtures is lost in transit) to make the replacement that does the explaining. Do collapse theories fare better?

According to collapse theories (e.g., Ghirardi, Rimini and Weber 1986; Gisin 1984; Pearle 1986; Shimony, Ghirardi and Pearle 1991), there is an initial strong coupling between object and apparatus corresponding to the transition represented by (4). That interaction results in the following state

$$a\,|\uparrow\,> \otimes |\,u > +\, b\,|\downarrow\,> \otimes |\,d > \tag{6}$$

which persists for a time that depends on the size or complexity of the composing partial systems. Since one of the systems (the luminescent screen) is macroscopic (as judged by size, degrees of freedom, or some other well-defined parameter), a collapse mechanism takes over in short order. Repeated applications of the collapse single out one of the two branches of the superposition in (6), renormalized – at least approximately. In the GRW theory (Ghirardi, Rimini and Weber *op. cit.*), for instance, the collapse mechanism multiplies one branch by a Gaussian that becomes sharply peaked over time. The collapse concentrates on that branch with a probability that approaches the norm-squared of the branch itself as collapses repeat and time goes on. In the limit, then, the overall transition produces the mixture $\rho_F$, or a close approximation. The physical story that goes with collapse is this. In the beginning the coupling entangles the spin components of the deflected particle with the fluorescent electrons in the screen, according to the usual dynamics. There is a time limit, however, on the life of the excited electron states and when that limit is reached they spontaneously collapse to a near-point which glows, due to the photons emitted during collapse. Because of the initial entanglement, that collapse is also centered on one of the spin components and occurs at a rate that depends on its norm.

Despite the story, this account has a number of unsatisfactory features. First off, it is really not so clear that the collapse need be to a small region that coincides with a particular spin component. For, as we saw above in connection with the replacement theory, the collapse is actually a change in the energy state, which need not be localized at all. The GRW theory is open to criticism on just these grounds (Albert and Vaidman 1988). Moreover, the concurrence of collapse probabilities with those derived from the coefficients in (5) has no physical foundation. It is simply put it in by hand in order to get the right probability distribution. Even worse, if the collapse only approximates the eigenstates ($|\uparrow > \otimes| u >$) or ($|\downarrow > \otimes| d >$) and hence the desired final mixture $\rho_F$, then in interpreting the actual superposed state of the object-apparatus system as yielding the registration of a definite value we breach the Rule of Silence. (See Albert and Loewer 1990, for a related criticism.) Thus collapse theories (like GRW) that only approximate the desired mixture are a kind of replacement theory, where the replacement mechanism consists of an on-going stochastic process.

On the other hand, 'exact' collapse theories (i.e., ones that actually achieve the right mixture $\rho_F$ and not merely an approximation) may seem able to explain why interference between the terms in the initial pure state does not show up in the final mixture. On each measured system only one term survives, so eventually there is nothing with which to interfere. If we recall the explanatory task, however, this answer may not seem very responsive. For we wanted to understand why measurements proceed just *as if* the interactions always start out from an initial mixture, regardless of whether the initial state is pure. Information that would enable one to differentiate between initial pure states and mixtures is scattered among the various terms of an evolving superposition. Exact collapse theories propose a series of spontaneous transitions that, in time, lop off all but one of those terms. Thus exact collapse theories simply postulate the loss of information that needed explaining. In terms of providing a satisfactory explanation, this account of the loss of information, even were it strictly correct, would not differ very much from that provided by a replacement theory.

The combination of strong coupling followed by collapse amounts to the transition

$$|\psi >< \psi| \otimes \alpha \Rightarrow (\rho_F + \rho_I) \Rightarrow \tilde{\rho}_t, \qquad (4C)$$

where $\tilde{\rho}_t$ is the collapsed state at time $t$, and $\tilde{\rho}_t \to \rho_F$ as $t \to \infty$. Once again there may be a problem over the limit. More important, however, is the replacement of the entangled state ($\rho_F + \rho_I$) by the collapsed one $\tilde{\rho}_t$, for that is where much of the excess information is discarded. Let me put the problem here in physical terms. We know from experience that (typically) quantum systems decay. There is no way to obtain the decay from basic quantum theory without somewhere invoking the collapse of the wave packet. Collapse theories do this systematically. In providing a systematic rule for discarding information, collapse theories make a virtue of necessity. In doing so, however, they thereby forego the possibility of explaining why the information is lost. According to the details of the particular assumed stochastic process, it just is. As in the rule provided by replacement theories, this codifies the problem of loss of information without providing further physical insight into its occurrence.

Perhaps we are asking too much, however, for quantum phenomena teach humility. They teach us to look critically at the sources of our puzzlement and at our needs for explanation. They suggest the wisdom of aligning our demands for insight with the character of the phenomena themselves. In the quantum theory we learn to be rigorous in our thinking but pragmatic in our expectations. This is the lesson many have taken from the investigations of the Bell theorem, where the 'puzzling' correlations between measurement results on separated systems can be seen as basic to the physics, not necessarily in conflict with relativity or local causality, and not in need of further explanation (Fine 1989). Maybe the lesson to learn from the measurement problem is that it is better to view the loss of information as a basic feature of interactions that do in fact produce definite results, than to treat it as a phenomenon for which we require an explanation. The correct response to the issue, then, would not be to 'explain' this feature but simply to give a general and reasonably precise account of it. Perhaps improved collapse theories, or replacement theories could eventually do just that. But I doubt it.

I doubt that we are ever likely to have a really clean and general account. To be sure, one might produce a general template that characterizes the form of a measurement interaction, and then fill in particular features to suit the circumstances of special applications, using an open catalogue of options. For collapse theories this would mean adjusting the relaxation times between collapses, the variable being collapsed upon (e.g., position or energy), the exact state that emerges from the collapse, and perhaps even the probability for collapse – all depending on the circumstances (Gisin 1989). For replacement theories the open catalogue would involve specific ways of eliminating extra variables, depending on the type of detector; e.g., depending on whether the registration of a result is internally induced or the product of external fields (Kobayashi and Ohmomo *op. cit.*). If we bear in mind that even the standard dynamics is not algorithmic (in the Schrödinger evolution a specific Hamiltonian has to be supplied for each separate case) and that the correspondence rules that associate physical quantities with operators are also open-ended, it seems to me that amendments to the dynamics or alterations in the rules, however general, are unlikely to be closed and context-free. That said, however, I do have a reasonably general and fairly definite proposal to make.

## IV. A PROPOSED SOLUTION

My proposal (Fine 1987 and 1992; Stairs 1992) also makes a virtue of necessity. The necessity is set by the fact that the object-apparatus system evolves during a measurement just *as if* the initial state were mixed rather than pure. My solution is to suggest that this is so because, from the perspective of the measuring instrument, the initial object state really is mixed; there is no 'as if' about it. Thus I suggest that we rethink how to apply the interaction formalism initially, rather than look for replacements or collapses further down the line. The fundamental starting principle is that if system I is in (pure) state $\psi$ and system II in (pure) state $\underline{\alpha}$, then the composite system evolves from state $\psi \otimes \underline{\alpha}$. We know, however, that if the evolution is unitary (actually linear, or even deterministic will do; see Gisin *op. cit.*) then the two systems immediately become entangled, and we will be faced with inventing strategies for discarding information down the line in order to achieve the disentanglement necessary to produce a definite result.

Instead of worrying over how to discard information later, I suggest that we do it sooner, replacing $\psi$ by the corresponding mixture $\rho$ (from equation (2)). We can then let the interaction run according to the Schrödinger equation to produce the desired mixture $\rho_F$, as in (3). There is a physical rationale for this procedure. It is that in making a measurement we do not interact with all the variables of the measured object. We only observe the particular aspect of the object that corresponds to the variable being measured, say spin in the $z$-direction. If the initial object state is a superposition over eigenstates of that spin variable (e.g., an eigenstate of spin-up in the $x$-direction), then there is no initial 'value' of spin in the $z$-direction to observe, at least if we accept the Rule of Silence. That initial superposed state, however, does carry information about the $z$-spin; namely, its probability distribution. As emphasized in section I, determining that distribution is what counts as a spin measurement. So, I suggest that what the apparatus 'sees' in coupling to the object is only the probability distribution for the measured variable, as represented by the mixed state, and not the whole (pure) state of the object. So far as the measurement interaction is concerned, starting states that have the same distribution function for spin in the $z$-direction are identical. Thus the measuring instrument really couples to what is common to a whole class of equivalent states and not to a particular one (where two states are equivalent with respect to an observable if they have the same probability distribution for values of that observable). If this is a plausible story, then we make a mistake in applying the interaction formalism in a fine-grained way. We need to coarse-grain in order to respect the discriminatory capacities of the measuring instrument.

A $z$-spin measurement is characterized by the fact that the initial object $z$-spin distribution is transferred to the final apparatus indicator distribution, and by the fact that each measurement registers some one result. This characterization is purely physical. It makes no reference to observers, whether conscious or not. It is also perfectly general: any interaction that transfers probability and gets results is a measurement. To represent the interaction formally we need to take both conditions into account. We represent an interaction that transfers probability by means of a dynamical group generated from a suitable joint Hamiltonian. We represent the fact that a result is produced by the procedure of course graining; specifically, we replace the initial pure object state by the corresponding mixture, and start the Schrödinger evolution with that.

This is an 'exact mathematical formulation'. What it formulates is the idea of an interaction with just part of a quantum system, the part (or 'aspect') represented by the probability distribution for a particular observable. The usual way of deploying the interaction formalism enforces a quite unreasonable holism. By entangling the interacting state functions, that formalism makes virtually any interaction capable of reflecting every aspect of the whole system. As in the case of measurement, much less may be true, and when it is we need a way of representing the nonholistic interaction formally. My way represents probability distributions by mixed states. It retains the usual dynamics and it respects the Rule of Silence. It may be desirable to have a more general way of treating nonholistic interactions, in terms of a fuller account of what constitutes a part or aspect of a system, and how that is to be represented formally (see Fine 1987). For measurement interactions, however, the relatively simple scheme sketched above seems to suffice.

## V. COLLAPSE OF THE WAVE PACKET

My reformulation of the measurement problem asks why the measurement proceeds as if the initial object state were mixed rather than pure. The answer provided above is that the measuring instrument in fact only interacts with part of the object system, a part which is adequately represented, formally, by the mixed state. So the interaction that actually occurs in a measurement is with the mixed state, and not with the whole pure one – which is why it seems to be that way. My reformulation also emphasizes that information is lost in a measurement. My account of the loss is that it occurs because the interaction is nonholistic: the apparatus only couples to a particular aspect of the object, not to the whole. What is lost pertains to aspects of the object to which the measuring device does not respond. The usual formulation of the measurement problems asks how we can account for the fact that measurements have results. My answer is that "having a result" is part of what we demand of a probability-transfering interaction in order to count it as a measurement. We represent this formally by adjusting the starting states of interactions differently, according to whether they do or do not produce 'results'. Given the right deployment of the interaction formalism, it is then trivial to show (as in equation (3)) that measurements produce results. Thus the answer to the usual formulation of the measurement problem does not lie in deriving that measurements have results (using special approximations or non-standard dynamics). The answer is contained in understanding how to use the interaction formalism.

There is a third way of formulating the measurement problem, one that John Bell has emphasized:

> [S]o long as the wave packet reduction is an essential component, and so

> long as we do not know exactly when and how it takes over from the Schrödinger equation, we do not have an exact and unambiguous formulation of our most fundamental physical theory. (Bell 1987, p. 51)

This third concern asks when, exactly, does the state function collapse. The answer contained in the account of nonholistic interactions is, exactly, never; there is no collapse. The Schrödinger equation always applies. A collapse seemed to be required in order to destroy interference and lose information. Reversing the paradox of classical statistical mechanics, where sensitivity to initial conditions is equivalent to forgetfulness of them (see Bell *ibid.*, p. 103), in quantum mechanics sensitivity entails complete recall. Thus to achieve forgetfulness, which is to say the appearance of a collapse, requires loss of sensitivity to the initial conditions. Loss of sensitivity is exactly what the aspect-sensitive deployment of the interaction formalism achieves.

Formulations of quantum mechanics without collapse include the pilot wave theory and the many worlds interpretation. (See Bell *ibid.*, p.117, for an exposition and comparison.) The program I advocate here has little in common with either. It is not a hidden variables account, for it strictly respects the Rule of Silence. Unlike the pilot wave, it does not privilege position variables, or any other. It does not invoke any quantum potential, as a multi-dimensional guiding field for real particles, nor does it entail non-local effects that propagate across the field, but below the level of observation, with superluminal velocities. Unlike the many worlds interpretation, measurements analyzed as above entail no splitting of universes and hence there is no need to worry about transworld communications. Bell has characterized the many worlds conception as giving an account of present correlations with present phenomena, and hence as renouncing the association of a particular present with a particular past (*ibid.*, pp. 134–5; see also Geroch 1984). My proposal for nonholistic interactions is not like that at all. For according to my proposal, just as in the usual quantum theory of measurement, the significance of a measurement is that it tells us what the probability distribution was for the observable being measured in the initial state of the object. Thus measurements are inherently backward-looking. Although they are not sensitive to the whole past, they do look back to and reflect a particular aspect of it.

## VI. PUZZLE CASES

My suggestion for tailoring the interaction formalism according to the anticipated result may seem to encounter problems in cases where we can change our decision about the measurement after it is in progress. Two such puzzle cases come to mind; namely, a delayed choice double slit experiment and a Stern-Gerlach experiment where we recombine the two beams. These may seem problematic on my account since they both allow for an interference pattern to be displayed which, one might think, would have been precluded by the switch from pure case to mixture that I recommend. That thought, however, is not correct.

Consider a delayed choice experiment. A low intensity beam of particles falls on a barrier with two suitable small and separated slits. Behind the barrier, at a respectable distance, is a detecting screen. In between the barrier and the detecting screen are particle counters capable of registering whether a particle passes through the top slit or the bottom. We can turn these counters on or off at will. The decision to turn the counters on or off is made only after the particle (assuming that only one at a time enters the apparatus) has passed the slits. With the counters off the particles build up an interference pattern on the detecting screen. The usual puzzle here arises from the conception that the interference pattern requires each particle (somehow) to go through both slits. By delaying the choice of whether to switch on the counters, on this conception, we seem to be able to make the particle go through one slit, or both, after the fact. Puzzle indeed.

My treatment of the experiment goes like this. We can expand the state function of a typical particle as a superposition of eigenstates corresponding to 'passage through the top slit' and 'passage through the bottom slit'. In runs where the counters are on, we need to replace this evolved pure state (at the time just after the particle passes the barrier) by the corresponding mixture over eigenstates of passage through one slit or the other. This will yield the observed counting rate. In those runs where the counters are turned off (after passing the barrier) a result is obtained on the detecting screen. Hence on those runs we need to analyze the interaction between the particle and the screen by replacing the particle pure state, at the time it encounters the screen, by the right mixture. The mixture we want is obtained by expanding the pure state, at the time the particle encounters the screen, in approximate eigenstates of position on the detecting screen (i.e., as corresponding to a coarse-grained position operator). This yields the result that, in such runs, the particle is detected on the screen and that, overall, the interference pattern builds up there. This treatment nicely illustrates how the replacement of pure states by mixtures is tailored to the specific sensitivity of the instrument. The counters are sensitive only to position at one time (near the time when the barrier is crossed). With the counters off the detecting screen is sensitive only to the later position of the particle. That sensitivity is enough to produce the interference pattern.

What then of the puzzle? In the preceding analysis nothing is made to happen after the fact. We do not make the particle go through the slits either singly or doubly. We only measure position at one time or another. The appearance of making things to have happened arises from a decision to treat the occurrence of the interference pattern as a sign that particles reaching the detecting screen have gone through both slits. This is one way of breaking the Rule of Silence, and the delayed choice experiment shows that it is not a very satisfactory way. The Rule of Silence says that we should not ask about passage through the slits in a run where we measure the position of a particle on the detecting screen. The approach to the measurement problem that I am sketching here respects that rule. Accordingly, it has nothing to say about passage through the slits in runs where the counters are off. This is orthodox Copenhagen non-speak. It too is unsatisfactory, but not because there is a problem about measurements and their results. With regard to solving problems in the quantum domain it is useful to treat one problem at a time, or at least to try. Here again, holism may not be the best basis on which to proceed.

The delayed choice experiment, however, can be given a further twist; namely, into the 'quantum eraser' (Scully and Drühl 1982). In this version, the counters are left on but after recording a particle's position on the detecting screen we erase the information contained in the counters concerning which slit the particle has passed through. With this erasure the interference pattern is observed on the screen, although not without it. Again we seem to have made something happen (the interference pattern) after the fact (i.e., after the particles have landed on the detecting screen). More importantly, in terms of my suggested analysis of measurement interactions, in an erasure experiment the measurement interaction may seem to be exactly the same as that in an experiment with the counters on but without erasure; namely, sensitive to position around the slits. But if that were the case, according to the preceding analysis, interference should not show up on the detecting screen. Consider, however, a time-reversed erasure. That is, suppose we first set the counters on, then we immediately erase that information, then we finally record the particle on the screen. The combination of turning the counters on and immediately erasing the count information amounts to an interaction sensitive only to position around the detecting screen, and hence in this case (according to the delayed choice analysis) we can demonstrate that the resulting pattern should show interference. If this is correct, then what difference does it make when the information is erased? When the whole measurement interaction is taken into account the net result is the same.

In an erasure experiment, just as in a time reversed erasure, we get a composite interaction with the measured object that is sensitive only to its position on the detecting screen. Thus the quantum erasure does not pose a difficulty for our analysis of measurement in terms of restricted sensitivity. It merely serves to highlight that in determining the range of sensitivity of an interaction we need to take into account the whole experimental arrangement. This is another feature of scientific practice in the quantum domain that Copenhagen has emphasized.

It is now straightforward to deal with a Stern-Gerlach experiment where the beams are combined after passage through the magnets but before a record is produced on the luminescent screens, for this experiment is similar to a double slit experiment with the counters turned off. We display interference between the two beams after recombination by measuring, say, spin in the $x$-direction. That measurement requires an interaction sensitive only to the $x$-spin distribution, and hence it is to be treated as starting from a state mixed over $x$-spin eigenstates. That mixture displays the interference between $z$-spin components that we get from the recombination of the beams, assuming that no $z$-spin has already been recorded. Nothing in the results of this treatment differs from the quantum theory, so nothing can go wrong with the analysis – that is, unless something is wrong with the quantum theory itself.

## VII. CONCLUDING REMARKS

My focus has been the measurement problem. As suggested in section I, it is usually thought that the problem arises out of a conflict between linear dynamics and the Rule of Silence. Linearity entangles the object and apparatus states, and the Rule of Silence applied to such entangled states forbids us from attributing a definite result to the interaction. I have shown that there is a significant third player in the genesis of the problem; namely, the application of the interaction formalism itself. Revising the rules for using that formalism provides a way out of the conflict, a way that respects the usual dynamics and the usual interpretive practices.

My way out trades on the idea that some interactions are sensitive only to certain aspects of a system, not to the whole thing. This nonholistic conception calls for a way of treating interactions with only part of a system. Where the part corresponds to the probability distribution for an observable, I suggest we represent it by a density operator over the eigenstates of the observable, one whose coefficients yield the probability distribution in question. This gives an objective, non-ignorance interpretation to mixed states for a single system. They represent 'parts' or 'aspects' of the system. This way of interpreting mixtures (in the context of an interaction with a system part), and a more general investigation of parts or aspects of a system, seems to me worth pursuing independently of its utility in reconceptualizing measurement interactions.

The basic scheme I have pursued is this: a measurement of an observable on a system in state $\psi$ is an interaction with the part of the system corresponding to the probability distribution for the observable that is given by state $\psi$. This conception of measurement is purely physical. It involves no 'observers'. When the interacting parts are represented by mixed states, this conception of measurement uses only the language of elementary quantum theory, where it can be given a general and precise mathematical treatment.

There is a long tradition that deplores the introduction of the concept of measurement as fundamental in the quantum theory. Einstein belongs to this tradition, as does Bell. The sticking point over measurement seems to relate to realism, and especially to concerns over objectivity. If the concept of measurement is what I make of it here, however, then I think that we need not worry about objectivity. No observers are required to make individual results definite. Interaction with part of a system makes results definite, although no particular one. Observers are not needed to collapse the wave packet, either; for the packet never collapses. The results of measurement, taken collectively, are not created by the measurement; for what measurements reveal (collectively) are aspects of the object already present in the initial, undisturbed state (namely, probability distributions). In all of these ways, the quantum theory is objective. My treatment of measurement helps to bring these objective features out in the open. Of course the quantum theory, like any other, has to be understood and applied by human beings. In focusing the treatment of measurement on the way the interaction formalism is applied, we highlight a pragmatic element present in all theories. This does not implicate special features of the quantum theory, with regard to objectivity, that need worry us.

It seems, then, that we can reconcile the quantum theory with the sort of observational correspondence discussed in section I; i.e., that the insolubility theorem need not stand in the way of a correspondence with the confirmed observational predictions of classical mechanics. As Heinz Post used to remind me, however, there is a conservation law for problems with the quantum theory. When we seem to dispose of one, another pops up; just as sweeping an object under the rug merely moves it from one place to another. Thus one may not be content with observational correspondence, requiring a more general and realist version instead. Such realist concerns relate to the Rule of Silence which, as we have seen, need not stand or fall with a solution to the measurement problem – our concern here. I will only say to Heinz that realism is a topic for another time (e.g. Fine 1988, Chap. 9).

## ACKNOWLEDGEMENTS

I am grateful for support provided by National Science Foundation Grant # DIR 8905571.

*Department of Philosophy*
*Northwestern University*

## BIBLIOGRAPHY

Albert, D. and Loewer, B. (1990) in A. Fine *et al.* (eds.), *PSA 1990*, Volume I, Philosophy of Science Association, E. Lansing, p. 277.

Albert, D. and Vaidman, L. (1988), *Physics Letters A*, 139, 1.

Bell, J.S. (1987), *Speakable and Unspeakable in Quantum Mechanics*, Cambridge University Press, Cambridge.

Bohm, D. (1952), *Physical Review*, 85, pp. 166, 180.

Born, M. (1953), *Scientific Papers Presented to Max Born*, Edinburgh, Oliver and Boyd.

de Broglie, L. (1956), *Tentative d'interpretation causale et non-linéaire de la méchanique ondulatoire*, Gauthier-Villars, Paris.

Fine, A. (1970), *Physical Review*, D 2, 2783.

Fine, A. (1987) in R. Kargon and P. Achinstein (eds.), *Kelvin's Baltimore Lectures and Modern Theoretical Physics*, M.I.T. Press, Cambridge, p. 491.
Fine, A. (1988), *The Shaky Game: Einstein, Realism and the Quantum Theory*, University of Chicago Press, Chicago, 1988.
Fine, A. (1989) in J. Cushing and E. McMullin (eds.), *Philosophical Consequences of Quantum Theory*, University of Notre Dame Press, Notre Dame, p. 175.
Fine, A. (1992), *Foundations of Physics*, 5, 125.
Fukuda, R. (1987a), *Physical Review*, A 35, 8.
Fukuda, R. (1987b), *Physical Review*, A 36, 3023.
Geroch, R. (1984), *Noûs*, 18, 617.
Ghirardi, G.C., Rimini, A. and Weber, T. (1986), *Physical Review*, D 34, 470.
Gisin, N. (1984), *Physics Review Letters*, 52, 1657.
Gisin, N. (1989), *Helvetica Physica Acta*, 62, 363.
Kobayashi, T. and Ohmomo, K. (1990), preprint of Tsukuba University, Japan, UTHEP-196.
Machida, S. and Namiki, M. (1988), *Foundations of Physics*, 18, 29.
Pearle, P. (1986) in R. Penrose and C. J. Isham (eds.), *Quantum Concepts in Space and Time*, Oxford University Press, Oxford, p. 539.
Post, H.R. (1971), *Studies in History and Philosophy of Science*, 2, 213.
Post, H.R. (1974), *Against Ideologies*, Inaugural Lecture, Chelsea College, University of London.
Scully, M.O. and Drühl, K. (1982), *Physical Review*, A 25, 2208.
Shimony, A. (1974), *Physical Review*, D 9, 2321.
Shimony, A., Ghirardi, G. and Pearle, P. (1991) in A. Fine *et al.* (eds.), *PSA 1990*, Volume II, Philosophy of Science Association, E. Lansing, p. 17.
Stairs, A. (1992), *Foundations of Physics*, 5, 105.
Wigner, E.P. (1963), *American Journal of Physics*, 31, 6.

SIMON SAUNDERS

# To What Physics Corresponds*

## I

In what follows I wish to reconsider certain ideas to be found in Post's (1971) defence of the 'retentionist' or 'accumulativist' view of science. In particular I shall focus on heuristics and methodology and will confine the discussion to physics, specifically to theories of *dynamics* (this, I hazard, is to be counted a constraint in principle: it seems unlikely that similar considerations will apply to any other branch of empirical science). Post's thesis (what he calls the "generalized principle of correspondence") is both historical and methodological; it may be simply put as the claim that what is taken over from preceding theories is not only those laws and experimental facts which are well-confirmed, but also 'patterns' and 'internal connections', that in this way the successor theory accounts for whatever success its precursor enjoyed, for it " . . . will in fact embody a good deal of the (lower) theoretical structure of the [precursor] theory." (1971, p. 229).

By 'dynamics' I mean to include statics and kinematics, as well as mechanics and field theory. The 'constraint in principle', as I understand it, is that in no other field does one see so powerful an interplay between mathematics and phenomenology, and only in mathematics has one the resources to elaborate a notion of 'patterns' and 'internal connections' that is something more than the generic concept of metaphor. For these reasons I shall further consider only those dynamical theories that achieved an internally consistent, systematic, and highly mathematical formulation, with a substantive and well-confirmed body of quantitative applications (with the exception of astronomy and statics, we are therefore limited to the Modern period). My principal target is the 'anti-accumulativist' (or 'anti-retentionist') consensus that has, by and large, replaced the traditional reductive account of inter-theory relationships that we owe to the positivists. This consensus appears a haphazard and perhaps temporary convergence of a number of themes in contemporary metaphysics and epistemology, ranging from social constructivism and his-

* Dedicated to Heinz Post on the occasion of his 75th birthday.



*S. French and H. Kamminga (eds.), Correspondence, Invariance and Heuristics*, 295–325.

toricist epistemology to linguistic holism and anti-realism. Correspondingly, those who have most vigorously championed the 'anti-retentionist' view of scientific progress come from widely different traditions (consider Kuhn, Feyerabend and Laudan). For convenience, however, I shall refer to this view as 'relativist'. It is my contention that relativists and realists alike have overlooked important and, in comparison to other disciplines, quite atypical features of dynamical physics[1]. It has long been apparent (to physicists if not to philosophers) that we do not really understand the role of mathematics in dynamical theory, that there is something more here than a 'mere' instrumental utility or economy of expression[2].

## II

Kuhn and Laudan acknowledge that mechanics is exceptional in the continuity of its development, but go on to isolate this 'branch' of science (but it is not any old branch) from quantum theory, spacetime theory, and gravity, as though it may be accounted some sort of recherché preoccupation of mathematicians in the late eighteenth and nineteenth centuries, outside of the mainstream of scientific advance. Newtonian theory, on the other hand, is treated on a par with the crude and figurative mechanical schemes of Galileo and Descartes. Thus Laudan (1981) cites "the theory of circular inertia" as a counter-example to the correspondence thesis (but this was not an internally consistent and systematic mathematical theory, still less one that achieved detailed empirical support), and considers the later history of mechanics a "rare occasion" on which the thesis of correspondence might prove justified.

Laudan's sights were set on more sweeping claims of 'convergence' of scientific theories, in the service of a still more resounding argument to successful reference, realism and the 'approximate truth' of contemporary theory. It is in connection with the more restricted claim that "in mature science" there is evidence that "mechanisms, models and laws" are preserved in theory-change, particularly within the developed theories of physics, that one is less than convinced. His examples (from physics) where (he claims) correspondence *fails* are these: Copernican astronomy vs. Ptolemaic astronomy, Newton's physics vs. Cartesian mechanics, astronomy, and optics, Franklin's electrical theory vs. that of Nollet, relativistic physics vs. the ether and "the mechanisms associated with it", statistical mechanics vs. "the mechanisms of thermodynamics", the wave theory of light vs. "the mechanisms of corpuscular optics".

One wants rather more detail (a curious failing, here and elsewhere, given that this is a historical critique), but surely Laudan is right to insist that one can always find some theorem, deduction, conjecture, or explanation that has no precise correlate in the successor theory (what Post calls 'Kuhn-losses'). But what are we to conclude from this? Clearly one needs desiderata as to which, and in what respect, such losses are to count as significant (one must make evaluative judgments). Alternatively, if one is not always to be compiling lists of examples, one wants a principled argument bearing one way or the other on the thesis at issue. Post's theory of heuristics is exemplary: given its emphasis on the heuristic importance of theoretical unity, and of the conservation of those features of past theories "which have been confirmed without exception", it is clear enough what 'Kuhn-losses' are to count as significant. Against this it may be said that significant theoretical innovation is only possible because scientists have been willing to tolerate Kuhn-losses, a point stressed by Feyerabend and Laudan.

Like Post, my concern is with heuristics; the point of view I shall sketch does, however, do justice to the latter objection. A part of the response is evaluative: that what are frequently called 'Kuhn-losses' are rather to do with high-level interpretations of dynamical theory (or with figurative schemata – usually precursors to the Newtonian synthesis – such as those of Galileo and Descartes). But this is only to say that some of these examples need not concern us; it cannot be denied that there is also radical innovation at a more substantive level, what is to be reconciled with the conservativism implicit in the thesis of correspondence.

To this end I want to focus on a certain kind of abstraction, distinctively mathematical: a level of 'pattern' or 'form' to a dynamical framework or application, the recognition of which (in existing theory) makes for the first phase of innovation (what Kuhn would include in 'normal' science)[3]. In what follows the term 'heuristic' will always be taken in this sense, i.e. that of an abstract principle, or a mathematical abstraction. As examples I would cite the action principle, the theory of Euler and Lagrange, Hamiltonian mechanics and the Hamilton-Jacobi theory, differential geometry, the theory of topological groups, Klein's 'Erlangen' program, the relativity principle of Poincaré and Einstein, the equivalence principle, the relationship between difference equations and the derivative, and the gauge principle (there are many more). Revolutionary developments, I suggest, occur when such a heuristic (embedded in extant theory) is subject to radical, and more or less autonomous development, with little or no regard to the 'high-level' interpretation of that theory.

What will emerge as central to the thesis of correspondence are questions to do with the 'plasticity' of such heuristics (and of abstract structures in general), and with criteria for their identity over time. It seems to me that there are many examples where this identity is not in doubt, despite dramatic changes, usually in the direction of greater abstraction, over the course of their development (we shall encounter some presently). There are also phases in which heuristics become more or less stable sub-structures of dynamical physics. Given this sort of 'entrenchment' (the parallels with Goodman's account of 'projection' are obvious) one does better to characterize these abstractions differently (I shall use the term *canonical*). It is, then, a part of the thesis of correspondence that such structures (heuristics and *a fortiori* canonical structures) are preserved or evolved in theory change.

I shall illustrate this account of heuristics in some detail in the context of the development of quantum mechanics (Sections V–VII); but first, and in a more philosophical context, let us consider Laudan's examples. Our criteria, as to what is to count as a dynamical theory, are exacting. Of those cited, only Ptolemaic and Copernican astronomy, statistical mechanics and thermodynamics, and relativity and the ether theories (presumably Laudan here means pre- and post-relativistic electromagnetic theory) come into consideration. Let me make some brief (and therefore sketchy) remarks on the first and third, the only *prima facie* cases of the wholesale *abandonment* of entrenched dynamical theory[4].

The 'Copernican Revolution' (a phrase due to Kant) is a much-cited example of the elimination of previous theory, but it also illustrates the cumulative methodology just summarized. To begin with let us view the introduction of epicycles (by Appolonius of Perga) into the geocentric astronomy as an example of a "first phase of innovation" (the autonomous development of the basic concept of Eudoxus and Plato, that celestial bodies move in circles with constant angular velocity). It was, therefore, a heuristic (and became – and remains – canonical). In particular, it was abundantly clear that one could represent the relative motion of the planets (including the Earth) and the Sun with respect to a system of epicycles centered on the Sun rather than the Earth. The obstacle to such an innovation was, of course, the conflict with Aristotelian *physics*. The contribution of Copernicus was precisely that he valued the parsimony and simplicity of the conceptual scheme of Plato, Eudoxus and Appolonius above that of Aristotelian physics and was prepared to modify the latter for these reasons. Considerations of simplicity then favored a heliocentric scheme. Of course something of the Aristotelian *Gestalt* (statics and astronomy to one side, here one has *only* a high-level interpretive scheme), had to be given up in this process.

In its details, the Copernican system made frequent use of epicycles and eccentrics, and could easily demonstrate the equivalence of heliocentric and geocentric motion[5]. What was altogether missing from the new system was the equant, introduced by Ptolemy, by means of which the deferent was assigned *non-uniform* angular velocity.

The controversies which surround this development are remarkable. Kuhn and Feyerabend focus on the abandonment of the Aristotelian world-view; what was involved was a wholesale change of paradigm. Laudan points out that since the Copernican theory did not preserve a 'mechanism' of the Ptolemaic theory, namely the equant, the latter cannot be understood as a limiting case of the former. Glymour argues on independent grounds (i.e. according to his 'bootstrap' methodology) that the Copernican theory is much superior to that of Ptolemy, and thereby defends a thesis of theory progression independent of correspondence. But I take my cue from Dijksterhuis (1961, p. 288):

> When at the end of his years [Copernicus] reviewed his life's work once more, he considered the greatest gain it had brought astronomy was not the changed position of the sun in the universe and the resulting simplifi-

cation of the world-picture, but the abolition of the *punctum aequans*, the atonement for the sin against the spirit of Platonic philosophy which Ptolemy had committed in an evil hour. Really to understand the workings of Copernicus' mind, one can no more overlook this statement than in studying Goethe one can ignore the fact that in his old age he appeared to attach more value to the theory of colours than to his literary achievements[6].

Einstein once remarked that each of us has our own Kant; it seems we each have our own Copernicus too. This does not seem the stuff of revolutions in the philosophy of science. But one point needs clarification. I remarked that the central heuristic – that of Appolonius – became and remains canonical. By this I mean that it was and still is an essential conceptual structure of dynamical theory. The technique can be simply illustrated for periodic motion in two dimensions. With the usual isomorphism onto the complex plane, motion using a single epicycle is an expansion of the form $c_1\exp(i\omega_1 t)$ + $c_2\exp(i\omega_2 t)$ (where $|c_1|$ is the radius of the deferent, $|c_2|$ that of the epicycle, and $\omega_1$, $\omega_2$ the angular velocities of the deferent and epicycle respectively). An astronomy based only on epicycles (in particular which does *not* make use of equants) corresponds to an expansion of the form $\sum_i c_i\exp(i\omega_i t)$ (with the Earth chosen as origin). That the innovation of Appolonius is recognizably a primitive version of Fourier analysis is an example of the 'plasticity' of a heuristic[7].

The relationships between pre- and post-relativistic theories of electromagnetism, and between Galilean and Einsteinian relativity, appear, on the face of it, much more problematic. The elimination of ether is certainly the most significant case of referential failure in dynamics. But reference is not what is here at issue, and the distinction between the theory proper, and an interpretive superstructure, must be handled with care. For example, in Helmholtz's (influential) synthesis of continental potential theories and Maxwellian field theory, the ether was treated as a dielectric substance in exactly the same way as any (ordinary) material medium; evidently there is still a sense in which we think of the vacuum state in this way (we still talk of the 'electric permitivity and magnetic permeability of free space'). Following Helmholtz, there was scarcely an application of ether theory by Continental theorists that cannot be taken over to the electrodynamics of material media. What appears to be at issue is the distinction between the abstract framework of continuum mechanics, and its application to electromagnetic phenomena *in vacuo*. The former is eminently canonical (as such it survives in contemporary dynamics); the latter application has been abandoned. It is not, of course, that the generic notion of *some* substructure to radiation has been given up – nor, for that matter, that this substructure has a *dynamical* description (it has a *quantum mechanical* description). But there is an important difference between these dynamical frameworks: the former, but not the latter, implies the existence of a privileged frame of reference[8].

The continuum mechanics proved remarkably successful in the formulation of Maxwell theory, but implied something more – that motion with respect to a privileged frame should be observable. The repeated failure of attempts to detect such effects then indicated that Maxwell theory could not, after all, be viewed as a species of this theory. But quite independent of these developments, Maxwell theory had been largely isolated from any detailed model of the underlying medium (and therefore also from mechanics), just because no viable and consistent model had been developed. Its relationship to mechanics was left hanging[9]. The subsequent development of relativity concerned a quite different canon, the concept of inertial frames (grounded in Newton's laws) and their covariance group, isolated and subjected to autonomous development as a heuristic (that is, independent of the specific form that it took in Newtonian theory), yielding the relativity principle. Once again what is at issue is heuristic plasticity. In the context of Maxwell theory (more specifically, to a certain consequence of this theory, the independence of light-speed from the speed of the source), Einstein was led to a wholesale reappraisal of the concept of simultaneity; from this the Lorentz group followed.

This history is therefore complex: two distinct heuristics were isolated within mechanics, wave equations (partial differential equations), and the relativity principle, both of which were subject to autonomous development. Their reconciliation (but also the reconciliation of the relativity principle with electromagnetic *phenomenology*) then led to a new theory of space and time, and thereby a new mechanics. But this did not bring about a *rapprochement* of continuum mechanics and Maxwell theory (i.e. through a relativistic mechanics of media), for the very concept of a medium (as opposed to more wide-ranging notions of some sort of geometric or dynamical substructure) defines a privileged frame of reference.

One could say that we have the abandonment of a canonical structure (Galilean relativity), the abandonment of an application of another (the continuum mechanics), and correspondingly, the elimination of ether *qua* absolute resting frame, and ether *qua* mechanical explicandum of Fresnel and Maxwell theory *in vacuo*. Should we not count these as instances of Kuhn-losses, of mechanical concepts, eminently successful in their time, subsequently consigned to oblivion? But this evaluation is too quick. Quite apart from the self-evident importance of Galilean theory to contemporary physics (and every other branch of empirical science), and the countless applications of continuum mechanics to the electrodynamical properties of material media, no ether which provided a mechanical reduction of Fresnel and Maxwell theory *in vacuo* proved satisfactory. Had the attempt to reduce Maxwell theory to mechanics been successful, one would, on the abandonment of ether (and with it, one supposes, the reduction), have a *prima facie* case of Kuhn-loss: but there was no such reduction.

III

How is it that Galilean theory co-exists with Einsteinian theory? One sort of relativist response (favored, for example, by Feyerabend) is that there is every reason to simultaneously pursue quite different theories, that in any case there is no real sense to the *incompatibility* of these theories (since they are 'incommensurable'). Of course this is not quite what happens (as though applications of Galilean relativity are somehow made in a spirit of rivalry, or mutual incomprehension, with their Einsteinian counterparts). But let us concede that the onus is on those who would maintain that dynamics is progressive and cumulative to provide an account of the curious 'peaceful coexistence' of the two theories. Evidently they must be brought into relation with one another, but here, it is claimed, one has neither invariance of meaning, nor of reference, with respect to concepts such as space, time, energy and mass. In Kuhn's original formulation of the objection, meaning and reference are lumped together: because the concepts have different meanings, " . . . we have had to alter the fundamental structural elements of which the universe to which they apply is composed" (Kuhn 1962, p. 102)[10]. It is not enough, in Kuhn's view, to recover the laws of Newtonian theory in a suitable approximation (relative velocities small in comparison to that of light); the positivist thesis that that, in any case, is the only validity that should have ever been claimed for Newton's laws (for the only data available concerned this regime) is rebutted on the grounds that proceeding in this way, one could equally insist that the claims of phlogiston theory, restricted to the concrete data then available, were not and cannot be challenged by successor theories.

That the *mere* recovery, in a suitable regime, of formal equations of a subordinated theory (what I shall call *analytic reduction*), is an inadequate grounding for the progressive and accumulative view of scientific progress, is a point well-taken. The argument runs: (1) if the precursor theory is taken to make claims restricted to the domain of experiments actually performed, it collapses into a mere catalog of observations; (2) if the equations are taken to express *conceptual* truths (or hypotheses), then these concepts will depend on the successor theory and in any case differ from those of the precursor; (3) for the latter reason, such formal equations cannot even be taken to have the same referents.

As it happens, on quite independent grounds, the notion that the *meanings* of 'theoretical terms' (as they were then called) were somehow fixed once and for all, had already come under attack from a quite separate quarter (as part of the fall-out of the Quinean critique of the analytic-synthetic distinction). Putnam's (1962) notion of 'law-cluster concepts' (and the issues there explored) largely anticipated the problematics of 'meaning invariance' subsequently posed by Feyerabend (e.g. that because 'energy' means something different in relativity and Newtonian theory, the one theory cannot contradict the other). Evidently this sort of line (that kinetic energy $E$ does not *mean* $p^2/2m$) appears necessary to rebut the charge (2), but on the other hand (of course Putnam was *not* responding to the relativist challenge), just because the meaning of a law-cluster concept is something more nebulous, it would seem that no response is possible to (3).

To be more precise, it seemed that if such concepts have no precise meaning, they have no precise denotation either. Israel Scheffler (who *was* responding to Kuhn and Feyerabend) made tentative steps (Scheffler 1966) towards freeing the reference (what he also called 'descriptive meanings') of such concepts from their theoretical meaning (the former, in contrast to the latter, he supposed stable under theory-change; note that even if the theoretical meaning determined reference, *at most* one could conclude that the reference *may* differ in (3)). A similar approach is evident in Dudley Shapere's notion of 'trans-theoretic terms' (Shapere 1969). Supposing the theoretical meaning of a term is what is 'grasped' in the use of a term (so that it is, roughly speaking, its Fregean sense), these strategies amount to making the sense/reference distinction carry much the same burden as the positivist distinction between theoretical and observational terms. But if they do not in fact *presuppose* that distinction (in which case reference *per se* is no longer what is at issue) it was scarcely clear how such a notion of reference could be made out.

The denouement to this story, what is probably the most popular contemporary response to the relativist challenge, we owe to Putnam and – with a somewhat different motivation – to Kripke. It is, essentially, that one allows 'natural kind' terms to provide a limited Fregean sense, a pointer to reference, and leave the rest to the *world*. The electron of J.J. Thomson, of Bohr, and of Dirac, was one and the same, not because they conceived of electrons in the same way, but because the same entity (whatever in fact it is – for of course the story does not stop with Dirac) was causally involved in all of the experiments to which their theorizing was directed (Putnam 1973, p. 197). Putnam's later elaboration of this doctrine (Putnam 1975) shared with Kripke a focus on (observable) natural kinds – but it was clear enough that failing a Fregean *pointer* to reference such causal theories cannot possibly be of much use in the case of unobservable objects or properties, for how else are we to know there is any one entity (or kind) causally responsible for a class of experimental phenomenology[11]?

In short, there had better be enough meaning invariance across theory change to ensure that we are talking about the same thing, not in the sense that 'that thing' is conceived of in exactly the same way (for of course it is not), but in the sense that whatever it is, we can be confident (because this much is invariant across theory change) that there is one sort of thing (or property or magnitude or whatever) causally involved in the relevant class of experiments. Or so it is claimed.

One sort of problem with this account (but I think it is one of many) is that it does not, after all, work very well in the context of relativity theory. The difficulty was thrown into sharp relief by Hartry Field in 1973 (Field 1973; although it was more or less implicit in Feyerabend's (1968) arguments for meaning-variance): accepting that relativistic concepts of mass (total and rest mass) do, in fact, refer, he asked: to which of these did the *Newtonian* concept of mass refer? The answer, according to Field, is that there is no very good answer, but neither is it the case that Newton referred to a thing sharing some of the properties of total relativistic and rest mass ('Newtonian mass'), because there *is* no such thing (and what would it be to say there is an *approximate* thing of this sort?). Field did not specifically address the 'causal' theory of reference, but it should be evident that it is not going to resolve this sort of conundrum: in this case the best 'expert' stereotype in the pre-relativistic corpus did not single out one referent rather than another, and supposing that the *real* referent is somehow singled out as whatever *in fact* enters into causal relationships with the experiments with which classical theorists were concerned is not much help, for on that account (supposing our present theories in some sense 'true') we will arrive at different referents depending on the particular experiment involved, and even worse, depending on how we reconstrue that experiment. More fatal still, to say even *this* (if we can say even this) we must make use of something more than the Fregean sense of a 'law-cluster' concept – we must make use of the *detailed* concepts of contemporary theory.

To suppose that on each tokening of the term 'mass', relative to a specific phenomenology and laboratory procedure, there is some determinate reference, some 'fact of the matter', as to the entity or magnitude causally responsible for the phenomenology at issue, on some notion of 'cause' neither so strong that it hinges on contemporary theory, nor so weak that one has referential indeterminism, is to whistle in the dark. For the moment I suggest the more pertinent response is to recognize that if the thesis of general correspondence can be sustained, *we need nothing more.* We can do no other than suppose that contemporary theory (our 'best theory') is referentially successful (that is, construing reference in terms of Fregean sense; no easy task, needless to say). And by such standards, we may attempt to find referents for applications of precursor theories. But if we can explain the success of precursors on the grounds of their correspondence with successors (and have good warrant to suppose our present theory superior), that will suffice. The notion of reference is a red-herring; we must meet the relativist challenge head-on.

A better perspective, I suggest, goes something like this: Putnam was much closer to the mark with the notion of 'law-cluster concepts', but there need be nothing vague in this notion; on the contrary, the 'clustering' of concepts may be as complex and precise as we care to make it. Putnam, contrasting the status of truisms of the form 'all bachelors are unmarried' with the principle '$E = p^2/2m$', remarks that the former " . . . cannot be rejected unless we change the meaning of the word 'bachelor' and not even then unless we change it so radically as to change the *extension* of the term 'bachelor' ". He continues:

> In the case of the terms 'energy' and 'kinetic energy', we want to say . . . that the meaning has not changed enough to affect "what we are

> talking about"; yet a principle superficially very much like "All bachelors are unmarried" has been abandoned. What makes the resemblance only superficial is the fact that if we are asked about what the meaning of the term 'bachelor' is, we can only say that 'bachelor' means 'unmarried man', whereas if we are asked for the meaning of the term 'energy', we can do much more than give a definition. We can in fact show the way in which the use of the term 'energy' facilitates an enormous number of scientific explanations, and how it enters into an enormous bundle of laws. (Putnam 1962, p. 53)

It might seem overly ambitious to ask: How, in detail, does the term 'energy' enter into an enormous bundle of laws? It might seem that to answer such a question, one would have to more or less recapitulate the basic structure of dynamical physics. But that is *exactly* what we must do. Only then can we address the question of how, precisely, the Newtonian concept of 'energy', (or 'mass', or 'momentum') compare with their relativistic analogs.

Of course it is not a matter of listing the equations of dynamics. It is their structure which is at issue, in particular the heuristics and canonical forms that we have already encountered. To begin with, we have the concept of spacetime as a topological group (Lie group), here the semi-direct product of a non-abelian group (containing the rotations and boosts) with the abelian group of translations on $\mathbb{R}^4$. The energy and momentum appear as the generators of transformations on the latter group. In the relativistic case the non-abelian group is the Lorentz group (and the semidirect product is the inhomogeneous Lorentz group or IHLG); in the non-relativistic case we have the inhomogeneous Galilei group or IHGG. The energy, which in both cases is the generator of time translations, is in the relativistic case what is also called the total relativistic mass. The non-relativistic mass, in contrast, has a quite different interpretation (as does also the relativistic rest-mass), bound up with more detailed properties of the respective Lie algebras: in the case of the IHGG, to the 'neutral elements' of the algebra (it therefore defines the momentum and the energy in conjunction with the velocity); in the relativistic case to the Casimir invariants (a function of elements of the Lie algebra, not a separate element). In both cases these quantities have vanishing Lie bracket with every element of the Lie algebra; they are therefore conserved. One has a quite reasonable understanding of their inter-relationships as provided by the theory of group contractions[12].

Of (perhaps) even greater importance is the structural characterization of the dynamical theory at the level of the spacetime manifold. The essential basis for comparison – the 4-dimensional manifold structure of Galilean spacetime – was laid down by Elie Cartan; this development, like those in pure group theory just discussed, followed the elaboration of the analogous ideas in special and general relativity by Minkowski and Einstein, ultimately going back to the theory of Gauss and Riemann. The differential geometry was itself an extension of the analytic 3-geometry of Descartes, an essential component of the Newtonian synthesis. That is, the 4-dimensional Galilean theory was in many ways the *last* theory to be properly investigated (this remark is even more pertinent when one comes to its Hilbert-space representation theory). We see here how canonical forms associated with contemporary theory are extended back to reformulate its precursors (or are recognized as implicit in precursor theory; cf. Post's notion of 'footprint'). In this way we understand that Newtonian theory makes use of a timelike congruence of spatial frames with associated degenerate (signature 3 and 1) space and time metrics. In comparison to the pseudo-Riemannian spacetime of the special theory, what is involved is the deformation of the light-cone structure into a set of spacelike hyperplanes (one is then left with only the time metric; in terms of the theory of group contractions, this is the zero momentum limit).

All of this is to give an abstract framework; we are a long way from writing down any *formal equations* common to the two theories (i.e. solving the analytic problem of reduction; the formal agreement between the two theories in a certain limiting regime, which relativists seem to regard as a foregone conclusion). In fact here it is the latter which is problematic, whereas the *conceptual* relations between the two theories are readily surveyable. The reason why formal agreement, at the level of specific equations of motion, is so difficult to establish is that the *mere* existence of a formal limit (taking $c$ as infinite) tells us nothing about how fast, and with respect to what class of interactions, limiting agreement between the solutions of equations of motion may be found; the equations themselves do not even always become identical in the limit (at *best* they coincide in the limit). What is needed is 'hard' analysis, the establishment of concrete bounds and rates of convergence in the parameter $v/c$, in comparison to the 'soft' analysis given above (compare Berry 1991). There are fortunately some results of this kind (for smooth and slowly-varying external fields), in both the quantum and classical case[13].

One moral that we can draw is that Field's dilemma – to what does the Newtonian mass correspond? – is problematic just because this quantity is *not* a generator of any 1-parameter subgroup of the IHGG. It shares this distinction, embarrassingly enough, with the quantity *position* in the case of the IHLG: the position is the generator of boosts in the Galilean case (this is why it is a self-adjoint operator, which satisfies canonical commutation relations, as also a Lie bracket, with the momentum in quantum mechanics), but since the boosts are *not* an abelian subgroup of the IHLG (because of the Wigner rotation; thus the Lorentz group is semi-simple, unlike the Galilean), their generators do not have vanishing Lie brackets with one another (hence they do not commute in quantum mechanics). As a result there are no covariant position operators in relativistic quantum mechanics, and no covariant Born interpretation either.

On the other hand, to strengthen a point I have already made, the IHLG and IHGG have substantial similarities. Most importantly, they both contain the Euclidean group as a subgroup, hence the generators of this subgroup have precisely the same group-theoretic meaning: the angular-momentum, energy, and momentum, are the generators of angle, time, and space translations, respectively, and are therefore conserved in any inertial frame. Correspondingly, that in Minkowski space one can pick out differently oriented spacelike hypersurfaces (with different associated energy and momentum, generating translations along the associated orthogonal vector fields), is directly responsible for the four-vector character of the combined energy and momentum (the 'combination' of energy and momentum conservation into a single conservation principle).

But it should be clear that all of this is not quite to specify the 'law-cluster concepts' of mass, energy, momentum, space and time. We have, to begin with, altogether left out of consideration the role of gravity, and more important still, there is no account here of concrete phenomenological applications. I do not suggest that these things can be completely codified (especially the latter). My point is that the *attempt* at codification (or integration) is precisely what is, and what has always been, the essence of the enterprise of dynamics. And in this process, the existence of heuristics and canonical forms by means of which precursor theories may be understood in terms of successor theories (here, most especially, the concept of dynamical variables as generators of symmetry transformations, thereby the connection between symmetry and conservation laws, the concept of Lie group and its use in the characterization of space and time, the concepts of spacetime manifold and differential geometry) have an importance that it would be hard to over-estimate[14]. The process is of course two-way: the precursor theory is used to interpret the successor equally as the successor interprets anew the precursor.

Here I have concentrated on the recognition that heuristics, developed in the pursuit of relativity theory, were present (and are now seen as canonical) in non-relativistic theory, that in this way the two theories can be reconciled. I have said little of the context of discovery of the Lorentz transformations. Here one finds a remarkable circumstance: with only small modifications of Einstein's procedure, one can in fact deduce the form of the coordinate transformations as functions of a single parameter, an invariant velocity; if it is infinite one obtains the Galilean transformations, and otherwise those of Lorentz. This fact is quite exceptional; not only is one fragment of the 'soft' reduction of these theories guaranteed, but in a certain sense the two theories are exhibited as the only possibilities consistent with the existence of global inertial frames[15]. As we shall see, something similar can be said of quantum and classical representations of probability, but in the case of relativity the procedure or perspective by which it and its precursor may be derived is close to that actually employed in the context of discovery.

## IV

The tenor of my argument is somewhat unorthodox; I have spoken of the *mathematical* structure of dynamics as a conceptualization of the world, leaving reference (if and when it can be made out) to Fregean sense. Let me now make some concessions.

Clearly there is a somewhat different practice of reference at the level of everyday (and not so everyday) experience. In particular, we refer to laboratory apparatus, and to systematic and detailed phenomenological properties to do with the manipulation of laboratory instruments and their construction, with a certain independence from theory. Just how 'deep' this sort of reference goes, the extent to which, indeed, one has a genuine autonomous basis for the interpretation of physical theory, has been well-illustrated in Hacking's accounts of microscopy and his subsequent elaboration (Hacking 1983) of 'entity realism' (as Allan Franklin has emphasized, "There are no anti-realists in the laboratory."). It is, I hazard, just such an appeal to 'common-sense' reference that underlay Scheffler's assumption of some sort of autonomy of reference from Fregean sense (if so, that it is parasitic on a theory/observation distinction is evident).

This sort of reference is surely what underlies talk of 'discovery' of elementary particles, of the 'real existence' of e.g. neutrinos (in some weak sense – cf. Grover Maxwell's arguments – these are indeed observable). But there is always a gap between what is observed (in Hacking's sense) and the putative referents of theoretical concepts; what is at issue appears more bound up with some sort of process of 'triangulation' by different methods (in each one of which there is *always* a gap). At the last, it seems we must make sense of some sort of coherence theory of truth, however unfashionable the strategy.

The appeal to any one method of observation, no matter how flexible a line one takes, is not going to be of much use when it comes to more general concepts and more structural features of dynamical theory; and the difficulty now (as was exhibited in connection with Newtonian 'mass', by most accounts an *observational* term), is that we do not know how changes at this level will force a re-evaluation of what we have been taking as 'observation' (as Putnam once remarked, we have " . . . the almost untouched problem, in thirty years of writing about 'theoretical terms' [as to] what is *really* distinctive about such terms." – and *that* comment was made thirty years ago). In other words, however plausibly we may suppose we make reference to objects and properties in laboratory manipulations, there can be no guarantee as to the stability of this sort of referential success (no guarantee that we will not learn new ways of thinking according to which we are *not* successful).

The most striking example is at the level of *everyday* experience. It *seems* that we make reference to *colours*, for example, that colours are *out there*, in the world, but 'in fact' (according to current theories of perception), colour words as ordinarily used (in particular as used to characterize objects) refer, if they refer to anything, to equivalence classes of relative spectral reflectancies of surfaces and boundaries, and only then subject to normative constraints on the range of permissible lighting. And what defines the equivalence class, is the structure of the *particular* retina and visual cortex which sees the colour (in particular, there is only a loose correlation with the wavelength of light incident on the retina; see Hardin (1988) for a comprehensive review).

Evidently we are in deep waters. It seems to me one has no real option but to approach the concept of reference in the context of cognitive science and the philosophy of mind (I take it that this is what a naturalized epistemology demands of us). For our present purposes I suggest that we may forgo talk of reference and 'entity realism' at any level deeper than that countenanced by Hacking. Such a strategy might best be termed 'methodological antirealism': by this I do not mean that we foreclose on issues of truth or reference. (The position is motivated by a *naturalized* epistemology; in a sense we *are* committed to a realism. There is an evident analogy with the notion of 'methodological solipsism'.) My claim is only that we may understand significant components in methodology, heuristics, and the notion of 'progress' more or less independently of questions of reference over and above the everyday (this is not Hacking's position, who – and something similar may be said of Laudan's more recent writings – countenances talk of 'progress' and 'success' in much the same sense that the terms might be applied to engineering and technological advance).

With this said, let me turn to quantum mechanics. There are three elements of the previous discussion that I wish to consider by turn. These are:

(1) *Heuristics*: Innovation proceeds by isolation and independent development of structural features of extant theory. Once entrenched, such heuristics (or canonical forms) are preserved in subsequent developments, and previous theory reformulated in their terms.

(2) *Unity*: Among these, one may provide over-arching abstract frameworks (Lie group, spacetime manifold) within which one dynamical structure may be embedded in another.

(3) *Analytic Reduction*: experimentally well-confirmed formal equations of the subordinate theory may be recovered in a well-defined approximation to the dominating theory.

I take it (3) is necessary to the thesis of 'zero Kuhn-loss'. Evidence that the problem of analytic reduction may be solved is provided by the existence of singular limits ('soft analysis') according to which formal equations of the one theory reduce to that of the other. Progress will in general depend on (2) being satisfied. (2) is, in any case, essential if the conceptual structure of the precursor theory is to be 'surveyable' within its successor. In turn, (2) (hence also (3)) depend on satisfaction of (1) (I take this to be self-evident). The conjunction of all three is what I take as the thesis of correspondence. The following three sections are somewhat more technical in nature and illustrate each component in turn in the context of quantum mechanics. I begin with (1) and the historical development of the theory.

V

By 1925 the conflict between the Bohr theory and classical electrodynamics had reached an impasse. According to the first, the frequencies of emitted radiation (the Bohr transition frequencies) could be related to the energy difference between the stationary states involved in the atomic transition; according to the second, the intensities were related to the amplitudes occurring in the Fourier expansion of dynamical variables (specifically the electric polarization). A partial, but obviously unsatisfactory, compromise was reached by modeling the atom as a multiply periodic system – despite the fact that one was concerned with alkaline metals, with only one optically active electron – for thereby one had two numbers available to parameterize each amplitude: the number of the harmonic, and the number of the degree of freedom involved. But there was a second problem: in this expansion each term represents a harmonic of the oscillatory motion of the electrical moment; according to classical electrodynamics, this frequency is also that of the radiation emitted or absorbed. Therefore it was necessary to identify these harmonics with the Bohr frequencies, *not* the orbital frequencies of the stationary states (those frequencies at which electrons actually vibrated, according to the Bohr theory). Correspondingly, these harmonics could not be simple multiples of a fundamental frequency (for the Bohr frequencies, given by differences in energy levels, displayed a quite different structure, except in the special casc when these energy levels were equally spaced, as in simple harmonic motion). Therefore this series expansion could not properly be understood as a Fourier expansion at all, and even if it could, it could not describe the optically active electron alone.

What were these objects, the oscillatory motions of which were described by the 'Fourier series'? It seemed they were the transition processes themselves, but how was it that more than one such transition process could be brought into play by the coupling of the atom with the radiation field? These objects were first introduced by an experimentalist, driven entirely by a pragmatic concern for a serviceable mathematical algorithm[16]. Ladenburg's 'dispersion electrons' – the objects vibrating at the transition frequencies – provided a key input to the Bohr-Kramers-Slater theory (where they were known as 'virtual oscillators'). Their 'number' could be directly measured via Ladenburg's algorithm, which was derived by a straightforward application of classical dispersion theory; but it had no relationship to the electron number in the sense of the Bohr theory and quantum chemistry.

Heisenberg thus contemplated a mathematical formalism with no reasonable physical or mathematical foundation: it seemed necessary to associate a large (actually infinite) number of dynamical variables with a single dynamical variable – describing the motion of a single electron – and for each of these develop a second infinity of labels of state through the Fourier expansion. But here the various frequencies associated with the Fourier coefficients are all multiples of the fundamental; one also had to break the strict proportionality of harmonics.

His response, an act of unqualified genius, was to consider the series expansion for each degree of freedom as simply a collection of terms, but to take over from the mathematics of Fourier series the rule for combination of terms, *modified* so that frequencies combined as given by the Bohr frequency condition. He ended up with a two-fold infinity of terms, a multiplication rule, and an association of each term with a Bohr transition: the basic elements of matrix mechanics. Let us look at the steps.

First, one has the classical Fourier expansion for the position coordinate (proportional to the electric dipole moment) for the kth degree of freedom of a multiply periodic system:

$$x_k(t) = \sum na_n(k)\exp(i\omega(k)nt)$$

The $\omega(k)$'s are the fundamental frequencies of the multiply periodic motion (one for each degree of freedom); the expansion is over the harmonics, frequencies which are integral multiples of $\omega$, as determined by the integer $n$. Second, a similar expansion for the square of the dipole moment (introducing a new set of fundamental frequencies $n(k)$ and Fourier coefficients $b(k)$) yields:

$$x_k^2(t) = \sum_r b_r(k)\exp(i\nu(k)rt)$$

By direct calculation from the original expansion this quantity should equal

$$\sum_{mn} a_n(k)a_m(k)\exp(i(\omega(k)n+\omega(k)m)t)$$

On writing $n + m = r$, it follows that $\nu = \omega$, since clearly:

$$\omega(k)n + \omega(k)m = \omega(k)(n+m) \qquad (1)$$

Evidently the fundamental frequency of any polynomial in the dynamical variables is unchanged, and only integral multiples of this fundamental occur in the oscillatory motion. As a consequence of Eq. (1), it follows that:

$$b_r(k) = \sum_n a_n(k)a_{r-n}(k) \qquad (2)$$

To see how these formulas must be modified in order to interpret the frequencies as Bohr transition frequencies, Heisenberg introduced the new notation (I depart slightly from Heisenberg's text):

$$\omega(k)n \leftrightarrow \omega(k,n)$$

$$a_n(k) \leftrightarrow a(k,n)$$

making clear that both degree of freedom and harmonic were to be treated on a par, both interpreted in terms of Bohr energy levels. Eq. (1) now takes the form (writing $m = r - n$):

$$\omega(k,n) + \omega(k,r-n) = \omega(k,r)$$

to be compared with the addition rule which follows from the Bohr frequency condition, namely $\omega(k,n) = (E_k - E_n)/h$, i.e.

$$\omega(k,n) + \omega(n,r) = \omega(k,r)$$

Imposing the latter, in place of Eq. (2) we obtain $b(k,r) = \sum_n a(k,n)a(n,r)$, i.e. the matrix multiplication law. In the original notation, this gives:

$$b_r(k) = \sum_n a_n(k)a_r(n) \tag{3}$$

This is, of course, just the beginning. Heisenberg then proceeded to apply this new calculus to a concrete dynamical problem (the anharmonic oscillator), essentially using *classical* dynamical theory amended in accordance with Eq. (3). The justification for this was starkly stated:

> If one seeks to construct a quantum-mechanical formalism corresponding as closely as possible to that of classical mechanics, it is very natural to take over the equation[s] of motion directly into quantum theory. (Heisenberg 1925, p. 267)

This is what one does if one wants to maintain the thesis of correspondence. At every step, Heisenberg first wrote down the classical equations (for a perturbative solution to the equation of motion $x + \omega_0 x + \lambda x^3 = 0$, expanding $x$ in a Fourier series), and then the analogous quantum mechanical quantities, employing Eq. (3) in place of Eq. (2).

A final point: I have suggested that the heuristic component of the correspondence thesis is bound up with the isolation and autonomous development of a mathematical abstraction embedded in previous theory. Here it is clear enough that the Fourier analysis was used as a heuristic in this sense; it is not so clear that the Bohr frequency condition, which evidently played a crucial role in the 'autonomous development' of the Fourier analysis, can be understood similarly. Its origins are in fact multi- faceted. The condition was hit upon by appeal to the more or less intuitive pictures of energy quantization, the concept of stationary states, and energy conservation, but it was Bohr's great genius to see in this condition a deeper connection with classical theory, in particular to the concept of the derivative and the action principle. The former is defined as the limit of a difference equation:

$$\delta_{J\to 0}^{\lim} \frac{1}{\delta J} \{ W(J + \delta J) - W(J) \} = \frac{dW}{dJ}$$

If, in particular, $W$ is the total energy and $J$ the action variable for a periodic motion one has, from Hamilton's equations, $dW/dJ = d\theta/dt$, where $\theta$ is the canonical conjugate to the action. When $\theta = \nu t$, with $\nu$ the frequency, one obtains $\delta W = \nu\delta J$, or $W(J+\delta J) - W(J) = \nu\delta J$. For $\delta J = h$, the 'minimum' change in action, the correspondence-limit of the Bohr frequency condition follows. This condition emerges as the product of the autonomous development of one of the canonical forms of classical dynamics.

The frequency condition was not, therefore, first formulated as a heuristic, but it was subsequently understood in this sense. Indeed, its autonomous development as the Bohr correspondence principle was, together with the quantization condition, the basis of the 'old' quantum theory. Meanwhile the quantization principle and the Hamiltonian mechanics emerged as so deeply interwoven – here I have in mind Ehrenfest's adiabatic theorem and the relations between separability, adiabatic invariance, action-angle variables and the Hamilton-Jacobi equations, as formulated by Schwarzchild, Levi-Civita, Epstein, Ehrenfest and Burgers – that one has a system of real structural integrity; the fundamental lacuna concerned the change of quantum number (quantum dynamics proper), which was related via the correspondence principle to the classical derivative and to classical dynamical continuity. A relationship between these heuristics – for that is what they clearly are – and probability theory was meanwhile established by Einstein (the theory of 'A and B coefficients'), and was further developed by Bohr (what became known as the Einstein-Bohr electrodynamics): symbols and coefficients classically given by derivatives in the action were taken to represent the probabilities of transitions between stationary states, and were written as difference equations in the Heisenberg-Kramers dispersion theory. These probabilities were in turn related to the radiation intensities (which returns us to Heisenberg's problem situation).

Schrödinger's route to the wave mechanics reveals a similar strategy, although here he was also guided by a new 'high-level' interpretation of previous modifications to dynamical theory (the wave-particle theory of Einstein and de Broglie). This in turn (specifically in de Broglie's contributions) hinged on the formal analogy between the principles of Hamilton and Fermat, thereby linking mechanics to optics (and hence wave theory). The crucial step was to seek modifications to the Hamilton-Jacobi theory. This procedure is rather better known than that of Heisenberg, since it is reviewed in a number of introductory texts, and I shall speak no more of it here.

I take it that the heuristic component to (1) finds overwhelming confirmation in the genesis of quantum theory. As for structural similarities with classical theory revealed by subsequent investigation, let us note: (i) Von Neumann's systematization, by means of which states in both classical and quantum theory are measures over the state space (phase space and Hilbert space respectively), in the one case given by a phase space integral of real functions on phase space, and in the other by the trace of self-adjoint operators on Hilbert space (the distinction between pure and impure states being formulated in exactly the same way in the two theories, with the same intended interpretation); (ii) Subsequent abstraction of this framework in terms of lattice theory; (iii) the Dirac correspondence, whereby the Poisson bracket algebra of classical observables is (formally) related to the commutator-bracket algebra (Heisenberg algebra) of quantum observables; (iv) the Weyl form of the canonical commutation relations, whereby the latter express (in infinitesimal form) the existence of additive groups of transformations on phase space (translations in space and boosts) as in classical theory; (v) the various Hamiltonian and Lagrangian function(al)s, identical in quantum and classical theory for electromagnetic couplings.

Once again these results are well-known; it is now clear that many of these structural inter-relationships of NRQM and classical mechanics hinge on properties of the IHGG (see e.g. Varadarajan 1970, Mackey 1962). Others, such as the Dirac correspondence, have been extensively investigated in the context of 'geometric quantization' and the synthesis of differential geometry and Hamiltonian and Lagrangian dynamics (see e.g. Woodhouse 1980). The lattice-theoretic structure is the object of the logico-algebraic approach (and also more abstract algebraic approaches to quantum mechanics and field theory); here one has the hope of a more generic, over-arching framework in the sense of (2) (see, e.g. Hooker 1975–9, Beltrametti and Cassinelli 1981, Saunders 1988). Indeed it is surprising, with such extensive interplay between quantum and classical concepts, that this remains only a hope. As we shall see, paradoxical though it sounds, there is actually *too much* interplay.

## VI

The difficulty is, of course, the problem of measurement. Here it is not so much that one cannot accommodate both quantum mechanics and classical theory in one and the same scheme, but rather, that we do not know how to formulate quantum mechanics *as a universal theory* in the first place. In a certain sense, the situation is actually made *worse* when viewed in a unified context which includes classical theory as a special case (for then we see that what should function as the fundamental level of description in quantum theory *makes no sense* when interpreted probabilistically, if the quantum mechanical description is to be 'complete').

As illustration, consider the Mackey theory (1963). In this framework the notions of 'state', 'observable', and (most importantly) 'probability' occur as primitives. We suppose we are given a prescription, for any state $f$, for assigning a probability to the proposition that the value of an observable $A$ lies in some subset of the reals, (for convenience we suppose these to be Borel, a very mild constraint). Since the values assigned are probabilities, we require $f(A, \emptyset) = 0$, $f(A, \mathbb{R}) = 1$, $f(A, U_i\, E_i) = \Sigma_i f(A, E_i)$ for $E_i$ pairwise disjoint Borel sets. Next, we impose some natural requirements of parsimony: if $f(A, E) = f(B, E)$ for all Borel sets $E$ and states $f$ then $A = B$. If $f(A, E) = g(A, E)$ for all observables $A$ and Borel sets $E$ then $f = g$. It is necessary to consider functions of observables: if $g$ is such a function, then the probability that $g(A)$ has value in $E$ should equal the probability that $A$ has value in $g^{-1}(E)$. To make this well-defined we require that $g$ is a Borel function. Armed with this notion, we can define 'properties' as special sorts of observables, namely those for which $f(a,\{0,1\}) = 1$ for every state $f$ (I use lower case letters to denote properties). $f(a,\{1\})$ (which may be written $f(a)$) is then naturally interpreted as the probability that the system 'has' property $a$. It follows that for any observable $B$ and Borel set $E$, $\chi_E(B)$ is a property (the property that the observable $B$ has value in $E$; here $\chi$ is the characteristic function of the set $E$). There is a further natural requirement: call properties $a$, $b$ *disjoint* when $f(a,E) + f(b,E) \leq 1$ for every Borel set $E$ and state $f$ (if a physical system has property $a$ with certainty, it is certain that it does *not* have property $b$, and *vice versa*). Require that for any countable set $\mathfrak{D}$ of pairwise disjoint properties, there is some property $c$ such that $f(c) = \sum_{a\in\mathfrak{D}} f(a)$ for every state $f$ (this is the appropriate modification of the $\sigma$-additivity of probability measures for disjoint outcomes, familiar from standard probability theory). In exact parallel to the familiar theory, require that for any finite set of real $\lambda_i$ in the interval (0,1) which sum to unity, and set of states $f_i$, then $\sum_i \lambda_i f_i$ is also a state. Armed with this structure (the set of states is now a convex set) we may define the pure states as extremals, i.e. those states which *cannot* be written as the convex sum of other states. Such states provide the most detailed or maximal description possible (possible, that is, in accordance with the present framework).

With this we have the basis of a generalized theory of probability. In fact, from these requirements, it follows that the set of properties is a $\sigma$-orthocomplete, orthomodular, orthocomplemented poset. The partial ordering is defined as follows: if $f(a,E) \geq f(b,E)$ for every state $f$ and Borel set $E$ then we write $a \geq b$. If in particular the system has property $b$ with certainty then it also has property $a$ with certainty. If we also require that this poset is a lattice we (almost[17]) have the conditions of the Piron theorem: such a system *must* have a representation as the projection lattice of a Hilbert space (over one of the reals, complex numbers, or quaternions). If, however, we insist that the pure states assign probability 0 or 1 to *all* properties (a probabilistic version of bivalence), then there also exists a representation of the lattice as a classical phase space[18].

The Mackey theory is commonly regarded as one of the 'cleanest' and most elegant abstract characterizations of classical and quantum theory, but, in a sense I now want to make plain, it also highlights the sense in which quantum mechanics is deeply problematic. The point is this: the abandonment of bivalence is, in a certain sense, incompatible with the notion that the quantum mechanical state is or can be regarded as a *complete* description of a system at a given time. And just because the abandonment of bivalence (virtually) *forces* a quantum mechanical representation of state (and is clearly a necessary condition given the standard probabilistic interpretation of NRQM), one has precious little room for manoeuvre. (One might suppose that the Hilbert space structure could be traced to some other generic viewpoint on the nature of microscopic phenomena; that it appears to follow from elementary notions of probability together with the abandonment of bivalence tells against such a view).

The conflict can be formulated very simply. One wants to assert *both* that a pure state description is possible for every system at all times, *and* that such a description is *complete* (i.e. that it cannot be supplemented by a more detailed description). But if that is so, there can be no sense to the notion that, given that a property $p$ of the system is assigned probability $f(p) = \lambda \notin \{0,1\}$, the system 'in fact' has or does not have that property with probability $\lambda$ and $1-\lambda$ respectively; for if it did, 'in fact', have property $p$, then it would surely have that property with probability 1, and therefore the system should be described by some other state $g$ such that $g(p)=1$ (i.e. f would *not* then be a complete description of the system).

The appeal to the notion that there is some *fact* to the matter as to whether or not a system has property $p$ is evidently crucial; I for one am happy to renounce any *a priori* appeal to intuition in this regard, *with the exception of those properties that are directly observed* (macroscopic pointer-positions and the like). It is this that makes this difficulty specifically a problem of *measurement*, the one case where we *must* interpret quantum mechanical probabilities at the level of what is directly observed.

At this point one might feel that what is at issue is some sort of 'emergence' of value-definiteness through measurement. It may be that it is only when one performs an experiment to determine whether or not the system has a given property, that such probabilistic statements come into play, and that one must then deal with a modified state (i.e. that describing the system coupled to the measurement device) – that in this modification resides the transition from the state $f$ to the state $g$ (cf. the terminology 'transition probability', for the probability that a system in the state $f$ 'will be found' in the state $g$). But so far we have made no mention of *dynamics*; the statement of the difficulty applies equally to the state describing the apparatus (and object system) *after* the measurement process is completed.

Alternatively one might suppose (in view of the fact that a pure state description of a composite system does *not* imply that each subsystem can be described by a pure state) that the difficulty is rather that *no* system (considered in isolation from the environment) *has* a pure state description. This point is particularly telling in consideration of the hypothetical (pure) descriptions of macroscopic objects, since every such object is certainly coupled to the environment (in contrast to microscopic systems). But it is hard to see how the same argument can fail to apply to the environment *together* with the macroscopic measuring system (and microscopic system).

One may, perhaps, make progress with the latter line, including 'the observer' as part of this environment (which appears to lead to some variant of the Everett 'relative-state' interpretation), but only insofar as the properties of systems, *as perceived by the observer*, are in fact always dispersion-free when described by the state attributed to the system by the observer. The generic solution to the problem of measurement appears to require that at least some of the properties of a system (those described at the classical level) are in fact bivalently described by some pure state (in the hidden-variable approach, of some *other* theory) at all times. In particular the supposition that 'wave-packet reduction' is an *objective* physical process, common to a variety of proposals, is committed to this strategy, as is also the supposition that one must look to dynamical evolutions which transform pure states into impure states (whether through non-linear modifications to the Schrödinger equation, or by appeal to super-selection rules or the existence of unitarily inequivalent representations for macroscopically distinguishable properties).

The present state of this problem may be briefly summarized: despite intensive debate over more than six decades, no satisfactory resolution of this dilemma has emerged. In particular, in any application of quantum mechanics to finite systems in finite times, pure states are preserved under unitary evolutions, and yield non-bivalent probability assignments to macroscopic properties.

## VII

There is another viewpoint. One can suppose that quantum theory is not, properly speaking, an independent theory of dynamics at all (that it is only quantum theory *together with* classical theory that can provide a consistent description of phenomena). So much is implicit in the usual text-book formulations of the 'measurement postulate' (the pragmatic 'operational' interpretation of quantum mechanics by means of which it is actually applied), where the distinction between quantum and classical descriptions is *sui generis*. In a certain sense, this appears the orthodoxy within the physics community (it is the viewpoint of Bohr), although it is not clear that more than lip-service is paid to it. The implications of this point of view are so revolutionary (both to philosophy and to physics) that one scarcely knows how to proceed; the most *conservative* approach would be to embrace a thoroughgoing instrumentalism (one recalls Bohr's statement: "there is no quantum reality").

But if this is the right sort of response to the measurement problem, it would seem that the relativist has been completely outflanked. One cannot say that classical theory has been superseded at all, for on this scenario *there is no such thing as quantum mechanics*, as an autonomous theory (distinct from and independent of classical mechanics). It would then seem that there could be no such thing as Kuhn-losses. If classical mechanics is only 'supplemented' by the quantum algorithm (let us now call it that), it is far from clear how any instance of successful application of classical mechanics could be jeopardized (in this sense it would seem that the analytic problem of reduction does not exist).

If, on the other hand, we suppose that the measurement problem must eventually submit to an autonomous resolution within or beyond quantum mechanics, then it would also seem that an analytic reduction will depend on the sort of resolution on offer. In particular, if we suppose the measurement problem points to the existence of a more detailed level of description of microscopic phenomena (that quantum mechanics is incomplete), it remains unclear whether the classical and quantum descriptions are to be considered as both 'approximating' this more detailed level of description, or whether the chain proceeds from this *via* quantum mechanics to classical theory.

These points notwithstanding, even though we do not know the circumstances under which the quantum description of phenomena (supplemented with the 'minimal' operational interpretation) *should* go over to a purely classical analysis, there is nonetheless the clear-cut question of whether, where classical analysis has been successful, consideration of quantum-mechanical effects modifies or otherwise calls into question those very applications (and here one must use the operational interpretation, disregarding awkward questions of whether or not the latter depends on the validity of classical mechanics).

The logical structure of such an approach to the analytic problem of reduction appears somewhat curious. It is by no means clear what conclusion should be drawn were the outcome to be unfavorable. But as it happens it seems we do not have to deal with this difficulty. Progress (of a particular sort) is already substantial.

The foregoing comments are brought into focus when we consider the strategy employed. In brief, the idea is to equate the equations of motion, and/or the solutions to the equations of motion, for classical pure states, to the expectation values for the corresponding quantum mechanical equations (or solutions), in some suitable pure state. At the crudest level we have Ehrenfest's theorem. This shows that for time-dependent Hamiltonian functions on $\mathbb{R}^6$ of the form

$$H = \frac{1}{2m}(p - A)^2 + V$$

the classical equations of motion are formally similar to the expectation values of the Heisenberg-picture quantum mechanical equations of motion for the Hamiltonian

$$\hat{H} = \frac{1}{2m}(\hat{p} - \hat{A})^2 + \hat{V}$$

for any state $\psi \in L^2(\mathbb{R}^3)$. Post (1971, p. 233) is, of course, quite right to point out that at best one has a correspondence with respect to 'mean-values' of dynamical quantities, and not with respect to 'higher-moments' of these quantities, but the situation is in fact worse than that. First, there remains a discrepancy

$$\frac{ih}{2}\langle \Delta \hat{A} \rangle_\psi$$

(evidently one must take the limit $\hbar$=0); second, one has not thereby proved (not even in a formal sense) that the expectation values of quantum operators satisfy the classical equations of motion, for there occur quantities of the form:

$$\langle \nabla \hat{V} \rangle_\psi$$
$$\langle \nabla \hat{A}.\hat{p} \rangle_\psi$$

etc., rather than:

$$\nabla \langle \hat{V} \rangle_\psi$$
$$\nabla \langle \hat{A}.\hat{p} \rangle_\psi$$

(we cannot even make sense of the latter expressions). But one point should be stressed: not even this qualified and formal result would be attainable were there not a sense in which dynamical variables *have the same meaning* in the two theories. And this, as we have seen, is a consequence of the fact that both theories define representations of the IHGG (it is for this reason that we have been able to employ the same symbols in the foregoing).

Coming back to Post's point, on reflection it is far from clear whether one can make sense of the notion that there be a correspondence between higher moments of quantum-mechanical dispersion and some sort of classical dispersion; the latter would require that we formulate a correspondence between quantum mechanical *pure* states and classical *impure* states (a sort of approximate 'hidden-variable theory'?), or else that we formulate a correspondence between unitary quantum evolution and a stochastic classical evolution. On reflection, given the 'curious logic' of the pragmatic problem of reduction posed above, such maneuvers do not even come into consideration. For at the pragmatic level, all that matters are successful applications of classical theory, and if we are considering classical *mechanics* (and not classical statistical thermodynamics or stochastic theories of e.g. Brownian motion), the classical variances are all exactly *zero*. In effect, we suppose we are given some phenomenon, and analyze that phenomenon (i) within classical theory, by whatever means prove most effective (ii) within quantum theory (presumably in a more integrated and controlled sense), and then seek an analytic basis for comparison of the two models. So much would be a principled approach if one supposed quantum theory incomplete, and both quantum and classical theory approximations to some unknown and deeper level of description[19].

This applies in particular to the detailed and more general approach to the analytic problem of reduction as developed by e.g. Hepp, Hagadorm, Heller, Simon, Taylor, Voros, and Yajima. These authors are concerned with the establishment of precise error bounds and controlled approximations to expectation values of the momentum and position coordinates parameterized by the time $t$ for some finite interval $t \in [0,T]$ (what may be called 'finite propositions'). Following Taylor (1984), one may state the strategy in the following terms: given classical parameters $(\Omega,H,T)$ (where $\Omega$ is a set of classical pure states) and acceptable error $\varepsilon$, for each $\alpha(0) \in \Omega$ prove that there exists $\psi \in \Psi$, where $\Psi$ is a set of quantum pure states, such that for all $t \in [0,T]$, one has

$$|\alpha(t) - \langle\hat{\alpha}(t)_\psi\rangle| < \varepsilon$$

Progress with the strategy has been considerable, but it should be clear that it does not provide a basis to a resolution of the problem of measurement (but rather seems to presuppose the incompleteness of quantum mechanics). A similar formulation in the case of higher moments of dynamical variables will beg all the interesting questions at issue: since the classical pure-state variances are all zero, the quantum mechanical variances will all have to be bounded by $\varepsilon$. The gap between the 'objectivity' of outcome in the classical case (variance zero) and the 'outcome-indeterminacy' in the quantum-mechanical case is now assimilated to the notion of 'error' (and what does that mean in this context?).

## VIII

What I contend should be concluded with respect to this decidedly unsatisfactory state of play is that there is a great deal more to be learnt about the relationship of quantum and classical mechanics, but that what makes this relationship so problematic is precisely the fact that we do not, as yet, have distinct and autonomous theories.

Given, further, the thesis (1) – that in dynamical physics one has no option but to build upon the structures laid down by pre-existing theory – and given further that the development of quantum theory unequivocally bears out this contention, it is by no means surprising that we should find ourselves in this position. Indeed, *regardless* of whether or not there are cases of 'Kuhn-losses', Kuhn *et al.* appear (to put the matter bluntly) to have *missed the point*. What emerges as the *central* characteristic of dynamical physics, from its inception with Newton to the present day, is the slow evolution of a mathematical and conceptual structure, in which changes introduced at one level are subject to complex, and only partially understood, inter-relationships at virtually all other levels. It is, in other words, exactly the conceptual and mathematical integrity of the entire monolithic framework which is the subject of sustained investigation, and here the problem of measurement – the most significant case of *failure* – appears to be a consequence of the *lack* of a sharp separation between quantum and classical descriptions of observable phenomena.

Of course the relativist case is not (or not primarily) made out in the context of the history of dynamical physics. Further, my defense of Post's position – given the powerful constraint that we attend only to mathematically highly developed and empirically well-supported theories – has perhaps only limited philosophical interest, in view of its studied indifference to the question of reference. It might seem that one can then find no purchase on the task of *explaining* the success of dynamical theory, in whatever sense we are to understand the term.

As it happens, I believe the latter conclusion would be unduly pessimistic. Just as (or so I have argued) variance in meaning is not incompatible with a cumulative and progressive view of dynamical theory, so too (as I have hinted), if one is prepared to adopt a more 'structuralist' view of physical ontology, one could say that in parallel, variance in reference is not incompatible with a cumulative and progressive view of the constitution of dynamical phenomena. It seems that what is required is that we take the mathematical conception of the world rather more *at face value*; that is, I suggest we allow that the 'furniture of the world' be rather more structural, more *mathematical*, than is ordinarily permitted within the nominalist tradition. In any case, it has long been apparent that no workable account of nomological necessity can be made out at the level of unstructured particulars (except in the context of an unfathomable and antiquated notion of the 'rule of law'). There may indeed be a further implication for the general account of inter-theoretic relations – that one has to do not only with different regimes of scale and energy, but with different *types* of description, the difference between describing what a system *does*, as opposed to what it *is* (this seems to me to apply most particularly to the relationship between the thermodynamical description of state versus that of dynamics, but also to the wider context of the application of mathematics to other disciplines).

These remarks are tentative; I wish only to indicate lines of future development. But within the narrower confines of the thesis of correspondence, I suggest that the relativist has only two options: *either* one must acknowledge that dynamical physics is somehow immune to the well-documented vagaries of other disciplines, *or* one must argue that the constraint to highly developed and empirically well-supported dynamical theory is too strong – that, in some sense, the thesis becomes vacuously true.

I have already remarked that one finds some acknowledgment, among historians and philosophers of widely differing persuasions, of the fact (I think one must call it a fact) that dynamics is somewhat anomalous. This (which would seem to deny the charge of circularity) may be considered a tolerable exception to those of the persuasion of Laudan and Kuhn, but only if one fails to consider that as a matter of record dynamical physics *preceded* the development of virtually every other branch of science (certainly as 'mature' disciplines). This in turn leads to the obvious question: what are the implications for the *general* account of scientific epistemology, methodology, and heuristics, (that is, an account directed to other disciplines), if every advance was made by scientists who had at their disposal the (essentially stable and unique) concepts of dynamics? And here it would seem that the epistemological situation is altered across the board in virtually every other discipline. This option (that dynamics is an 'exceptional case') leads, willy-nilly, to a wholesale re-evaluation of conventional historiography and philosophy of science.

Is it, then, that the 'triviality' charge may be well-founded? Laudan (*op. cit.*) appears to adopt this line in his attack on the much broader claim of 'convergent realism'. Adherents of this claim (the early Putnam, Harmann, Boyd, Watkins) typically restrict to 'mature' branches of science, and it is with this that Laudan takes issue. In consideration of Krajewski, for example, who insists that the thesis is nevertheless not rendered vacuous (and is subject to historical evaluation) because "every branch of science crosses at some period the threshold of maturity", Laudan retorts that since the threshold is not specified, it can always be shifted to counter any conceivable unfavorable historical observation. As it stands, this response is not particularly relevant to the thesis at issue here. The nearest analog would run along the lines that what is to count as *mathematically highly developed* and *well-confirmed* are shifting goalposts.

But this is no argument at all, except (perhaps) in connection with the Newtonian theory of light. For what is at issue are *accepted* theories of dynamics, beginning (astronomy and statics apart) with Newton. There have only ever been the theories of mechanics, electromagnetism, relativity, and quantum theory, and these we have just reviewed. And if the relativist insists that one consider the rich and extensive pre-history of dynamics, then we may simply adopt a normative stance: let modern science be judged to begin with Newton (and what is *ad hoc* about this judgment?)[20]. Might the relativist instead object that we have simply separated off that part of physics (indeed, that part of empirical science) that resists the relativist critique? The objection might have some force if 'that part' were some sort of heterogenous collection of results, those that happened never to have been subsequently challenged; but the whole point is that that which has been separated off displays an internal and structural integrity of quite fabulous proportions. Taken in a different spirit the observation is exactly correct: we have isolated some part of science that resists the relativist critique. It is that part which is distinctively mathematical, which employs an ill-understood but characteristically progressive heuristics and methodology, the emergence of which constituted the beginning of the Modern period and which has provided its most central and unifying framework. That this part exists – could there be a more damaging critique of the relativist thesis?

Post, I suggest, was right, and Kuhn was wrong. Let relativists defend their thesis, for if it fails in so central a field as this then it would seem wholly misconceived. There are passing fashions in philosophy, of which this has surely been most pernicious: the important tasks of understanding better the role of mathematics in science, of developing a theory of truth and reference appropriate to the thesis of correspondence in dynamics, which do justice to its vast and exponentially-increasing range of phenomenological applications, lie still before us.

*Dept. of Philosophy,*
*University of Harvard*

## NOTES

1. The so-called 'scientific realism' of recent decades has proved largely ineffectual in the face of the challenge from relativist and pragmatist writers as diverse as Fine, Laudan, Cartwright, Hacking, Rorty, and Putnam. Laudan's (1981) onslaught against 'convergent realism' appeared particularly damaging to the arguments, based on a retentionist view of theory change, to be found in (Boyd 1973), (Putnam 1978), (Watkins 1978), and (Newton-Smith 1981). McMullin's (1984) otherwise admirable riposte to Laudan explicitly *excluded* dynamical physics from his critique (and indeed, his methods are scarcely applicable outside of the macroscopic realm).
2. Among physicists, the writings of Dirac, Wigner and Penrose are exemplary; see Mickens (1990) for a recent collection of papers devoted to the issue. Among philosophers, Steiner (1989) and Zahar (1980) are exceptions. The examples there summarized illustrate the thesis of correspondence, especially its heuristic component (but Steiner may not agree). These issues are important to contemporary methodology; see Cao and Schweber (1993) for current debates concerning renormalization theory, effective field theory, and superstring theory.
3. There are parallels here to Post's conception of the 'footprint' of a successor theory, present in a precursor, a further heuristic in theory innovation. Much of what follows may be considered an elaboration of this device, although I shall not adopt his terminology.
4. It should not need to be stated, but apparently does after all need to be stated, that the investigation of quantum and classical statistical mechanics and field theory, in relation to equilibrium and non-equilibrium thermodynamics, constitutes the major part of condensed matter physics and hence one of the most active areas in contemporary research (for a truly exotic example, consider Hawking radiation and the Hawking-Bekenstein equations).
5. See e.g. Dijksterhuis (1961, p. 291) for an immediate demonstration.
6. For an evaluation of shifting interpretations of Osiander's preface to *De Revolutionibus*, see Dijksterhuis (*ibid.*, pp. 296–7). The reader should also bear in mind Ptolemy's fictionalism, in contrast to the serious commitment of peripatetic (and of course Platonic) philosophy to perfectly uniform motion.
7. There is some evidence that the full generality of the theory of epicycles (that any periodic motion may be so represented) was already conjectured by Appolonius and Hipparchus (the latter proved that any eccentric motion could be replaced by a simple epicyclic motion, the basis of the equivalence between geocentric and heliocentric systems). I do not know who was the first to recognize the relationship between epicycles and Fourier analysis, but it may be found in Torretti (1978, p. 19).
8. I make this remark with some reservations, in view of the difficulties over the definition of local measurable quantities in relativistic quantum theory. See, e.g. Bacry 1988, Fleming 1988. For relationships between the ether and quantum field theory, see Saunders and Brown 1991.
9. As a result, certain features of Maxwell theory – e.g. radiation pressure – remained contentious. The price to be paid for the relative autonomy of Maxwell theory from detailed models of the ether was, roughly speaking, widespread uncertainty with regard to the proper formulation of energy and momentum conservation in radiative interactions, for it was just here that the ether was thought to have an important substantive role (as the bearer of properties). For a detailed evaluation of these functions of the ether I refer to Stein (1981). My claim is that with relativity these matters were resolved, and that a strategy – but no previous resolution – was abandoned.
10. As will become clear, I am sympathetic to the blurring of the distinction between sense and reference for such abstractions.
11. Kripke makes a similar concession in the context of homonymous names, where one has, so to speak, the same 'phenomenology', but distinct causal chains to quite different objects (Kripke 1980, pp. 8–10). Compare also Putnam (1990).
12. For an excellent introduction see Sudarshan and Mukunda 1974; for more background on

the quantum mechanical case see Varadarajan 1970. For the theory of group contractions see Inonu and Wigner 1953.

13. There are very difficult, though perhaps not insuperable, problems when one comes to covariant (relativistic) theories of inter-particle forces. The use of retarded potentials is certainly inadequate (there are, indeed, 'no-go' theorems which indicate that no non-trivial theory of this sort may be formulated consistent with 'manifest' covariance – see Sudarshan and Mukunda 1974). On the other hand, the inverse-square law and its applications to gravity are both understood in the context of field theory (and in particular General Relativity).
14. These canonical forms make up the nearest thing we have to inviolable laws; but we see that they depend on the existence of spacetime isometries, and that we must dispense with them in general relativity for spacetimes of low or vanishing symmetry. How this is to be done, if, indeed, there is any sense to a global dynamics within such spacetimes, is an unresolved problem particularly when addressed in the context of quantum theory (where symmetries play an even more important role in defining the basic dynamical variables).
15. This insight is due to Ignatowsky. A number of qualifications should be made; see Torretti 1983, pp. 76–82 and Brown, this volume. I trust it is clear that there are deep connections between the relativity principle, Klein's Erlangen program, and the symmetry properties of dynamical equations.
16. Ladenburg was fortunately undeterred by the implication of the Bohr theory that there must be infinitely many such electrons, because infinitely many possible transition frequencies, for he could measure but few of them. In this way he cut the Gordian knot that had held up development of the Bohr-Einstein electrodynamics for several years (see Van Der Waerden 1973, for a detailed commentary and original papers).
17. That additional (more technical) requirements are needed to secure the Piron representation theorem (see, e.g. Varadarajan (*op. cit.*) suggests that there may be some more general structure than Hilbert space which satisfies Mackey's axioms in the absence of bivalence, on which a dynamics can be formulated. In this way one would obtain a generalization of quantum mechanics. This is, indeed, a principal motivation of the quantum logic program. However, no such structure has been found, for the more general lattices or posets (that do not satisfy the Jauch-Piron axioms) do not appear powerful enough to support dynamical applications.
18. The essential step uses a generalization (due to Loomis) of the Stone representation theorem for complete Boolean algebras. One ends up with phase space as it should be, a $C^\infty$-manifold in a metric topology, endowed with a Borel $\sigma$-algebra of sets generated by this topology. The quantum mechanical case (where one gives up bivalence) is in certain respects more simple.
19. There is an equivocation in the case of radiation theory. In semiclassical theory one has a rather good idea of how one can formulate a correspondence between classical impure states and quantum pure states (using the zero-point fluctuation and coherent states, respectively). One also knows more or less precisely when the semi-classical description will fail (through the Sudarshan correspondence, Sudarshan 1963; see also Glauber 1963). The equivocation arises because the phenomenology associated with coherent states can be understood in terms of the notion of 'intensity' (in a sense one does not need bivalence in the 'wave limit' of such an ensemble).
20. Evidently there is a great deal to be said on this score; so too, concerning the pre-history of mechanics, and the general relationship between mathematical innovation and the systematization of experience (the experimental method). These topics are so large that I can do no more than gesture to them.

## BIBLIOGRAPHY


Bacry, H. (1988), *Localizability and Space in Quantum Physics*, Lecture Notes in Physics, Vol. 308, Springer-Verlag, Heidelberg.

Beltrametti, E.G., and Cassinelli, G. (1981), *The Logic of Quantum Mechanics*, Addison-Wesley, Reading.

Berry, M. (1991), 'Asymptotics, Singularities and the Reduction of Theories' in D. Prawitz, B. Skyrms, and D. Westerstähl (eds.), *Proceedings of the 9th International Congress of Logic, Mathematics, and Philosophy of Science, Uppsala*, Studies in Logic and the Foundations of Mathematics, North Holland, Amsterdam.

Boyd, R. (1973), 'Realism, Underdetermination, and a Causal Theory of Evidence', *Noûs*, 7, pp. 1–12.

Cao, T. and Schweber, S. (1993), 'The Conceptual Foundations and Philosophical Aspects of Renormalization Theory', *Synthese*, forthcoming.

Dijksterhuis, E.J. (1961), *The Mechanization of the World Picture*, Princeton University Press, Princeton.

Feyerabend, P.K. (1968), 'How to be a Good Empiricist' in P.H. Nidditch (ed.), *The Philosophy of Science*, Oxford University Press, Oxford.

Field, H. (1972), 'Theory Change and the Indeterminacy of Reference', *Journal of Philosophy*, 70, pp. 462–81.

Glauber, R. (1963), 'Coherent and Incoherent States of the Radiation Field', *Physical Review*, 131, pp. 2766–88.

Hacking, I. (1983), *Representing and Intervening*, Cambridge University Press, Cambridge.

Hardin, C.L. (1988), *Color for Philosophers*, Hackett Publishing Co., Indianapolis.

Heisenberg, W. (1925), 'Über quantentheoretische Umdeutung kinematischer und mechanischer Beziehungen', *Zeitschrift für Physik*, 33, pp. 879–93. English translation in Van der Waerden (1967).

Hooker, C. (1975–1979), *The Logico-Algebraic Approach to Quantum Mechanics*, Vols. 1 & 2, University of Western Ontario Series in Philosophy of Science, University of Western Ontario, London.

Inonu, E., and Wigner, E.P. (1953), 'On the Contraction of Groups and their Representations', *Proceedings National Academy of Science*, 39, pp. 510–24.

Kripke, S. (1980), *Naming and Necessity*, Blackwell, Oxford.

Kuhn, T. (1962), *The Structure of Scientific Revolutions*, International Encyclopedia of Unified Science, Vol. 2, University of Chicago Press, Chicago.

Laudan, L. (1981), 'A Confutation of Convergent Realism', *Philosophy of Science*, 48, pp. 19–48.

Mackey, G.W. (1963), *Mathematical Foundations of Quantum Mechanics*, Benjamin/Cummings, Reading.

McMullin, E. (1984), 'A Case for Scientific Realism' in J. Leplin (ed.), *Scientific Realism*, University of California Press, Berkeley.

Mickens, R.E. (1990), *Mathematics and Science*, World Scientific, Singapore.

Newton-Smith, W. (1981), *The Rationality of Science*, Routledge and Kegan Paul, Boston.

Post, H. (1971), 'Correspondence, Invariance and Heuristics: In Praise of Conservative Induction', *Studies in History and Philosophy of Science*, 2, pp. 213–55.

Putnam, H. (1962), 'The Analytic and the Synthetic' in H. Feigl and G. Maxwell (eds.), *Minnesota Studies in the Philosophy of Science Vol. 3*, University of Minnesota Press, Minneapolis. Reprinted in *Philosophical Papers*, Volume 2, Cambridge University Press, Cambridge.

Putnam, H. (1973), 'Explanation and Reference' in G. Pearce and P. Maynard (eds.), *Conceptual Change*, D. Reidel, Dordrecht. Reprinted in *Philosophical Papers*, Vol. 2, Cambridge University Press, Cambridge.

Putnam, H. (1975), 'The Meaning of 'Meaning'' in K. Gunderson (ed.), *Language, Mind, and Knowledge*, Minnesota Studies in the Philosophy of Science, VII, University of Minnesota Press, Minneapolis. Reprinted in Philosophical Papers, Vol. 2, Cambridge University Press, Cambridge.

Putnam, H. (1978), *Meaning and the Moral Sciences*, Routledge & Kegan Paul, Boston.
Putnam, H. (1990), 'Is Water Necessarily $H_20$?' in *Realism With a Human Face*, Harvard University Press, Cambridge.
Saunders, S.W. (1988), 'The Algebraic Approach to Quantum Field Theory' in H. Brown and R. Harré (eds.), *The Philosophical Foundations of Quantum Field Theory*, Clarendon Press, Oxford.
Saunders, S.W. and Brown, H. (1991), 'Reflections on Ether' in S. W. Saunders and H. Brown (eds.), *The Philosophy of Vacuum*, Clarendon Press, Oxford.
Scheffler, I. (1966), *Science and Subjectivity*, Hacking Publishing Co., Indianapolis.
Shapere, D. (1969), 'Towards a Post-Positivist Interpretation of Science' in P. Achinstein and S. Barker (eds.), *The Legacy of Logical Positivism*, Baltimore.
Stein, H. (1981), "Subtler Forms of Matter' in the Period Following Maxwell' in G.N. Cantor and M.J.S. Hodge (eds.), *Conceptions of Ether: Studies in the History of Ether Theories, 1740–1900*, Cambridge University Press, Cambridge.
Steiner, M. (1989), 'The Application of Mathematics to Natural Science', *The Journal of Philosophy*, 76, pp. 449–80.
Sudarshan, E.C.G. (1963), 'Equivalence of Semiclassical and Quantum Mechanical Descriptions of Statistical Light Beams', *Physics Review Letters*, 10, pp. 277–9.
Sudarshan, E.C.G. and Mukunda, N. (1974), *Classical Dynamics: A Modern Perspective*, John Wiley, New York.
Taylor, P. (1984), The Relationship Between Classical and Quantum Mechanics, *D. Phil. Thesis*, University of Oxford, Oxford.
Torretti, R. (1978), *Philosophy of Geometry from Riemann to Poincaré*, D. Reidel, Dordrecht.
Torretti, R. (1983), *Relativity and Geometry*, Pergamon Press, Oxford.
Van Der Waerden, B. (1967), *Sources of Quantum Mechanics*, North Holland, Amsterdam.
Varadarajan, V. (1968), *Geometry of Quantum Theory*, Vol. 1, Van Nostrand, Princeton.
Varadarajan, V. (1970), *Geometry of Quantum Theory*, Vol. 2, Van Nostrand, Princeton.
Watkins, J. (1978), 'Corroboration and the Problem of Content-Comparison' in G. Radnitzky and G. Anderson (eds.), *Progress and Rationality in Science*, D. Reidel, Dordrecht.
Woodhouse, N. (1980), *Geometric Quantization*, Clarendon Press, Oxford.
Zahar, E. (1980), 'Einstein, Meyerson and the Role of Mathematics in Physical Discovery', *British Journal for the Philosophy of Science*, 31, pp. 1–43.

MICHAEL REDHEAD

# Is the End of Physics in Sight?*

One of the central concerns of Heinz Post's work in the philosophy of science has involved problems of intertheory relations, both with respect to the diachronic correspondence between successive theories and the synchronic reduction of one theory to another. The general correspondence principle articulated by Post (1971) demands that successor theories and reducing theories should conserve the successes of replaced theories and reduced theories, respectively, while overcoming their perceived flaws. Post's thesis is that unification by means of theory reduction and 'conservative induction' in the formulation of new theories, providing that the correspondence principle is satisfied, are progressive procedures and, hence, goals worth pursuing in science. The aim of this paper is to evaluate what the prospects are for a final 'end-product' for these unifying processes of theory evolution and theory reduction.

## I. INTRODUCTION

In 1980 Stephen Hawking delivered his inaugural lecture as Lucasian Professor at Cambridge with the title 'Is the End in Sight for Theoretical Physics?'[1]. Hawking was cautiously optimistic that by the end of the twentieth century the answer might have turned out to be 'yes'. Since we have now reached the half-way mark, I thought it might be appropriate to take stock. I have deliberately drawn the title rather more widely than Hawking, since I want to say something also about the experimental side of the subject.

The view I am going to discuss is that physics, and more generally experimental science, has developed at an increasingly rapid rate since its inception in the seventeenth century, and will simply burn itself out in a final frenzy

* A version of this paper was delivered as the Presidential Address to the British Society for the Philosophy of Science on 15th October 1990. I dedicate it to my friend and teacher Heinz Post. Some of the material was subsequently included in my Tarner Lectures, 'From Physics to Metaphysics', delivered in Cambridge in February 1993.



*S. French and H. Kamminga (eds.), Correspondence, Invariance and Heuristics*, 327–341.

of intellectual effort. It is rather like the exploration of the physical Globe – there is only one Globe and once all the continents, mountains, rivers, etc. have been discovered, the work is essentially finished, there is just detail to fill in. Although it may be granted that there are infinitely many exact locations on the Globe there is a real sense in which the job of map-making on any 'interesting' scale is finite, completable and has in fact been completed over a total span of less than five hundred years.

Towards the end of the nineteenth century Lord Kelvin is famously supposed to have claimed that all that remained in physics was the job of filling in the next decimal place[2]. Matter, and energy in its diverse forms, were governed by mechanical laws and the principles of thermodynamics. Essentially there was nothing more to be said . . . and then, well, physics really started to take off with Einstein's relativity theories and, in the 1920s, with the supreme discovery of quantum mechanics. As Max Born remarked, "Physics . . . will be over in six months"[3]. Kelvin had been a little premature, but now, thought Born, physicists, after some false starts, had got it right. There was still mopping up to be done, chemistry and biology had got to be worked out in detail, but with the Dirac equation, to which Born was actually referring in this remark, the ultimate basic law had been discovered. There was only one sort of matter composed of corpuscles of positive and negative electricity, and only one sort of interaction, electromagnetism, aside from gravitation, and that had been sorted out by Einstein in his General Theory of Relativity.

But almost at once this synthesis dissolved itself. First, there were more and more new particles, and new types of force involved in nuclear physics. But gradually new order emerged. There were six sorts of lepton (generalizing the notion of electron) and there were six sorts of quark (generalizing in a sense the notion of proton) and then there were all the interactions, strong, weak, electromagnetic and gravitational and their associated particles, the W-particles, the photon, the gluons and the graviton. So what had seemed simple in 1927 looked very messy forty years later! But then things began to fall into place, the Weinberg-Salam electroweak unification in the late sixties became the model for a unification of all the particles and forces. One particular such scheme, known technically as $N = 8$ extended supergravity theory, was the one that Hawking was particularly excited about ten years ago[4].

But I am sorry to disappoint you: $N = 8$ supergravity has fallen by the wayside and has been replaced by a totally different sort of TOE (Theory of Everything), viz. superstrings. The particles are not massive points at all, but excitations of linear 'entities', the superstrings. But once physicists go for 'extended' objects as fundamental, why restrict oneself to strings? So already the two-dimensional membrane and more generally higher-dimensional $p$-brane are under active investigation. So where will it all end? The theoretical physicists working in this area feel that the Holy Grail is within their grasp, it just eludes them, but one more effort and it will all be done!

## II. THE PHILOSOPHER'S VIEW

Let me now turn from the heady speculations of the theoretical physicists to the more carefully argued and, dare I say, down to earth views of philosophers of science.

Firstly, what could one mean by the end of physics? Ideally, I suppose, that one could explain the occurrence of every event past, present and future. Already problems crowd in. What counts as an explanation, and what counts as an event? Again pursuing the ideal, I suppose we would want to be able to *deduce* the occurrence of events from a simple set of interlocking principles, a unified theory, together with boundary conditions and initial conditions, and 'event' would be the translation into a canonical physicalist language of *every* 'natural' happening. But that sounds very much like a nineteenth-century ideal. Twentieth-century physics has been in certain senses both less and more demanding. Firstly, modern physics has broadly embraced the possibility of indeterminism, that it may not be possible to deduce the occurrence of individual events at all. So, can one accept as the end of physics a principled limitation on what can be explained, a limitation set by physics itself? But the very notion of event as the ascription of values to physical variables is itself very problematic, from the perspective of quantum mechanics. The fact of the matter is, one couldn't accept that the end of physics has arrived until the interpretational problems of quantum mechanics have been sorted out[5]. But what about the initial and boundary conditions? There has been a cherished hope in modern physics that it may turn out, in the context of a TOE, that nothing happens as a matter of pure contingency, that there is a reason for everything, and this is where modern physics essays to be *more* demanding than classical physics. The hope here is that logical consistency, together with empirical adequacy, may dictate a *unique* way the world is. Philosophers, fed underdeterminism with their mother's milk, may be amazed at such presumption. The reason that physicists even entertain such an idea is that it is extraordinarily difficult to develop consistent TOEs at all. TOEs have been plagued with problems of infinities and so-called anomalies. The present interest in superstring theory arises from the belief that such theories are finite and anomaly-free. This is, at present, a much more significant criterion than empirical adequacy. In fact, superstring theories predict a host of phenomena that no-one has seen, lots of unobserved particles, and in most versions, the instability of the proton. But there is still enough flexibility in the theory that these non-observations may be explained away. But that does not quite square with the ideal of a unique self-consistent scheme! Furthermore, TOEs are not purely logical theories, but are constrained by very general physical principles. In the 1960s there was a great deal of talk of bootstrap physics[6], in which everything would be mutually self-supporting, but a genuine unique bootstrap is scarcely a credible possibility[7]. In practice bootstrap physics starts with very general principles, and then consistency may possibly fill in the details uniquely, but that is not the same thing as a purely rational physics, in which everything would arise from pure reason, and experiments would not be necessary at all as a check on the adequacy of the theory.

Next, we should refer to the imperialistic aspect of physics, that all the other sciences are *reducible* to physics, and ultimately to elementary particle physics at that. It is quite true that physics has encroached on chemistry and biology, and chemists and biologists themselves pay lip service to the view that their subjects, in ultimate ontological commitment, are just a part of physics. There are not many vitalists or dualists around among scientists these days. But that does not mean that *methodologically* the best way to tackle problems in chemistry or biology is by solving the *N*-particle Schrödinger equation or its more modern equivalents. In fact, methodologically speaking, chemistry and biology are in many respects autonomous disciplines. Does this mean that the end of physics could only truly be said to have been reached when, per impossibile, the present way of doing chemistry and biology had been totally replaced by a fundamental physics approach? When we turn to the social and human sciences, psychology and so on, the possibility of a *practical* reduction to physics is patently absurd, so we are left with a dilemma: either in principle physics cannot explain the 'soft' sciences, a view which most scientists and philosophers would deny, or in principle it can, but the infeasibility of actually doing so in practice means that physics will never be completed, in the sense that there are tasks which in principle it could perform, but in practice it is inconceivable that it ever will.

But, in point of fact, we do not have to look to the other sciences to make this point. In most areas of physics recourse is not in practice made to the fundamental laws of elementary particles. Physics is a collection of approximations, theoretical models that cover a limited range of conditions, and so on (Redhead, 1980b). Indeed, Nancy Cartwright (1983) has argued in her book *How the Laws of Physics Lie*, that no fundamental analysis for complex phenomena, such as the operation of a laser, is possible even in principle. I certainly would not want to go as far as that myself. But the practice of physics is certainly much closer to Cartwright's description than the neat, tidy discipline laid out in some of the textbooks.

But it may be objected that I have been concerned with practical limitations on completing physics, subject, as we have observed, to constraints that are themselves part of physics on what a completed physics could even in principle achieve. But are there theoretical as distinct from practical reasons why physics cannot be completed? This depends essentially on one's views about the scientific enterprise generally. The first point to get out of the way: even if from a God's eye point of view we hit on a totally correct TOE, we would certainly never know it. The infinity of its empirical content could never be surveyed by human scientists, and even the most determined inductivist would allow that, as far as we can tell today, the best confirmed theory may turn out tomorrow to be in error. So, even in principle, it would never be possible to announce: it is all over, you can hang up your galvanometers, lads, and go home. But let us play at being God for the moment, is it possible for human scientists to hit, even by chance, on the final TOE? If you are a realist about scientific claims, there are two possibilities: (1) There is no TOE, because there are infinitely many levels of structure that can be unpacked, like an infinitely descending sequence of Chinese boxes, or to use the more colloquial expression: it is boojums all the way down[8]. If this is the case, the ideal of a simple unified scheme must be relativized to levels – what you see depends on how closely you look, and there is no reason in principle why things should not get more complicated, rather than simpler, beyond a certain level. (2) There is a TOE, and however miraculous it may seem, we might hit on it with superstrings or something of the kind. I can see no argument that can possibly tell us which of these options is correct. Or, to put it another way, a TOE is in principle possible.

Of course, to an anti-realist, or a social constructionist, this realist discussion looks patently absurd. There is no way the world is that can be captured by the deductive closure of a TOE. Science is an open-ended sequence of attempts to represent aspects of an objective external world to human enquirers where no objective external world exists to be represented. But suppose we give up the idea of a TOE as being true and judge it only by its empirical adequacy in the sense, say, of Van Fraassen (1980). Then again, in principle there could be such a TOE. But this position is fraught with all the usual difficulties of identifying a 'neutral' empirical content.

The Popperian image of scientists endeavouring fallibly at both the theoretical and observational level, of improving their cognitive grip on a real external world, and in principle never-to-be-completed sequence of revisions of how we conjecture the world to be, motivates the view of an incompletable *practice* of physics.

## III. PROGRESS IN SCIENCE

But is physics getting nearer to completion? That raises the whole thorny question of progress in science to which I will now briefly turn. There is an obvious sense in which applied science, i.e. technology, has progressed in a linear cumulative fashion over the past five hundred years. With the possible exception of 'lost' skills, like crafting violins in the manner of Stradivarius, there has been a steady expansion of our control over the natural world; the television set and the word processor did not exist until comparatively recently. But if this cumulative model is granted for applied science, how are we to square this with the situation in pure science, where the cumulative model is standardly challenged? My first comment is that the links between pure and applied science involve very much a two-way traffic. Indeed, as one commentator put it, "thermodynamics owes much more to the steam engine than ever the steam engine owed to thermodynamics" (Price 1986, p. 240). Applied science itself has a certain autonomy – ground once gained never has to be relinquished, whereas in pure science, or so it is claimed, there is to-oing and fro-ing rather than linear progress.

Let us look at some of the arguments adduced for this state of affairs. In a scientific revolution, it is claimed, there are always 'Kuhn losses', problems which the old paradigm (to pursue Kuhn's famously elastic terminology) can handle, but which the new paradigm cannot. And, of course, the very idea of comparing problems solved across scientific revolutions is anyway denied by the ardent incommensurabilist. But am I, locked in my quantum-mechanical worldview, really supposed not to be able even to understand what it was like to be a classical physicist? Of course that is all nonsense. I find no difficulty in reading Newton's *Principia* or Maxwell's *Treatise*. I may not agree with it, but I can certainly understand what Newton and Maxwell were on about[9]. Does classical mechanics and electromagnetism account for the discrete lines of the hydrogen spectrum? I do not believe it does, and that is a problem which, in some sense, was solved by quantum mechanics.

## IV. THE CLASSICAL LIMIT

So, can quantum mechanics do everything that classical physics could do and then some more? Well, I must admit that there are all sorts of problems in constructing a classical limit for quantum mechanics. It is not just a question of putting Planck's constant equal to zero. The quantum-mechanical solutions actually display complicated singularities at the origin, when considered as functions of variable $h$. There is an abrupt discontinuity between the physics of an infinitesimal $h$ and a vanishing $h$, if I can express the matter somewhat crudely. But the presence of such a discontinuity does not mean that, qua mathematician, I cannot fully understand what a world with $h$ strictly zero is like, and how it differs from a world with infinitesimal $h$. The same is true of relativity theory. As $c \to \infty$, the metric of space-time becomes singular in a precise mathematical sense[10]. Many of the nice properties of relativistic space-times depend on the existence of a non-singular metric, and these properties simply disappear in a discontinuous way in classical space-times. But that does not mean that everything is discontinuous – the dependence of mass on velocity,

$$m = m_0/\sqrt{1 - v^2/c^2}$$

behaves quite smoothly as $c \to \infty$, so there is a well-defined sense in which some classical phenomena are perfectly well captured as approximately true in relativistic theories, while others have an all-or-nothing aspect. Let us take a crude example. At a distance $x$ the time difference between events declared simultaneous relative to a stationary frame $K$, becomes $vx/c^2$, relative to a moving frame $K'$. However large $c$ may be, $x$ can always be chosen large enough to reveal the relativity of simultaneity. But if $c$ is actually infinite (i.e. $1/c = 0$), the relativity of simultaneity has simply been eliminated.

So the mathematical reflection of incommensurability is just that some aspects of the classical world are not 'approximately' true in the new theories, but are simply and misleadingly false. But that looks like cumulative progress: everything that classical physics is still useful for corresponds to smooth limiting behaviour in the relevant parameters; where classical physics goes definitely wrong is either as we move smoothly *away* from the limit in parameter space, or as we move across a discontinuity. Schematically we illustrate the situation in Figures 1–3.

In Figure 1, the value of a quantitative physical property is shown as $X$, when the parameter $p$ is zero, with error bars indicated by the lines AB, CD. The actual value of $p$ is $p_0$ and the dotted curve shows how the property varies with $p$. It only moves outside the error bars of classical behaviour for $p > p_0$.

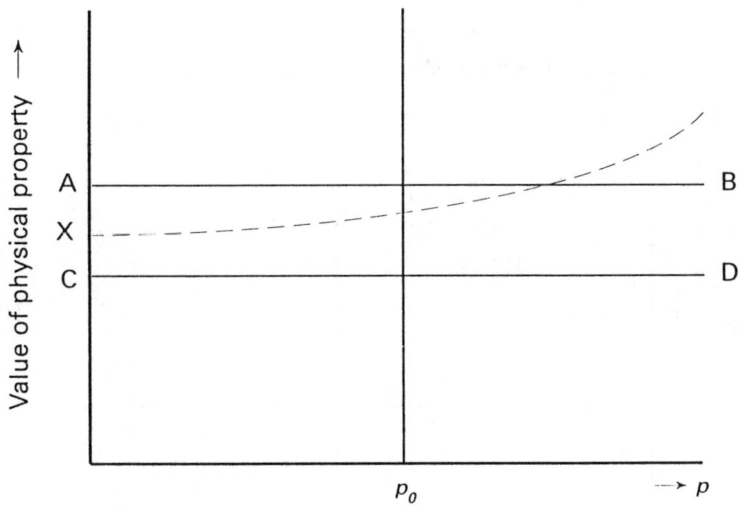


*Fig. 1.* Smooth limiting behaviour. The value of some physical property is $X$ when the parameter $p$ is zero, with error bars indicated by the lines AB and CD. The actual value of $p$ is $p_0$ and the dotted curve shows how the property varies with $p$. It only moves outside the error bars of classical behaviour for $p > p_0$.

In Figure 2 we have a situation in which non-classical behaviour is already revealed for $p = p_0$.

Finally, in Figure 3 we show a situation of singular discontinuity: even for infinitesimal $p$ the new theory is quite different from classical physics.

Of course, in respect of Figure 3, if the error bars are sufficiently widely drawn we may not be able to distinguish Figure 3 from Figure 2 or even Figure 1. The point really is that by refining accuracy, non-classical behaviour can always be revealed, however small $p$, counterfactually, might be.

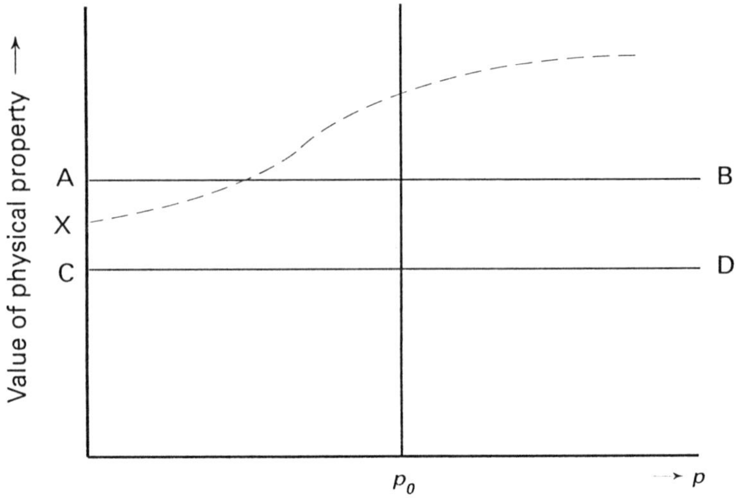


*Fig. 2.* As Figure 1, but here non-classical behaviour is already displayed for $p = p_0$.

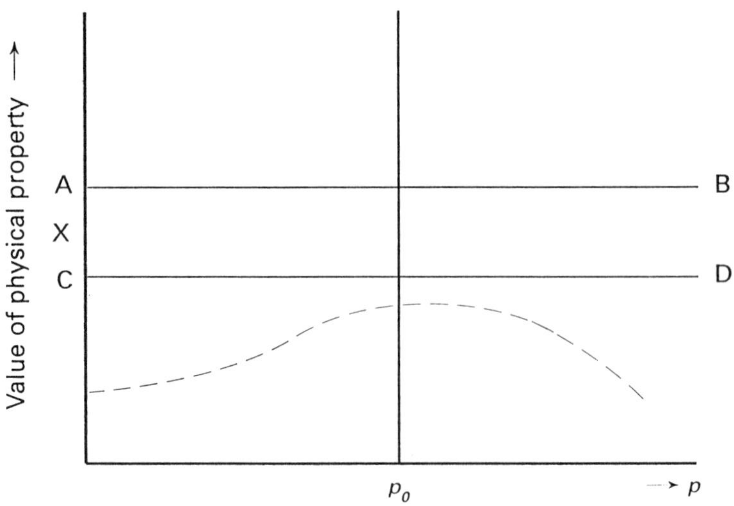


*Fig. 3.* Singular discontinuity. Here the behaviour of the property in the new theory is quite different from that in classical physics even for infinitesimal $p$. Symbols as in Figures 1 and 2.

As another simple illustration which is relevant to the quantum-mechanical case, consider the classical limit of the wave equation

$$\frac{\partial \varphi}{\partial x^2} - \frac{1}{c^2}\frac{\partial^2 \varphi}{\partial t^2}$$

as $c \to \infty$.

The equation itself becomes

$$\frac{\partial^2 \varphi}{\partial x^2} = 0$$

Suppose $\varphi$ is just the displacement of a stretched string, of length $L$, with fixed end points. The solution of (2) is then $\varphi = 0$, but now look at the solution of (1), subject to the same boundary conditions, as $c \rightarrow \infty$. For any finite $c$ we have pulses travelling to right and left, reflected with 180° change of phase at each fixed end. If $l$ is the length of a pulse assumed to be of unit amplitude, the time average of displacement over a resolution time $T$ is bounded by $l/Tc$. For fixed $T$, as $c \rightarrow \infty$, the time average goes to zero, but for any $c$, however large, we can always make $T$ small enough to reveal the criss-crossing pulses on the string. In the same way, diffraction phenomena always exist on an appropriate scale in the neighbourhood of a geometrical shadow.

These considerations are important in understanding Post's General Correspondence Principle. In the case of what he calls inconsistent correspondence[11], the types of behaviour illustrated in Figures 1 and 2, as contrasted with Figure 3, are to be sharply distinguished. Our remarks should also be relevant to Worrall's structural realism programme (Worrall 1989). The role of discontinuous limiting behaviour is important here in assessing the cumulative aspect of scientific progress.

But how does this talk of cumulative approach to the truth square with the verisimilitude problem broached famously by Tichy (1974) and Miller (1974). We are committing ourselves in truth in this whole discussion to essentialism, the view that the behaviour of certain variables is what a theory is 'really' about, that the physicist should not be interested in Minnesotan weather, or adding the number of planets to the number of days in the week, or whatever. But to make a principled approach to ruling out Miller-type predictions as counting in the assessment of false theories is, I agree, a still unresolved problem.

We have so far discussed motion in a parameter space scaffolded by what might be called fixed parameters, e.g. the examples of Planck's constant and the velocity of light illustrate that they are not under our control. But, already, in drawing the distinction between Figures 1 and 2, we have in mind physical properties identified by exploration in a variable parameter space. Consider the paradigm example of velocity. For small velocities the variable mass cannot be identified for physical $c$, with appropriate error bars as in Figure 1, but for higher velocities, the physical $c$ does allow the new type of behaviour to be revealed (cf. Figure 2). Let us use $\nu$ as a generic symbol for variable parameters, and retain $p$ as a symbol for fixed parameters. As we explore larger and larger regions of $\nu$, discrepancies with observation open up, but are restored by appropriate movement in $p$-space, as illustrated in Figure 4.

Along OA, BC, DE we are moving parallel to the $\nu$-axis, opening up

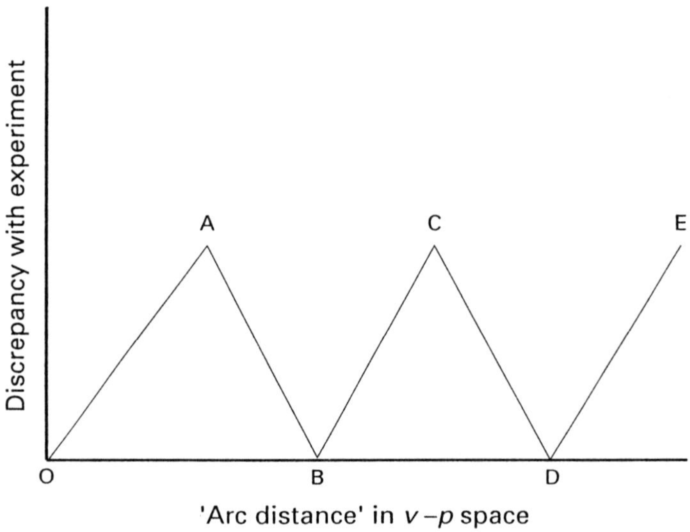


*Fig. 4.* Discrepancy with experiment.

successive disagreements with experiment, which are corrected by movement along AB, CD, etc. parallel to the $p$-axis. The track in $\nu$-$p$ space is shown in Figure 5.

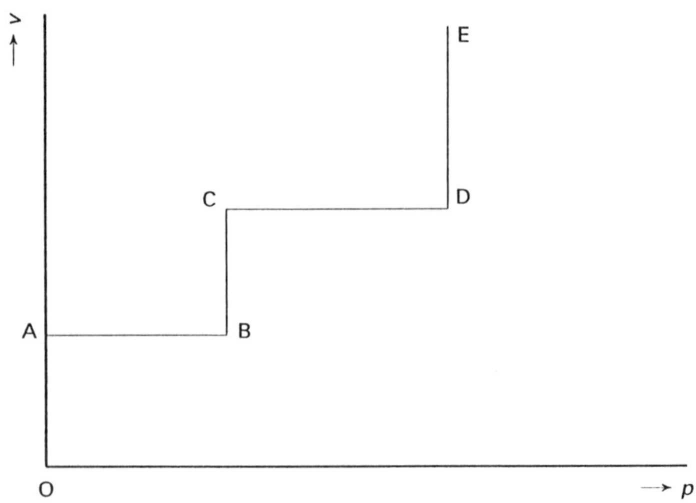


*Fig. 5.* Schematic progression of physics in $\nu$-$p$ space.

## V. THE END OF PHYSICS?

The question, does physics come to an end, can now be posed in the form: does the sawtooth behaviour sketched in Figure 4 continue indefinitely or does it terminate in a TOE, as illustrated in Figure 6?

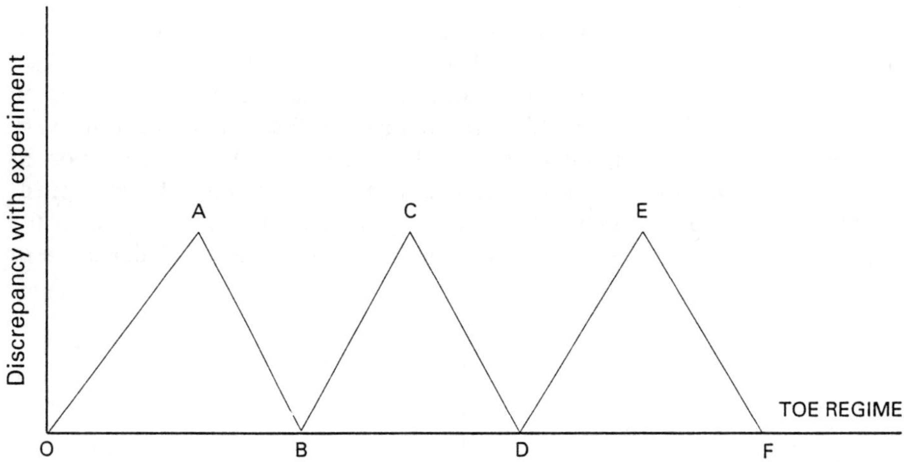


*Fig. 6.* Termination of discrepancy with experiment.

The answer to the question 'could physics come to an end?' is then that in principle the behaviour of Figure 6 could obtain rather than an indefinite prolongation of Figure 5. But the question of whether the end is in sight, in the sense that we could rationally *anticipate* the future course of science is, of course, quite impossible to answer even in principle.

But is the end of physics in sight from a pragmatic point of view, that the experimental investigation of the ν-space is approaching limits imposed by practical economic considerations, for example? Let us look at some dimensions of the ν-space. Firstly, there is physical space and time, at very small scales involved in atomic and subatomic physics, and at very large scales dealt with in astronomy. Then there is complexity, the understanding of how complex systems concatenated of elementary systems behave. Finally, there is 'field strength' in a generalized sense, intense gravitational fields, low- and high-temperature regimes, extremes of pressure, and so on. The first point to notice is that ν-space and *p*-space do not operate independently. As soon as a new type of interaction is contemplated, a movement in *p*-space, then the question of variable intensities of these fields, a movement in ν-space, becomes possible for experimental investigation. This is an example of what Rescher (1984) refers to as Kant's Principle of Question Propagation in action. As Kant puts it in the *Prolegomena*, "Every answer given according to fundamental laws of experience always gives birth to a new question" (Kant 1783, section 57). So we can only be sure that all regions of ν-space are being explored if we are sure that there is no more movement in *p*-space, which, as we have seen, is impossible to establish.

But, even if we have established what parameters to investigate in ν-space, it is true that, in general, the exploration of any given dimension is subject to a law of diminishing returns. It is generally agreed that exponential growth in resources is required to maintain a linear growth in significant output, so far as exploration is concerned in ν-space. If, for economic reasons, resources are capped at some particular cut-off, then the output is subject to so-called logarithmic retardation. This behaviour is illustrated in Figure 7, adapted from Rescher (1978 p. 115). The application of these ideas become very clear in the context of high-energy particle accelerators, where the exponential growth regime in input resources is almost certainly about to be capped! Higher energies or correlatively smaller distances illustrate very well the question of exploring ν-space. Present machines reach down to distances of

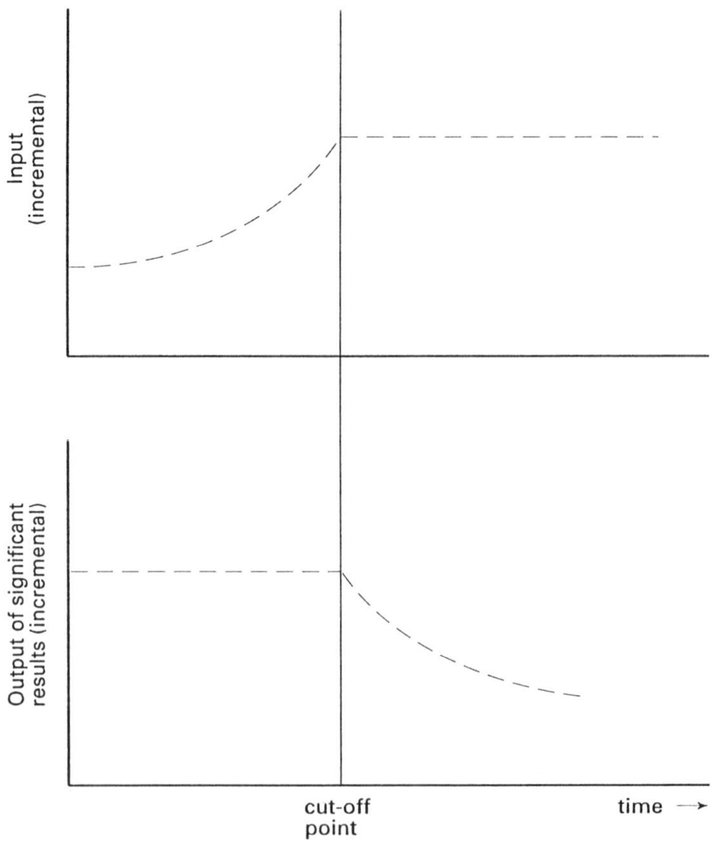


*Fig. 7.* The relation between the output of significant research results and the input of economic resources. (Adapted from Rescher 1978, p. 115)

the order of $10^{-16}$ cm. Beyond this there is a region often referred to as Asymptopia, where nothing interesting is supposed to happen until the grand unification regime is reached at a distance of about $10^{-29}$ cm, where the effective coupling constants for strong, weak and electromagnetic interactions become equal. Beyond this, at $10^{-33}$ cm, lies the Planck length where quantum gravity begins to make nonsense of any ordinary conception of space and time. But all this is really pure surmise, based on theoretical principles that have far outstripped the empirical basis except in the rarefied field of early-universe physics. But of course, that in itself is a curious sort of experimental science; we cannot run the universe over again to check how things really turned out! It is even arguable that modern theoretical physics has entered a regime of baroque decadence, and in trying to achieve everything is in danger of falling flat on its face and achieving nothing!

Before closing, I would like to make a few miscellaneous additional remarks. Firstly, there is an obvious difficulty in thinking about something, like the future course of science, which we cannot in principle know. All that we can do as we have seen is to extrapolate from our limited past experience.

Secondly, there is a real danger in scientism, trying to apply the methods of science to unsuitable areas of experience, such perhaps as the subjective content of human thought. As Tom Nagel (1986) has remarked, science, in providing the objective View from Nowhere, leaves out the ineliminable subjectivity of experience. But against this we must beware lest too much is banished *a priori* from proper scientific investigation. Fringe science, anything from polywater to cold fusion, should not be rejected out of hand, but only after error, fraud or charlatanism has been properly exposed.

Finally, there is the question of *insolubilia*, the question of whether there are *scientific* questions that can never be answered. This of course was famously debated in the last century by Emil Du Bois-Reymond and Ernst Haeckel[12]. In this debate I side squarely with Haeckel. Any question which, in principle, science cannot answer, such as the ultimate explanation of order and regularity in the world, is *ipso facto* not a *scientific* question. There are limits on science that arise not internally but externally. Any question that science can properly pose can be given a scientific answer.

## VI. CONCLUSION

It is often felt that physics has now dealt with everything that matters to us in any everyday sense, and is only concerned, at its research frontiers, with such extreme conditions of energy, temperature, etc., that from the point of view of intrinsic interest and significance, physics has already reached an effective limit or boundary. But questions of interest and significance are of course essentially subjective, and I would submit myself that the further development of physics, about whose future course we can admittedly only speculate, will provide the most stimulating intellectual challenge of the new millennium. In my view the end of physics is a receding horizon. To achieve the end of physics is as unrealistic as grasping the pot of gold at the end of a rainbow.

Returning to Post's general correspondence principle, I suggest that this principle does give content to some notion of progress in science and that it forms the basis for useful heuristic guidelines. Nevertheless, as I have argued here, the ultimate goal of a completely unified science will remain elusive. Of course, Post's thesis that, overall, progress in science is linear does not logically entail that there is some final state towards which science is progressing. Heinz Post would surely be excited by the prospect, not of some Theory of Everything, but of the continuing confrontation with novel surprises in science.

*Dept. of History and Philosophy of Science,*
*University of Cambridge,*
*Cambridge, U.K.*

## NOTES

1. See Hawking (1980) for the published text of this Lecture.
2. This view is attributed (supposedly to Lord Kelvin) in Maxwell's remark in his inaugural lecture at Cambridge:

   > The only occupation . . . left to men of science will be to carry on these measurements to another place of decimals.

   See Maxwell (1965, Vol.2, p. 244); compare also Badash (1972).
3. Quoted in Hawking (1980, p. 1).
4. For technical details of this and other recent developments in fundamental physics, see for example Collins, Martin and Squires (1989).
5. These problems are discussed, for example, in Redhead (1987).
6. See Redhead (1980a) and also the very detailed discussion in Cushing (1990).
7. I say scarcely credible, because there are some rationalist philosophers around who might believe this sort of thing.
8. Compare the title of David Mermin's book (Mermin, 1990) on atomic physics, *Boojums All the Way Through*.
9. I once discussed with the Professor of History of Science at Oxford the qualifications for being an Historian of Science. She explained that the most important qualification was not to be educated as a scientist!
10. A non-singular metric $g$ is one for which

    $$\forall X g(X,Y) = 0 \Rightarrow Y = 0$$

    for arbitrary vectors $X$ and $Y$.
11. Post refers to 'inconsistent correspondence' in cases where a successor theory degenerates into an *approximate* version of the replaced theory under the conditions under which the latter theory has been well confirmed (Post 1971, p. 230). Even in these cases, Post argues, the successor theory explains the successes of the old theory and, hence, the correspondence principle is satisfied. The types of behaviour illustrated in Figures 1 and 2

satisfy the correspondence principle (that in Figure 2 in the sense of inconsistent correspondence), but the behaviour shown in Figure 3 does not.

12. See Du Bois-Reymond (1884) and Haeckel (1899).

## BIBLIOGRAPHY

Badash, L. (1972), 'The Completeness of Nineteenth-Century Science', *Isis*, 63, pp. 48–58.

Cartwright, N. (1983), *How the Laws of Physics Lie*, Clarendon Press, Oxford.

Collins, P.D.B., Martin, A.D. and Squires, E.J. (1989), *Particle Physics and Cosmology*, John Wiley, New York.

Cushing, J. (1990), *Theory Construction and Selection in Modern Physics: The S Matrix*, Cambridge University Press, Cambridge.

Du Bois-Reymond, E. (1884), *Über die Grenzen des Naturerkennens: Die Sieben Welträthsel – Zwei Vorträge*, Veit, Leipzig.

Haeckel, E. (1889), *Die Welträthsel: Gemeinverständliche Studien über Monistische Philosophie*, Emil Strauss, Bonn.

Hawking, S.W. (1980), *Is the End in Sight for Theoretical Physics?*, Cambridge University Press, Cambridge.

Kant, I. (1783), *Prolegomena to Any Future Metaphysics That Will be Able to Present Itself as a Science*, Hartknoch, Riga.

Maxwell, J.C. (1965), *The Scientific Papers of James Clerk Maxwell*, ed. by W. D. Niven, Dover, New York.

Mermin, N.D. (1990), *Boojums All the Way Through: Communicating Science in a Prosaic Age*, Cambridge University Press, Cambridge.

Miller, D. (1974), 'Popper's Qualitative Theory of Verisimilitude', *The British Journal for the Philosophy of Science*, 25, pp. 166–77.

Nagel, T. (1986), *The View From Nowhere*, Oxford University Press, Oxford.

Post, H.R. (1971), 'Correspondence, Invariance and Heuristics', *Studies in History and Philosophy of Science*, 2, pp. 213–55.

de Solla Price, D.J. (1986), *Little Science, Big Science . . . And Beyond*, Columbia University Press, New York.

Redhead, M.L.G. (1980), 'Some Philosophical Aspects of Particle Physics', *Studies in History and Philosophy of Science*, 11, pp. 279–304.

Redhead, M.L.G. (1980), 'Models in Physics', *The British Journal for the Philosophy of Science*, 31, pp. 145–63.

Redhead, M.L.G. (1987), *Incompleteness, Nonlocality and Realism: A Prolegomenon to the Philosophy of Quantum Mechanics*, Clarendon Press, Oxford.

Rescher, N. (1978), *Scientific Progress*, Blackwell, Oxford.

Tichy, P. (1974), 'On Popper's Definition of Verisimilitude', *The British Journal for the Philosophy of Science*, 25, pp. 55–60.

Van Fraassen, B.C. (1980), *The Scientific Image*, Clarendon Press, Oxford.

Worrall, J. (1989), 'Structural Realism: The Best of Both Worlds', *Dialectica*, 43, pp. 99–124.

# Notes on Contributors

**Harvey Brown** is University Lecturer in Philosophy of Physics at Oxford University, and Fellow of Wolfson College, Oxford. He was introduced to the philosophy of science in 1971, when he took Heinz Post's course at Chelsea College, and went on to take a Ph.D. in Post's department. His primary research interests lie in the foundations of quantum mechanics and relativity theory, and he has published widely on these subjects in scientific and philosophical journals. Harvey Brown is co-editor of *Philosophical Foundations of Quantum Field Theory* (1988, 1990; with Rom Harré) and of *The Philosophy of Vacuum* (1991; with Simon Saunders).

**Alan Chalmers** is Associate Professor in the History and Philosophy of Science Unit at the University of Sydney. He completed a Ph.D. on Maxwell's electromagnetic theory at Chelsea College, which Heinz Post supervised. His research interests lie in the history and philosophy of physics, as well as general philosophy of science. He is the author of *What is This Thing Called Science?* (1980; second edition 1982) and *Science and its Fabrication* (1990).

**Newton C. da Costa** is Professor of Logic in the Department of Philosophy at the University of São Paulo, Brazil, and a member of the Institut International de Philosophie in Paris. He has published extensively in a wide variety of fields, including logic, mathematics, foundations of probability theory and the philosophy of science. He has edited or co-edited several collections of papers and is the author of five books. Professor da Costa is particularly renowned for his seminal work in paraconsistent logic, and his collaboration with Steven French on this subject is just one of many projects in which he is currently engaged.

**Elspeth Crawford** is Lecturer in Teaching Studies at the Moray House Institute of Education, Heriot-Watt University, Edinburgh. She was a research student in the Chelsea Department of History and Philosophy of Science from 1975 and completed her Ph.D. thesis, *The Ideas of Particle and Field in Michael Faraday's Work, 1831–1845*, under Heinz Post's supervision. An experienced



*S. French and H. Kamminga (eds.), Correspondence, Invariance and Heuristics*, 343–346.

teacher of physics, she has chaired the Education Section of the British Society of the History of Science and served on the Editorial Board of *Physics Education*. Her current interests are in the application of psychodynamic thinking to educational practice and philosophy.

**James Cushing** is Professor of Philosophy and Physics at the University of Notre Dame. He was a frequent visitor to the History and Philosophy of Science Department at Chelsea and contributed to its programme of seminars, both as participant and presenter. He is the author of *Theory Construction in Modern Physics: the S-Matrix* (1990), and co-editor of *Philosophical Consequences of Quantum Theory: Reflections on Bell's Theorem* (1989).

**Melvin P. Earles** was a member of the Academic Board at the time of the founding of the Department of History and Philosophy of Science at Chelsea College. At the invitation of Heinz Post, he lectured on the history of biology in the department for many years and supervised several successful Ph.D. candidates working on biological topics. He was subsequently appointed an Honorary Research Fellow in the department. Melvin Earles has published contributions to the history of pharmacy including a study of the first editions of the *Pharmacopoeia Londinensis* and a survey of early theories of the mode of action of drugs and poisons.

**Arthur Fine** is the John Evans Professor of Philosophy at Northwestern University, and past President of the Philosophy of Science Association. He was Visiting Professor in Heinz Post's department at Chelsea College, and for several years returned to visit, taking part in the various discussions and seminars. He has published extensively on issues in the history and philosophy of science, especially as they relate to the natural and social sciences. Recent works include *The Shaky Game: Einstein, Realism and the Quantum Theory* (1989).

**Allan Franklin** is Professor of Physics at the University of Colorado. His research interests are in the history and philosophy of science, particularly with respect to the roles of experiment, an endeavour in which he has received much encouragement from Heinz Post. He has been a regular visiting researcher, and then Honorary Research Fellow, in the Department of History and Philosophy of Science at Chelsea (King's) College since 1978. He is the author of *The Neglect of Experiment* (1986), *Experiment, Right or Wrong* (1990), and the forthcoming *The Rise and Fall of the Fifth Force.*

**Steven French** is an Assistant Professor in the Department of Philosophy and Religion at Southeast Missouri State University, but will be moving in 1993 to Leeds University in the U.K. He was first introduced to the philosophy of science by Heinz Post, who also suggested the topic for his doctoral research. In 1985 Steven French took up a position at the University of

Campinas, São Paulo, Brazil, where he met Newton da Costa. They have since collaborated on a variety of papers, focussing on the philosophical applications of paraconsistent logic and the accommodation of inconsistency within the model-theoretic approach in general.

**Giora Hon** is a lecturer in the Department of Philosophy, Haifa University, Israel. His principal research concerns the problem of experimental error, the subject of his doctorate which Heinz Post supervised. He has published widely in major journals of history and philosophy of science. His extensive study of Kaufmann's experiments, which were thought to have refuted both Lorentz's theory of electrons and Einstein's theory of relativity, is forthcoming in *The Autonomy of Experiment/the Sovereignty of Practice* (ed. J. Buchwald, with I. Hacking).

**Harmke Kamminga** is in charge of the modern biomedicine programme of the Cambridge Wellcome Unit for the History of Medicine. She joined Heinz Post's department at Chelsea College in 1975, where she obtained her Ph.D. for a thesis on theories of the origin of life. She later became a Lecturer at Chelsea/King's College, taking over Post's graduate teaching in philosophy of science – which was a hard act to follow, and all the more fun for that reason. She has published in both history and philosophy of science, mostly with reference to the biological sciences, and generally tries to subvert physical imperialism.

**Clive Kilmister** held the Chair of Mathematics at King's College, London, until his retirement in 1984. His work in mathematical physics has long been marked by a concern with philosophical issues, especially in relation to relativity theory. He regularly attended Heinz Post's weekly seminar and, upon his retirement, was elected Honorary Research Fellow in the Department of History and Philosophy of Science founded by Post. Professor Kilmister has played an active role in the British Society for the Philosophy of Science and is a past President of the Society. His books include *Special Theory of Relativity* (Pergamon, 1970), *General Theory of Relativity* (Pergamon, 1973) and *Philosophers in Context: Russell* (Harvester, 1984). He is currently preparing books on Arthur Eddington's philosophy of science and on the philosophy of combinatorial physics.

**Noretta Koertge** is Professor of History and Philosophy of Science at Indiana University. Her research interests in theories of scientific method and the development of science began with a doctoral dissertation on the General Correspondence Principle, supervised by Heinz Post. Although not always heeding it, she remembers Post's advice to "Beware of intellectual fads" and "Eschew flights to the meta-level".

**Avinash Puri** is a lecturer at Hackney College, London, where he teaches mathematics, philosophy and science – the last mostly to students of plumbing and construction crafts. He was a student of Heinz Post in the 1970s and remained an active participant in Post's weekly seminar at Chelsea College and later at King's College. He plans to present a collection of his 'tales from the classroom' in a forthcoming book.

**Michael Redhead** is Professor of History and Philosophy of Science at the University of Cambridge, a Chair which he has occupied since 1987. After obtaining his Ph.D. in Mathematical Physics, he attended Heinz Post's lectures in the early 1970s. He subsequently became a Research Associate in the Chelsea Department of History and Philosophy of Science and succeeded Heinz Post as Head of Department in 1983. He has published extensively on the philosophy of physics and in 1988 received the Lakatos Award for his book *Incompleteness, Nonlocality and Realism*. He was elected Fellow of the British Academy in 1991.

**Simon Saunders** is Assistant Professor in the Department of Philosophy, Harvard University. He obtained his Ph.D. from the Department of History and Philosophy of Science at Chelsea College, London. He has broad interests in metaphysics and epistemology, but his area of specialisation is the philosophy of physics, particularly the foundations of quantum theory. Simon Saunders co-edited, with Harvey Brown, *The Philosophy of Vacuum* (1991) and is currently working on a book on Dirac and the development of quantum field theory.

**Eric Scerri** was Heinz Post's last research student and he obtained his Ph.D. from King's College, London, in 1992. He continues to do research in what is now the Department of Philosophy at King's College, as well as teaching chemistry in a sixth form college. His interests are focussed on the philosophy of chemistry and physics, quantum chemistry and chemical education.

**Barrie Tonkinson** has a technology-based career in aerospace, specialising in vertical take-off and, more recently, flight simulation. He took a masters degree in mathematical physics at Chelsea College in 1981 and then joined the Department of History and Philosophy of Science at Chelsea, where he was taught by Heinz Post. He is a permanent student in history and philosophy of science where Heinz Post continues to offer friendship and guidance. His paper in this volume comes out of a long-standing collaboration with Clive Kilmister.

# Index of Names


Abney, K. 276
Abraham, P. 185
Adams, J.C. 100
Anandan, J. 228, 251, 253ff, 258
Anderson, J.L. 249ff
Appolonius 298, 299, 322
Archimedes xxi, 161, 176ff, 185
Aristotle 189, 193ff, 202
Aronson, S. 84, 86ff, 93ff, 99
Aston, F.W. 47

Baccheschi, E. 185
Bacon, F. 131
Barbour, J.B. 231ff, 255ff
Bell, A. 129ff
Bell, J.S. 85, 131, 228, 240ff, 245, 247, 268, 279, 281, 286, 288ff, 292
Beller, M. 275
Ben-Menahem, Y. 262
Bender, C.F. 57ff
Berenson, B. 185
Berkeley, G. 166
Bernstein, J. 85
Berzi, V. 256
Besso, M. 179
Bion, W.R. xx, 106ff, 118, 121ff
Boccaccio, G. 172ff, 185
Bock, G.J. 86ff
Bohm, D. xxii, 261ff, 265ff, 275ff, 280ff
Bohr, N. xxi, 48ff, 62, 137ff, 147ff, 153ff, 179, 261, 266, 273ff, 279, 302, 309ff, 316, 323
Bondi, H. 213
Borel, E. 313ff, 323
Born, M. 62, 273ff, 305, 328
Boyd, R. 321
Braginskii, V.B. 94
Brans, C. 85, 90, 98
Broglie, L. de 281, 272ff, 312
Brown, B. 148ff, 156
Brown, H.R. xxii, 229ff, 254ff, 323
Brush, S. 204
Bucherer, A.H. 245, 248
Budden, T. 256ff
Bunge, C.F. 56, 59
Burgers, J.M. 52, 312
Buridan, J. 193

Cabibbo, N. 85
Cantor, G. 123
Capek, M. 181
Carnap, R. xiv
Carnot, S. 183
Cartan, E. 304
Cartwright, N. 322, 330
Casimir, H.B.G. 304
Chalmers, A. xxi, 204
Christenson, J.H. 86
Chuaqui, R. 141
Cimabue 173
Clausewitz, K. 134
Clifford, W.K. 224
Cohen, L.J. 139ff, 144ff, 150
Colella, R. 85ff, 98ff
Copernicus, N. 193, 198, 204, 298ff
Costa, N.C.A. da xx, 141, 148
Cotton, F.A. 63
Coupal, D.P. 99
Crawford, E. xx
Crick, F. 78
Curd, M. 135
Curie, M. 79
Cushing, J.T. xxii, 276, 340

Dalton, J. 48
Dante 172




*S. French and H. Kamminga (eds.), Correspondence, Invariance and Heuristics*, 347–351.

Danton, G. 68
Darden, L. 80
Darwin, C. 71ff, 77, 81
Davidson, E.R. 57, 59
De Rujula, A. 100
DeBouard, X. 86
Democritus 179
Denbigh, K. 186
Descartes, R. 79, 203, 296ff, 304
Dicke, R.H. 85, 90, 94, 98
Dijksterhuis, E.J. 202, 298, 322
Dingle, H. 180
Dirac, P.A.M. 62, 148, 303, 312ff, 322, 328
Doppler, C.J. 209, 247
Dresden, M. 277
Du Bois-Reymond, E. 339
Duhem, P. 67, 262ff

Earman, J. 186, 228, 248ff, 257
Eddington, A.S. 237, 256
Ehlers, J. 228, 242
Ehrenfest, P. 52, 262, 312, 317
Einstein, A. xxiff, 62, 79, 89ff, 100, 147, 178ff, 183, 207ff, 218ff, 223ff, 227ff, 233ff, 239ff, 243, 246ff, 250ff, 255ff, 263ff, 269, 272ff, 279ff, 297, 299ff, 304, 306, 312, 323, 328
Eötvös, R. 90, 93ff
Eudoxus 298
Euler, L. 297
Everett, H. 315

Fackler, O. 99
Fajans, K. 46ff, 62
Faraday, 229
Faraday, M. xx, 105ff, 112ff, 123, 229
Fekete, E. 94
Feyerabend, P. xiii, xvii, 66, 122, 204, 296ff, 301ff
Field, H. 302ff, 305
Fine, A. xxii, 269, 276, 322
Fischbach, E. 84ff, 93ff
Fizeau, A.H.L. 230, 246
Fock, V.A. 55, 57
Fourier, J. 299, 310ff, 322
Franklin, A. xx, 99, 307
Franklin, B. 296
French, S. xx, 81
Fresnel, 210
Fresnel, A.J. 210, 230, 233, 237, 246, 300
Freud, S. 107
Frey, D. 171
Friedman, M. xxii, 228, 237ff, 247ff, 253, 256ff
Frontov, V.N. 91ff
Fujii, Y. 90ff, 93ff, 97ff

Galbraith, W. 86
Galileo xxi, 176, 189ff, 196ff, 231ff, 234, 255, 296ff
Gauss, K.F. 304
Gerlach, W. 274, 281ff, 289, 291
Geymonat, L. 191
Gibbons, G.W. 91ff, 98
Gigerenzer, G. 126
Giotto 172ff, 185
Glashow, S.L. 101
Glymour, C. 298
Goe, G. 185
Goethe, J.W. 299
Gooding, D. 123
Goodman, A. 297
Gorini, V. 256
Greenaway, F. 48
Grossman, N. 89

Hacking, I. 307ff, 322
Haeckel, E. 72, 74, 76ff, 81, 339
Hamilton, W.R. 297, 311ff
Harding, S. 134
Harman, G. 149ff, 156
Harmann, 321
Hartmann, L. xiv
Hartree, R. 55, 57
Hawking, S. 322, 327ff, 340
Haxton, W. 100
Heilbron, J. 274
Heisenberg, W. 48, 51, 164, 264, 266ff, 273ff, 309ff, 317
Heller, 318
Helmholtz, H. 299
Hempel, C. 215
Henry, J. 117ff
Hepp, 318
Heracleitus 174, 179
Hertz, H.R. 67, 182, 248
Hevesey, G. von 46ff
Hilbert, D. 265, 305, 312, 314, 323
Hinshelwood, C.N. xiv, 62
Hipparchus 322
Hodges, A. 256
Hodgson, J.R. 257
Hoffman, B. 230
Holding, S.C. 94
Hon, G. xxi
Horwich, P. 139
Hull, D. 133
Hume, D. 139

Hund, F. 51
Huxley, T. 72
Huygens, C. 227, 236

Inonu, E. 323

Jacobi, K. 297, 312
James, F.A.J.L. 123
Jaskowski, S. 143, 151
Jauch, J.M. 323
Jones, J. 135
Jordan, E.P. 273
Jørgensen, C.K. 62ff

Kaluza, 251
Kamminga, H. xix, 80, 277
Kant, I. xiv, 71ff, 77, 298ff, 337
Keats, J. 114
Kelvin, Lord 328, 340
Kilmister, C. xxi
Klein, C.F. 252ff, 297, 323
Klein, M. 107
Koertge, N. xvii, xx, 59ff
Koyré, A. 201
Kragh, H. 52
Krajewski, 321
Kramers, H.A. 274, 309, 312
Kripke, S. 302, 322
Krotkov, R. 90, 94
Kuhn, T.S. xviii, 45, 67, 146, 296ff, 300ff, 308, 316, 319ff, 332
Kyburg, H. 156
Kyritsis, C. 256

Ladenburg, R.W. 309, 323
Lagrange, J.L. 297
Lakatos, I. xvii, 67, 132, 184
Landau, L.D. 190
Langevin, P. 245, 248
Laplace, P. 71ff, 77
Laudan, L. 276, 296ff, 308, 320ff
Laue, M. von 246
Lee, T.D., 85, 100
Leibniz, G.W. 212, 221
Leonardo (da Vinci) 173ff, 182, 185
Leplin, J. 276
Leverrier, U.J.J. 100
Levi-Civita, T. 312
Lie, M.S. 238, 304, 306, 308
Lifshitz, E.M. 190
Lipton, P. 81
Long, D.R., 91, 100
Longino, H. 128ff
Loomis, E. 323
Lorentz, H.A. xix, xxii, 100, 208, 210ff, 213ff, 220, 222ff, 227, 228ff, 233ff, 236ff, 243ff, 251, 255, 257ff, 300, 304ff
Lorenz, K. xvi
Lucas, J.R. 257

Mach, E. 176ff, 179, 182, 185
Machover, M. xv
Mackey, G.W. 313ff, 323
MacKinnon, E.M. 50
Magnus, Albertus 197
Maia, A. 254ff
Martindale, A. 185
Marzke, R.F. 242, 245
Maxwell, G. 307
Maxwell, J.C. 67, 115, 123, 180, 210ff, 227, 229, 240, 243ff, 248, 251, 255, 257, 279, 300, , 322, 332, 340
McMullin, E. 322
Mendeleef, D.I.: see Mendeleev, D.I.
Mendeleev, D.I. 48, 60, 63, 131, 135
Mermin, D. 340
Meyer, S.C. 69, 80
Meyerson, E. 185
Michelson, A.A. 148, 229, 240, 246ff
Mikenberg, I. 141
Mikkelsen, D.R. 91ff
Miller, A.I. 254, 257
Miller, D. 335
Milne, E.A. 213
Minkowski, H. xxii, 216, 220, 227ff, 238ff, 241, 249ff, 254, 257, 304, 306
Morley, E.W. 148, 229, 240, 246ff
Moseley, H.G.J. 60
Mukunda, N. 322ff

Nagel, T. 339
Nägeli, C. 72
Nernst, H.W. 47, 62
Newman, M.J. 91ff
Newton, I. xxi, 78ff, 84, 93, 98ff, 185, 189ff, 204, 207ff, 221, 223, 227, 231ff, 234ff, 244, 250, 255, 296, 300, 301, 303, 319, 321, 332
Nollet, J.A. 296
Norton, J. 149, 156, 250, 251
Nugaev, R.N. 255

O'Hanlon, J. 90ff, 93
Oliveira, M.B. de 145ff, 154, 156
Oparin, A.I. 74ff, 81
Oresme, N. 193
Oseen, C. 62
Osiander, A. 322

Ostwald, F.W. 184
Overhauser, A.W. 85ff, 98ff

Paneth, A.F. xiii, 46ff
Paneth, J. xiv
Panov, V.I. 91ff, 94
Parmenides 179, 181
Pascal, B. 178
Pasteur, L. 78
Pauli, W. xxii, 51, 54, 227ff, 230, 240ff, 273ff
Pekar, D. 94
Penrose, R. 254ff, 322
Perring, J. 85
Pflüger, E. 73ff, 76ff, 81
Pickering, A. 99
Pirani, F.A.E. 228, 242
Planck, M. 49, 147, 149, 156, 231, 255, 263, 332, 335, 339
Plato, 172, 298
Poincaré, H. 112, 297
Poisson, S.D. 312
Popper, K. xvii, 65, 79, 123, 127
Post, H.R. xiiiff, 45ff, 48, 59ff, 65, 66, 68, 71, 77ff, 83, 98ff, 105ff, 121ff, 125ff, 131, 135, 137ff, 147, 153, 155ff, 168ff, 174ff, 179ff, 183ff, 189ff, 203, 207, 223, 227, 256, 261ff, 276, 279ff, 293, 295ff, 305, 317ff, 327, 335, 340
Priest, G. 148, 153, 156
Prigogine, I. 171, 183, 186
Ptolemy, 298ff, 322
Puri, A. xxi
Putnam, H. 67, 78, 301ff, 307, 321ff

Quine, W.V.O. 67, 262ff

Rapp, R.H. 92, 100
Redhead, M.L.G. xxiii, 340
Reichenbach, H. xx, 125ff, 135
Rescher, N. 156, 337, 338
Richter, H.E. 81
Riemann, G.F.B. 304
Rindler, W. 218, 256
Ritz, W. 180, 181, 229
Roll, P.G. 90, 94
Rorty, R. 322
Ross, G. 63
Ross, W.D. 195ff
Rueger, A. 147
Rugerius, L. 197
Russell, B. 277
Rutherford, E. 47, 49

Salam, A. 84, 328
Saunders, S. xxiii, 230
Scerri, E. xvii, xix
Schaefer, H.F. 57ff
Scheffler, I. 302, 307
Scherk, J. 90
Schild, A. 228, 242
Schiller, F. 175
Schrödinger, E. 48, 54, 164, 265, 267, 271ff, 282, 283, 286ff, 289, 312, 330
Schwartz, C. 56, 59
Schwarzschild K., 218, 312
Scriven, M. 80
Shapere, D. 302
Sitter, W. de 255
Sklar, L. 185ff, 228, 239ff, 243
Slater, J.C. 309
Smith, G. 256ff
Smith, J. 149ff, 153ff
Snow, C.P. xv
Sommerfeld, A. 54, 273
Sperber, D. 143, 156
Spero, R. 92
Stacey, F.D. 92
Stachel, J. 255
Stalnaker, R.C. 144, 156
Stein, H. 322
Steiner, M. 322
Stern, O. 281ff, 289, 291
Stiegler, K.D. 256
Stocks, J.L. 204
Stokes, G.G. 148
Stoner, E. 54
Stradivarius, A. 331
Sudarshan, E.C.G. 322ff

Talmadge, C. 84
Tarski, A. 141
Taylor, P. 318
Thomson, J.J. 302
Tichy, P. 335
Tonkinson, B. xxi
Torretti, R. 255ff, 322ff
Tuck, G.J. 94
Tyndall, J. 72, 108, 110, 117, 122
Tzanakis, C. 256

Ullian, J.S. 151ff
Urey, H. 47ff

Vailati, G. 185
van der Steen, W.J. 80
van Fraassen, B. 185, 331

Varadarajan, V. 323
Verstraeten, G. 186
Vitelleschi, M. 197

Wagoner, R.V. 90ff
Wallace, W. 197
Watkins, J. 321
Weinberg, M. 129ff
Weinberg, S. 84ff, 99, 328
Weingard, R. 252, 256ff
Weinhold, F. 56, 59
Weismann, A. 72
Weisskopf, V.F. 51
Werner, S.A. 85ff, 98
Westfall, S. 99
Weyl, H. 172, 178, 181, 251, 312
Whewell, W. 116
Whiting, B.F. 91ff, 98
Wien, W. 149
Wigner, E.P. 305, 322ff
Wilkinson, G. 63
Will, C. 95, 100
Winnie, J. 237, 256
Worrall, J. 335

Yang, C.N. 100
Yukawa, H. 84

Zahar, E. 322
Zee, A. 90
Zeeman, P. 274
Zener, C. xiv

# Index of Subjects


*ab initio* calculations 54ff, 59
absolute objects 243, 250
accelerated frame 207
acceptance xx, 137ff
adiabatic principle 50, 52ff
antecedent conditions xx, 65ff, 75ff, 80ff
aperiodic systems 53
approximation 182ff, 224
art 171ff, 184
astronomy
– Copernican 196, 296, 298ff
– Ptolemaic 296, 298ff
asymmetry 171, 173, 178ff, 181ff, 185
Asymptopia 339
atomic structure 50
atomic theory 309ff
atomism xviii, 175, 179, 182
*Aufbauprinzip* 50ff, 53
auxiliary statements 67, 79
axioms 67ff, 72, 76, 79

belief
– factual 138, 143ff, 150
– representational 143ff, 152, 155, 156
Bell's theorem 131, 268
boundary conditions 61, 66ff, 76, 79, 181, 183, 329, 335
Brans-Dicke theory 85, 90, 98

causality 180, 185
change
– explanation of 172, 174ff, 179, 181ff, 185
– irreversible 172, 182ff
chronometric hypothesis 238, 240ff, 252
coherence 263, 272
collapse
– state function 284ff, 292ff
– wave function 266ff
collapse theories 283ff
complementarity 266, 273ff
completeness 261, 313ff
conservatism 168
contingent elements 175ff
continuum mechanics 299ff
coordinate transformations 232, 234ff, 245ff, 249ff, 256
Copenhagen interpretation xxii, 261, 265ff, 274ff
correspondence xviiiff, xxii, 45, 61, 65, 105, 138, 146ff, 154, 168, 183ff, 227, 237, 296ff, 303, 308, 311, 313, 317ff, 320
correspondence principle xixff, 50, 62, 261ff, 295, 312, 327, 340ff
correspondence relation xix, 280ff, 293
CP symmetry 85
CP violation 85ff, 98

deductive closure 138, 149ff, 156
delayed choice experiment 289ff
determinism 271, 273, 277
diagonalisation 175ff
discovery xviii, 45, 105ff, 111, 119ff, 159ff
– context of xixff, 65, 79, 125ff, 128, 130, 134, 137, 155
double slit experiment 289ff
Duhem-Quine thesis 67, 262ff
duration 222
dynamical objects 245, 249, 251

education 54, 61, 107, 113, 121, 159
Ehrenfest theorem 312, 317
electrodynamics 210, 231, 233, 240, 243, 249, 256ff, 299, 309, 312, 323
– classical 227ff, 243ff
electromagnetic induction 115, 123




*S. French and H. Kamminga (eds.), Correspondence, Invariance and Heuristics*, 353–356.

electron shells 51ff, 54
electrotonic state 106, 114ff, 120, 123
energy dependence 86ff, 93ff, 97ff
Eötvös experiment 90, 93ff, 96ff, 100ff
equivalence
- empirical 269
- observational 268ff
ether 227, 229ff, 232ff, 247, 255, 257, 296, 298ff, 322
etiological plausibility 69, 73, 76ff
evolution
- theory of 71ff, 75ff, 81
exclusion principle 54
experiment xviii, xx, 83ff, 90ff, 98ff
explanation
- historical 69ff, 80
explanatory import 68

Fifth Force xx, 84ff, 89, 91, 93, 97ff
footprint xviiiff, xxiff, 227, 305, 322

Galilean invariance xxi, 189ff
Galilean transformation 190, 220
general correspondence principle xviiiff, xxiii, 45, 65ff, 77ff, 80, 138, 146ff, 154ff, 207, 261, 303, 327, 335, 340
general relativity: see relativity
- general
gravitation 78, 84ff, 98, 100, 223, 328
gravitational constant *(G)* 90, 93
gravitational mass xviii, 94, 100
gravitational potential 90, 94, 98
gravitational red-shift 221
gravity
- centre of 185
group theory 297, 300, 304ff, 308, 312, 323
GRW theory 284ff

Hamilton's principle 312
Hartree-Fock method 55
heuristic fruitfulness 138, 153ff
heuristic objects 106
heuristic plasticity 297, 299ff
heuristics xviiiff, xxi, 65ff, 78ff, 105ff, 168, 183ff, 295, 297ff, 304, 306, 308, 311ff, 320ff
hidden variables 268, 280ff, 289
historical contingency 272, 275
historical plausibility 69, 73, 76ff
holism
- relational 268
homosexuality 129
hypercharge 88, 94, 96ff
hyperphoton 85ff, 94, 100

idealization 176
ideology xx, 126ff, 132, 134
imprinting
- process of 179
incompleteness xviii, 71ff, 77, 262, 319
inconsistency xviii, 65, 72, 138, 147ff, 152ff, 156
inconsistent correspondence xxi, 207, 335, 340ff
indeterminism 271, 273, 277, 329
inertia
- circular 199ff, 202ff
- Galileo's 202ff
- horizontal 196ff, 199ff, 202
- law of 202
- linear 199, 202
- Newtonian 199, 202ff
- principle of 193, 199, 223ff
inertial mass xviii, 94, 100
initial conditions 67, 76, 79, 178, 180, 329
insolubilities 339
insolubility theorem 279ff, 293
interaction formalism 286ff, 292ff
invariance xviiiff, xxi, 65, 105, 166ff, 176, 180ff, 183
- of meaning 301ff, 320
- of reference 301, 320
isotopes 46ff, 62

k-calculus 213
K mesons 85ff
kinematics/dynamics distinction 228, 239ff, 243ff, 247, 249ff, 253, 258
Kuhn loss xviii, 45, 296ff, 300, 316, 319, 332

law-cluster concepts 301ff, 306
laws 66ff, 74, 76ff, 172, 178, 180ff, 328, 330, 337
- as ontology 320
lever
- law of xxi, 176ff
lie
- Bion's concept of xx, 107ff, 116, 121ff
life
- origin of 71ff, 81
light postulate 227, 229ff, 233, 236, 241, 250ff, 254ff
logic
- Jaskowski's 143, 151
- non-adjunctive 149
- paraconsistent 142ff, 148ff, 151ff, 156
logical omniscience 150ff

Mackey axioms 313ff

many worlds interpretation 289
mathematics
– role of 295ff, 306ff, 309ff, 319ff, 323
matrix mechanics 310ff
Maxwell's equations 180
measurement problem xxii, 266, 270, 279ff, 283, 286, 288, 290, 292ff, 313, 315ff, 319
motion
– horizontal 192ff, 197, 199ff, 203
– natural 194ff, 201
– neutral 197

Nazi science 129
necessity 173ff, 182
negative capability 113ff, 118, 121
Newtonian gravitation 78, 85, 89ff, 93, 97ff
Newtonian mechanics xviii, xxi, 67, 209, 220, 296ff, 300ff, 303ff, 319, 321
nonlocality 268ff

ontology 48, 210, 212, 265ff, 270ff
orbital xix, 52, 54ff, 57, 61

partial structures xx, 138, 142, 152
periodic table 46, 48ff, 53, 59ff, 62ff
permanence 175
physics
– classical 179, 182ff
– modern 178, 182ff
plausibility xx, 69, 73, 76ff, 133ff
pragmatic circle, 207, 209, 222ff
prejudice 106ff, 110ff, 114, 116ff, 119, 121
process
– irreversible 172, 179ff, 182ff
– measuring 179, 183
– microscopic reversible 182
– reversible elementary 180, 182ff
psychoanalysis 106, 118
pursuit
– context of 84, 99, 125, 128, 133ff, 137, 155

quantum eraser 291
quantum mechanics xix, xxiiff, 48, 54, 57, 59, 61ff, 164, 261ff, 264ff, 296, 298ff, 305ff, 308ff, 321ff, 328ff
– classical limit of 332
quantum numbers 51, 53ff
quantum potential 266ff, 289
quantum theory xxiiff, 48ff, 179, 183, 279ff, 283, 285ff, 289, 292ff

rational decision making 130ff, 133ff
rational reconstruction 264, 275
rationality 175
realism 261, 269ff, 292ff, 308
– convergent 296, 321ff
– entity 307ff
– scientific 262ff, 270ff, 275, 322ff
– structural 335
reduction 46, 48, 61, 79, 300ff, 305ff, 308, 316ff, 327
reference 296, 299, 301ff, 307ff, 319ff
relativistic mass 301ff, 304, 306
relativity
– Einstein's 207, 209, 297ff, 304, 306, 321ff, 328
– Galilean xxi, 190ff, 200, 203ff, 300ff
– Galileo's xxi, 192ff, 199ff, 203
– general xviii, 85, 89ff, 207, 218ff, 239, 242, 244, 248ff, 252, 254, 256
– Newtonian 199
– perceptual 192ff, 200ff, 204
– principle of xxii, 227ff, 231ff, 241, 245, 247ff, 255ff, 297, 300, 323
– special xxiff, 207ff, 220ff, 227ff, 233ff
representations
– propositional 143
– semi-propositional 143ff, 152
revelation 105, 113ff, 120
rule of silence 279ff, 283, 285, 287ff, 292ff

Schrödinger equation 54, 265ff, 330
Schrödinger evolution 282ff, 286ff
semantic approach 141ff, 148, 152
short-range force 93ff
simplicity xviii, 98, 134ff, 178
simultaneity 208, 332
solvability 130, 132ff
space-time geometry
– Galilean 227, 242ff, 254, 257
– Minkowskian 227ff, 239, 241, 243ff, 249ff, 254, 257
– Newtonian 227ff, 243ff, 254
space-time theories 222, 228, 248ff
special relativity: see relativity – special
spin 281ff, 284, 287, 291ff
state
– mixed 279, 282ff, 290ff
– pure 282ff, 290
Stern-Gerlach measurement 281ff, 289, 291
symmetry xviii, xxi, 171ff, 176ff, 222
synchrony condition 209

teleology 70, 72, 81
theory confirmation 297, 308, 321
theory convergence 296

Theory of Everything (TOE) xxiii, 328ff, 337, 340
theory of light
- Newtonian 296, 321
thinking process 106, 108ff, 112, 120, 122
three-body problem 54
time xxiff, 179ff, 185, 207ff, 339
- arrow of 180, 182
- frame 208, 210
- global 208, 210ff
- metrical 221
- physical 179
- topological 221ff
true thought 111, 113, 118
truth xx, 107, 109ff, 121ff, 167
- correspondence 137ff, 140ff, 151ff
- partial 137ff, 141ff, 151ff, 155
- pragmatic 138, 141ff, 150
Tuskegee experiment 129

uncertainty relation 267, 274
underdetermination xxii, 262ff, 269ff, 276
uniformly accelerated observer 216
universal statements 68, 79ff

29. J.J. Wiatr (ed.): *Polish Essays in the Methodology of the Social Sciences.* [Synthese Library 131] 1979 ISBN 90-277-0723-5; Pb 90-277-0956-4
30. P. Janich: *Protophysics of Time.* Constructive Foundation and History of Time Measurement. Translated from the 2nd German edition. 1985 ISBN 90-277-0724-3
31. R.S. Cohen and M.W. Wartofsky (eds.): *Language, Logic, and Method.* 1983 ISBN 90-277-0725-1
32. R.S. Cohen, C.A. Hooker, A.C. Michalos and J.W. van Evra (eds.): *PSA 1974.* Proceedings of the 4th Biennial Meeting of the Philosophy of Science Association. [Synthese Library 101] 1976 ISBN 90-277-0647-6; Pb 90-277-0648-4
33. G. Holton and W.A. Blanpied (eds.): *Science and Its Public.* The Changing Relationship. [Synthese Library 96] 1976 ISBN 90-277-0657-3; Pb 90-277-0658-1
34. M.D. Grmek, R.S. Cohen and G. Cimino (eds.): *On Scientific Discovery.* The 1977 Erice Lectures. 1981 ISBN 90-277-1122-4; Pb 90-277-1123-2
35. S. Amsterdamski: *Between Experience and Metaphysics.* Philosophical Problems of the Evolution of Science. Translated from Polish. [Synthese Library 77] 1975 ISBN 90-277-0568-2; Pb 90-277-0580-1
36. M. Marković and G. Petrović (eds.): *Praxis.* Yugoslav Essays in the Philosophy and Methodology of the Social Sciences. [Synthese Library 134] 1979 ISBN 90-277-0727-8; Pb 90-277-0968-8
37. H. von Helmholtz: *Epistemological Writings.* The Paul Hertz / Moritz Schlick Centenary Edition of 1921. Translated from German by M.F. Lowe. Edited with an Introduction and Bibliography by R.S. Cohen and Y. Elkana. [Synthese Library 79] 1977 ISBN 90-277-0290-X; Pb 90-277-0582-8
38. R.M. Martin: *Pragmatics, Truth and Language.* 1979 ISBN 90-277-0992-0; Pb 90-277-0993-9
39. R.S. Cohen, P.K. Feyerabend and M.W. Wartofsky (eds.): *Essays in Memory of Imre Lakatos.* [Synthese Library 99] 1976 ISBN 90-277-0654-9; Pb 90-277-0655-7
40. B.M Kedrov and V. Sadovsky (eds.): Current Soviet Studies in the Philosophy of Science. (In prep.) ISBN 90-277-0729-4
41. M. Raphael: Theorie des geistigen Schaffens aus marxistischer Grundlage. (In prep.) ISBN 90-277-0730-8
42. H.R. Maturana and F.J. Varela: *Autopoiesis and Cognition.* The Realization of the Living. With a Preface to 'Autopoiesis' by S. Beer. 1980 ISBN 90-277-1015-5; Pb 90-277-1016-3
43. A. Kasher (ed.): *Language in Focus: Foundations, Methods and Systems.* Essays in Memory of Yehoshua Bar-Hillel. [Synthese Library 89] 1976 ISBN 90-277-0644-1; Pb 90-277-0645-X
44. T.D. Thao: *Investigations into the Origin of Language and Consciousness.* 1984 ISBN 90-277-0827-4

45. A. Ishimoto (ed.): Japanese Studies in the History and Philosophy of Science. (In prep.) ISBN 90-277-0733-3
46. P.L. Kapitza: *Experiment, Theory, Practice.* Articles and Addresses. Edited by R.S. Cohen. 1980 ISBN 90-277-1061-9; Pb 90-277-1062-7
47. M.L. Dalla Chiara (ed.): *Italian Studies in the Philosophy of Science.* 1981 ISBN 90-277-0735-9; Pb 90-277-1073-2
48. M.W. Wartofsky: *Models.* Representation and the Scientific Understanding. [Synthese Library 129] 1979 ISBN 90-277-0736-7; Pb 90-277-0947-5
49. T.D. Thao: *Phenomenology and Dialectical Materialism.* Edited by R.S. Cohen. 1986 ISBN 90-277-0737-5
50. Y. Fried and J. Agassi: *Paranoia.* A Study in Diagnosis. [Synthese Library 102] 1976 ISBN 90-277-0704-9; Pb 90-277-0705-7
51. K.H. Wolff: *Surrender and Cath.* Experience and Inquiry Today. [Synthese Library 105] 1976 ISBN 90-277-0758-8; Pb 90-277-0765-0
52. K. Kosík: *Dialectics of the Concrete.* A Study on Problems of Man and World. 1976 ISBN 90-277-0761-8; Pb 90-277-0764-2
53. N. Goodman: *The Structure of Appearance.* [Synthese Library 107] 1977 ISBN 90-277-0773-1; Pb 90-277-0774-X
54. H.A. Simon: *Models of Discovery* and Other Topics in the Methods of Science. [Synthese Library 114] 1977 ISBN 90-277-0812-6; Pb 90-277-0858-4
55. M. Lazerowitz: *The Language of Philosophy.* Freud and Wittgenstein. [Synthese Library 117] 1977 ISBN 90-277-0826-6; Pb 90-277-0862-2
56. T. Nickles (ed.): *Scientific Discovery, Logic, and Rationality.* 1980 ISBN 90-277-1069-4; Pb 90-277-1070-8
57. J. Margolis: *Persons and Mind.* The Prospects of Nonreductive Materialism. [Synthese Library 121] 1978 ISBN 90-277-0854-1; Pb 90-277-0863-0
58. G. Radnitzky and G. Andersson (eds.): *Progress and Rationality in Science.* [Synthese Library 125] 1978 ISBN 90-277-0921-1; Pb 90-277-0922-X
59. G. Radnitzky and G. Andersson (eds.): *The Structure and Development of Science.* [Synthese Library 136] 1979 ISBN 90-277-0994-7; Pb 90-277-0995-5
60. T. Nickles (ed.): *Scientific Discovery.* Case Studies. 1980 ISBN 90-277-1092-9; Pb 90-277-1093-7
61. M.A. Finocchiaro: *Galileo and the Art of Reasoning.* Rhetorical Foundation of Logic and Scientific Method. 1980 ISBN 90-277-1094-5; Pb 90-277-1095-3
62. W.A. Wallace: *Prelude to Galileo.* Essays on Medieval and 16th-Century Sources of Galileo's Thought. 1981 ISBN 90-277-1215-8; Pb 90-277-1216-6
63. F. Rapp: *Analytical Philosophy of Technology.* Translated from German. 1981 ISBN 90-277-1221-2; Pb 90-277-1222-0
64. R.S. Cohen and M.W. Wartofsky (eds.): *Hegel and the Sciences.* 1984 ISBN 90-277-0726-X
65. J. Agassi: *Science and Society.* Studies in the Sociology of Science. 1981 ISBN 90-277-1244-1; Pb 90-277-1245-X

66. L. Tondl: *Problems of Semantics.* A Contribution to the Analysis of the Language of Science. Translated from Czech. 1981
ISBN 90-277-0148-2; Pb 90-277-0316-7
67. J. Agassi and R.S. Cohen (eds.): *Scientific Philosophy Today.* Essays in Honor of Mario Bunge. 1982 ISBN 90-277-1262-X; Pb 90-277-1263-8
68. W. Krajewski (ed.): *Polish Essays in the Philosophy of the Natural Sciences.* Translated from Polish and edited by R.S. Cohen and C.R. Fawcett. 1982
ISBN 90-277-1286-7; Pb 90-277-1287-5
69. J.H. Fetzer: *Scientific Knowledge.* Causation, Explanation and Corroboration. 1981 ISBN 90-277-1335-9; Pb 90-277-1336-7
70. S. Grossberg: *Studies of Mind and Brain.* Neural Principles of Learning, Perception, Development, Cognition, and Motor Control. 1982
ISBN 90-277-1359-6; Pb 90-277-1360-X
71. R.S. Cohen and M.W. Wartofsky (eds.): *Epistemology, Methodology, and the Social Sciences.* 1983. ISBN 90-277-1454-1
72. K. Berka: *Measurement.* Its Concepts, Theories and Problems. Translated from Czech. 1983 ISBN 90-277-1416-9
73. G.L. Pandit: *The Structure and Growth of Scientific Knowledge.* A Study in the Methodology of Epistemic Appraisal. 1983 ISBN 90-277-1434-7
74. A.A. Zinov'ev: *Logical Physics.* Translated from Russian. Edited by R.S. Cohen. 1983 ISBN 90-277-0734-0
*See also* Volume 9.
75. G-G. Granger: *Formal Thought and the Sciences of Man.* Translated from French. With and Introduction by A. Rosenberg. 1983 ISBN 90-277-1524-6
76. R.S. Cohen and L. Laudan (eds.): *Physics, Philosophy and Psychoanalysis.* Essays in Honor of Adolf Grünbaum. 1983 ISBN 90-277-1533-5
77. G. Böhme, W. van den Daele, R. Hohlfeld, W. Krohn and W. Schäfer: *Finalization in Science.* The Social Orientation of Scientific Progress. Translated from German. Edited by W. Schäfer. 1983 ISBN 90-277-1549-1
78. D. Shapere: *Reason and the Search for Knowledge.* Investigations in the Philosophy of Science. 1984 ISBN 90-277-1551-3; Pb 90-277-1641-2
79. G. Andersson (ed.): *Rationality in Science and Politics.* Translated from German. 1984 ISBN 90-277-1575-0; Pb 90-277-1953-5
80. P.T. Durbin and F. Rapp (eds.): *Philosophy and Technology.* [*Also* Philosophy and Technology Series, Vol. 1] 1983 ISBN 90-277-1576-9
81. M. Marković: *Dialectical Theory of Meaning.* Translated from Serbo-Croat. 1984 ISBN 90-277-1596-3
82. R.S. Cohen and M.W. Wartofsky (eds.): *Physical Sciences and History of Physics.* 1984. ISBN 90-277-1615-3
83. É. Meyerson: *The Relativistic Deduction.* Epistemological Implications of the Theory of Relativity. Translated from French. With a Review by Albert Einstein and an Introduction by Milić Čapek. 1985 ISBN 90-277-1699-4

84. R.S. Cohen and M.W. Wartofsky (eds.): *Methodology, Metaphysics and the History of Science.* In Memory of Benjamin Nelson. 1984 ISBN 90-277-1711-7
85. G. Tamás: *The Logic of Categories.* Translated from Hungarian. Edited by R.S. Cohen. 1986 ISBN 90-277-1742-7
86. S.L. de C. Fernandes: *Foundations of Objective Knowledge.* The Relations of Popper's Theory of Knowledge to That of Kant. 1985 ISBN 90-277-1809-1
87. R.S. Cohen and T. Schnelle (eds.): *Cognition and Fact.* Materials on Ludwik Fleck. 1986 ISBN 90-277-1902-0
88. G. Freudenthal: *Atom and Individual in the Age of Newton.* On the Genesis of the Mechanistic World View. Translated from German. 1986 ISBN 90-277-1905-5
89. A. Donagan, A.N. Perovich Jr and M.V. Wedin (eds.): *Human Nature and Natural Knowledge.* Essays presented to Marjorie Grene on the Occasion of Her 75th Birthday. 1986 ISBN 90-277-1974-8
90. C. Mitcham and A. Hunning (eds.): *Philosophy and Technology II.* Information Technology and Computers in Theory and Practice. [*Also* Philosophy and Technology Series, Vol. 2] 1986 ISBN 90-277-1975-6
91. M. Grene and D. Nails (eds.): *Spinoza and the Sciences.* 1986 ISBN 90-277-1976-4
92. S.P. Turner: *The Search for a Methodology of Social Science.* Durkheim, Weber, and the 19th-Century Problem of Cause, Probability, and Action. 1986. ISBN 90-277-2067-3
93. I.C. Jarvie: *Thinking about Society.* Theory and Practice. 1986 ISBN 90-277-2068-1
94. E. Ullmann-Margalit (ed.): *The Kaleidoscope of Science.* The Israel Colloquium: Studies in History, Philosophy, and Sociology of Science, Vol. 1. 1986 ISBN 90-277-2158-0; Pb 90-277-2159-9
95. E. Ullmann-Margalit (ed.): *The Prism of Science.* The Israel Colloquium: Studies in History, Philosophy, and Sociology of Science, Vol. 2. 1986 ISBN 90-277-2160-2; Pb 90-277-2161-0
96. G. Márkus: *Language and Production.* A Critique of the Paradigms. Translated from French. 1986 ISBN 90-277-2169-6
97. F. Amrine, F.J. Zucker and H. Wheeler (eds.): *Goethe and the Sciences: A Reappraisal.* 1987 ISBN 90-277-2265-X; Pb 90-277-2400-8
98. J.C. Pitt and M. Pera (eds.): *Rational Changes in Science.* Essays on Scientific Reasoning. Translated from Italian. 1987 ISBN 90-277-2417-2
99. O. Costa de Beauregard: *Time, the Physical Magnitude.* 1987 ISBN 90-277-2444-X
100. A. Shimony and D. Nails (eds.): *Naturalistic Epistemology.* A Symposium of Two Decades. 1987 ISBN 90-277-2337-0
101. N. Rotenstreich: *Time and Meaning in History.* 1987 ISBN 90-277-2467-9
102. D.B. Zilberman: *The Birth of Meaning in Hindu Thought.* Edited by R.S. Cohen. 1988 ISBN 90-277-2497-0

# Boston Studies in the Philosophy of Science

103. T.F. Glick (ed.): *The Comparative Reception of Relativity.* 1987 ISBN 90-277-2498-9
104. Z. Harris, M. Gottfried, T. Ryckman, P. Mattick Jr, A. Daladier, T.N. Harris and S. Harris: *The Form of Information in Science.* Analysis of an Immunology Sublanguage. With a Preface by Hilary Putnam. 1989 ISBN 90-277-2516-0
105. F. Burwick (ed.): *Approaches to Organic Form.* Permutations in Science and Culture. 1987 ISBN 90-277-2541-1
106. M. Almási: *The Philosophy of Appearances.* Translated from Hungarian. 1989 ISBN 90-277-2150-5
107. S. Hook, W.L. O'Neill and R. O'Toole (eds.): *Philosophy, History and Social Action.* Essays in Honor of Lewis Feuer. With an Autobiographical Essay by L. Feuer. 1988 ISBN 90-277-2644-2
108. I. Hronszky, M. Fehér and B. Dajka: *Scientific Knowledge Socialized.* Selected Proceedings of the 5th Joint International Conference on the History and Philosophy of Science organized by the IUHPS (Veszprém, Hungary, 1984). 1988 ISBN 90-277-2284-6
109. P. Tillers and E.D. Green (eds.): *Probability and Inference in the Law of Evidence.* The Uses and Limits of Bayesianism. 1988 ISBN 90-277-2689-2
110. E. Ullmann-Margalit (ed.): *Science in Reflection.* The Israel Colloquium: Studies in History, Philosophy, and Sociology of Science, Vol. 3. 1988 ISBN 90-277-2712-0; Pb 90-277-2713-9
111. K. Gavroglu, Y. Goudaroulis and P. Nicolacopoulos (eds.): *Imre Lakatos and Theories of Scientific Change.* 1989 ISBN 90-277-2766-X
112. B. Glassner and J.D. Moreno (eds.): *The Qualitative-Quantitative Distinction in the Social Sciences.* 1989 ISBN 90-277-2829-1
113. K. Arens: *Structures of Knowing.* Psychologies of the 19th Century. 1989 ISBN 0-7923-0009-2
114. A. Janik: *Style, Politics and the Future of Philosophy.* 1989 ISBN 0-7923-0056-4
115. F. Amrine (ed.): *Literature and Science as Modes of Expression.* With an Introduction by S. Weininger. 1989 ISBN 0-7923-0133-1
116. J.R. Brown and J. Mittelstrass (eds.): *An Intimate Relation.* Studies in the History and Philosophy of Science. Presented to Robert E. Butts on His 60th Birthday. 1989 ISBN 0-7923-0169-2
117. F. D'Agostino and I.C. Jarvie (eds.): *Freedom and Rationality.* Essays in Honor of John Watkins. 1989 ISBN 0-7923-0264-8
118. D. Zolo: *Reflexive Epistemology.* The Philosophical Legacy of Otto Neurath. 1989 ISBN 0-7923-0320-2
119. M. Kearn, B.S. Philips and R.S. Cohen (eds.): *Georg Simmel and Contemporary Sociology.* 1989 ISBN 0-7923-0407-1
120. T.H. Levere and W.R. Shea (eds.): *Nature, Experiment and the Science.* Essays on Galileo and the Nature of Science. In Honour of Stillman Drake. 1989 ISBN 0-7923-0420-9

121. P. Nicolacopoulos (ed.): *Greek Studies in the Philosophy and History of Science.* 1990 ISBN 0-7923-0717-8
122. R. Cooke and D. Costantini (eds.): *Statistics in Science.* The Foundations of Statistical Methods in Biology, Physics and Economics. 1990 ISBN 0-7923-0797-6
123. P. Duhem: *The Origins of Statics.* Translated from French by G.F. Leneaux, V.N. Vagliente and G.H. Wagner. With an Introduction by S.L. Jaki. 1991 ISBN 0-7923-0898-0
124. H. Kamerlingh Onnes: *Through Measurement to Knowledge.* The Selected Papers, 1853-1926. Edited and with an Introduction by K. Gavroglu and Y. Goudaroulis. 1991 ISBN 0-7923-0825-5
125. M. Čapek: *The New Aspects of Time: Its Continuity and Novelties.* Selected Papers in the Philosophy of Science. 1991 ISBN 0-7923-0911-1
126. S. Unguru (ed.): *Physics, Cosmology and Astronomy, 1300-1700.* Tension and Accommodation. 1991 ISBN 0-7923-1022-5
127. Z. Bechler: *Newton's Physics on the Conceptual Structure of the Scientific Revolution.* 1991 ISBN 0-7923-1054-3
128. É. Meyerson: *Explanation in the Sciences.* Translated from French by M-A. Siple and D.A. Siple. 1991 ISBN 0-7923-1129-9
129. A.I. Tauber (ed.): *Organism and the Origins of Self.* 1991 ISBN 0-7923-1185-X
130. F.J. Varela and J-P. Dupuy (eds.): *Understanding Origins.* Contemporary Views on the Origin of Life, Mind and Society. 1992 ISBN 0-7923-1251-1
131. G.L. Pandit: *Methodological Variance.* Essays in Epistemological Ontology and the Methodology of Science. 1991 ISBN 0-7923-1263-5
132. G. Munévar (ed.): *Beyond Reason.* Essays on the Philosophy of Paul Feyerabend. 1991 ISBN 0-7923-1272-4
133. T.E. Uebel (ed.): *Rediscovering the Forgotten Vienna Circle.* Austrian Studies on Otto Neurath and the Vienna Circle. Partly translated from German. 1991 ISBN 0-7923-1276-7
134. W.R. Woodward and R.S. Cohen (eds.): *World Views and Scientific Discipline Formation.* Science Studies in the [former] German Democratic Republic. Partly translated from German by W.R. Woodward. 1991 ISBN 0-7923-1286-4
135. P. Zambelli: *The Speculum Astronomiae and Its Enigma.* Astrology, Theology and Science in Albertus Magnus and His Contemporaries. 1992 ISBN 0-7923-1380-1
136. P. Petitjean, C. Jami and A.M. Moulin (eds.): *Science and Empires.* Historical Studies about Scientific Development and European Expansion. ISBN 0-7923-1518-9
137. W.A. Wallace: *Galileo's Logic of Discovery and Proof.* The Background, Content, and Use of His Appropriated Treatises on Aristotle's *Posterior Analytics.* 1992 ISBN 0-7923-1577-4

# Boston Studies in the Philosophy of Science

138. W.A. Wallace: *Galileo's Logical Treatises.* A Translation, with Notes and Commentary, of His Appropriated Latin Questions on Aristotle's *Posterior Analytics.* 1992 ISBN 0-7923-1578-2
Set (137 + 138) ISBN 0-7923-1579-0
139. M.J. Nye, J.L. Richards and R.H. Stuewer (eds.): *The Invention of Physical Science.* Intersections of Mathematics, Theology and Natural Philosophy since the Seventeenth Century. Essays in Honor of Erwin N. Hiebert. 1992
ISBN 0-7923-1753-X
140. G. Corsi, M.L. dalla Chiara and G.C. Ghirardi (eds.): *Bridging the Gap: Philosophy, Mathematics and Physics.* Lectures on the Foundations of Science. 1992 ISBN 0-7923-1761-0
141. C.-H. Lin and D. Fu (eds.): *Philosophy and Conceptual History of Science in Taiwan.* 1992 ISBN 0-7923-1766-1
142. S. Sarkar (ed.): *The Founders of Evolutionary Genetics.* A Centenary Reappraisal. 1992 ISBN 0-7923-1777-7
143. J. Blackmore (ed.): *Ernst Mach – A Deeper Look.* Documents and New Perspectives. 1992 ISBN 0-7923-1853-6
144. P. Kroes and M. Bakker (eds.): *Technological Development and Science in the Industrial Age.* New Perspectives on the Science–Technology Relationship. 1992 ISBN 0-7923-1898-6
145. S. Amsterdamski: *Between History and Method.* Disputes about the Rationality of Science. 1992 ISBN 0-7923-1941-9
146. E. Ullmann-Margalit (ed.): *The Scientific Enterprise.* The Bar-Hillel Colloquium: Studies in History, Philosophy, and Sociology of Science, Volume 4. 1992 ISBN 0-7923-1992-3
147. L. Embree (ed.): *Metaarchaeology.* Reflections by Archaeologists and Philosophers. 1992 ISBN 0-7923-2023-9
148. S. French and H. Kamminga (eds.): *Correspondence, Invariance and Heuristics.* Essays in Honour of Heinz Post. 1993 ISBN 0-7923-2085-9
149. M. Bunzl: *The Context of Explanation.* 1993 ISBN 0-7923-2153-7

*Also of interest:*
R.S. Cohen and M.W. Wartofsky (eds.): *A Portrait of Twenty-Five Years Boston Colloquia for the Philosophy of Science, 1960-1985.* 1985 ISBN Pb 90-277-1971-3

*Previous volumes are still available.*

KLUWER ACADEMIC PUBLISHERS – DORDRECHT / BOSTON / LONDON